Molecular Mechanism of Alzheimer's Disease

Molecular Mechanism of Alzheimer's Disease

Special Issue Editor

Ian Macreadie

MDPI • Basel • Beijing • Wuhan • Barcelona • Belgrade

MDPI

Special Issue Editor
Ian Macreadie
RMIT University
Australia

Editorial Office
MDPI
St. Alban-Anlage 66
4052 Basel, Switzerland

This is a reprint of articles from the Special Issue published online in the open access journal *International Journal of Molecular Sciences* (ISSN 1422-0067) from 2018 to 2019 (available at: https: //www.mdpi.com/journal/ijms/special_issues/AD)

For citation purposes, cite each article independently as indicated on the article page online and as indicated below:

LastName, A.A.; LastName, B.B.; LastName, C.C. Article Title. *Journal Name* **Year**, *Article Number, Page Range.*

ISBN 978-3-03921-407-5 (Pbk)
ISBN 978-3-03921-408-2 (PDF)

Contents

About the Special Issue Editor

Ian Macreadie is a molecular biologist who has developed yeast to produce foreign proteins, including a viral subunit vaccine. He was a project leader at CSIRO for 24 years working on HIV, malaria, and Pneumocystis jirovecii. For the past decade, he has coordinated courses on industrial microbiology and protein technologies at RMIT University. His current research examines the molecular aspects of Alzheimer's disease, focusing on amyloid beta. In addition, he works on the gut microbiota of Australian animals to find out how they survive with limited diets. He is the Editor-in-Chief of Microbiology Australia, the official journal of the Australian Society for Microbiology

Preface to "Molecular Mechanism of Alzheimer's Disease"

The cause of Alzheimer's disease (AD) remains debated more than a century after its discovery. Amyloid beta remains a smoking gun at the scene, continuing to be associated with the disease. The genetics of familial AD clearly point to mutations within amyloid beta sequences or near the protease cleavage sites where amyloid beta is cut from the Alzheimer's precursor protein (APP). Likewise, in an extensive Icelandic genetic study, people protected from AD were found to have APP mutations leading to a 40% reduction in amyloid beta. Further, amyloid beta, in an oligomeric form, continues to be identified as being toxic to neurons, suggesting it is the cause of neuronal death, while amyloid beta in plaques is not toxic. Plaques can be likened to a graveyard—they have no opportunity to cause harm.

Evidence for amyloid beta being the cause has been sought through intervention with therapeutic antibodies that can remove amyloid beta from those with AD or those who are progressing towards AD. However, those antibody treatments, while removing amyloid beta, did not change AD outcomes, causing many pharmaceutical developers and researchers to abandon amyloid beta as a target. Because of the complexities, there are good reasons to look at how other agents play into AD, such as tau protein, ApoE, responses to oxidative stress, folate status, sleep status, RNA, etc.

It can be argued that previously therapeutic antibody interventions were unsuccessful because they were applied too late and the damage, such as neuronal death, that may have been caused by amyloid beta, was irreversible. Therefore, testing such interventions much earlier is more appropriate. While the early detection of AD has improved greatly, there is no evidence that therapeutic antibody interventions can stop AD progression from the time of early detection. It therefore seems that an intervention targeting amyloid beta should be tested on cohorts of asymptomatic individuals, a proportion of who are expected to develop AD, but there are ethical arguments against using this approach. Even if therapeutic antibodies were to succeed, they remain a highly expensive and invasive intervention. Future hopes appear to lie with less invasive interventions that may involve protective chemotherapies and nutraceuticals.

Large-scale, well-controlled epidemiology studies of prescription drug users have given some insights. For example, in the Veterans Administration study, millions of ex-service personnel on statins were monitored for AD progression. The study showed that simvastatin was unique in providing protection against development of AD (and, coincidentally, Parkinson's disease). While subsequent studies have shown that simvastatin cannot cure AD, it does give hope that additional therapeutic options can be found for prevention. The statins in cell culture reduce BACE prenylation, which reduces BACE activity, leading to less amyloid beta; but perhaps not all statins do this in the brain. Simvastatin is the most lipophilic statin, raising the possibility that it may have effects in the brain.

The insights into AD are today coming from nontraditional approaches, including yeast, although it was once considered that yeast, although a model eukaryote, had no role in assessing AD, which was thought to be caused by brain plaques. Decades later, we know better, and many yeast researchers are now using the "awesome power of yeast" to find what yeast can tell us about AD and to rapidly test compounds that affect specific targets such as amyloid beta and tau.

This book, "Molecular Mechanisms of Alzheimer's Disease", includes contributions that cover many aspects of current and ongoing studies in the early detection of AD and factors involved in AD.

Ian Macreadie
Special Issue Editor

International Journal of
Molecular Sciences

MDPI

Review

Imaging and Molecular Mechanisms of Alzheimer's Disease: A Review

Grazia Daniela Femminella [1], Tony Thayanandan [2], Valeria Calsolaro [1], Klara Komici [3], Giuseppe Rengo [4,5], Graziamaria Corbi [3] and Nicola Ferrara [4,5,*]

[1] Neurology Imaging Unit, Imperial College London, London W12 0NN, UK;
    g.femminella@imperial.ac.uk (G.D.F.); v.calsolaro@imperial.ac.uk (V.C.)
[2] Imperial Memory Unit, Charing Cross Hospital, Imperial College London, London W6 8RF, UK;
    tony.thayanandan@nhs.net
[3] Department of Medicine and Health Sciences, University of Molise, 86100 Campobasso, Italy;
    klara.komici@unimol.it (K.K.); graziamaria.corbi@unimol.it (G.C.)
[4] Department of Translational Medical Sciences, Federico II University of Naples, 80131 Naples, Italy;
    giuseppe.rengo@unina.it
[5] Istituti Clinici Scientifici Maugeri SPA—Società Benefit, IRCCS, 82037 Telese Terme, Italy
*   Correspondence: nicola.ferrara@unina.it; Tel.: +39-081-7463786; Fax: +39-081-7462339

Received: 5 October 2018; Accepted: 14 November 2018; Published: 22 November 2018

Abstract: Alzheimer's disease is the most common form of dementia and is a significant burden for affected patients, carers, and health systems. Great advances have been made in understanding its pathophysiology, to a point that we are moving from a purely clinical diagnosis to a biological one based on the use of biomarkers. Among those, imaging biomarkers are invaluable in Alzheimer's, as they provide an in vivo window to the pathological processes occurring in Alzheimer's brain. While some imaging techniques are still under evaluation in the research setting, some have reached widespread clinical use. In this review, we provide an overview of the most commonly used imaging biomarkers in Alzheimer's disease, from molecular PET imaging to structural MRI, emphasising the concept that multimodal imaging would likely prove to be the optimal tool in the future of Alzheimer's research and clinical practice.

Keywords: Alzheimer's disease; positron emission tomography (PET); magnetic resonance imaging (MRI)

1. Introduction

Alzheimer's disease (AD) is a neurodegenerative disease that is responsible for 60–80% of all cases of dementia worldwide. Recent epidemiological data indicate that approximately 5.7 million Americans of all ages are living with AD in 2018 and 10.5 million people were suffering with dementia in Europe in 2015. The prevalence of dementia in Europe ranges from 4.7% to 6.8% [1]. Estimated projections suggest that by 2025, the number of people over 65 with AD will reach 7.1 million in the U.S., which is almost a 29 percent increase from the 2018 prevalence, and by 2050 the population affected will grow to 13.8 million, posing a great burden on health systems [2]. Clinically, AD is typically characterised by impairment in short-term memory to such an extent as to interfere with activities of daily living, while later symptoms include impairment in the other cognitive domains, such as language, orientation, judgment, executive functions, behavioural changes, and, ultimately, motor difficulties.

The first criteria proposed for AD diagnosis were developed in 1984 and focused only on clinical symptoms. However, the exceptional amount of research conducted since has helped clarify that the phase of dementia in AD is preceded by a long preclinical phase of several decades that evolves through a continuum, with the prodromal stage of mild cognitive impairment (MCI), and ultimately leads to dementia [3]. In this long preclinical phase, an early diagnosis can be made with the help of biomarkers. Based on this evidence, the National Institute on Aging (NIA) and the Alzheimer's Association in 2011 published new guidelines incorporating biomarker tests in addition to clinical symptoms, moving from a symptom-based definition to a biology-based definition of AD [4].

The biology of AD is characterised by two major protein abnormalities in the brain of affected individuals: the extracellular accumulation of amyloid β (Aβ) plaques and intraneuronal deposits of neurofibrillary tangles (NFTs). Insoluble Aβ plaques are formed of aggregated Aβ peptides that derive from the abnormal cleavage of the amyloid precursor protein (APP) into hydrophobic Aβ peptides. Aβ is thought to be the trigger or the driver of the disease process, mainly based on evidence from familial AD cases, leading to the amyloid hypothesis of AD [5]. NFTs are composed of hyperphosphorylated tau protein aggregates which accumulate in the neuron cytoplasm, leading to destabilisation of microtubules and axonal transport [6]. Both proteinopathies can trigger oxidative stress, microvascular dysfunction, and blood–brain barrier (BBB) disruption, and can induce the activation of an inflammatory response within the brain, ultimately resulting in neuronal damage and consequent neurodegeneration [3].

All these pathological changes that manifest at earlier or later phases of the AD continuum can now be explored with the use of biomarkers, some of which are still only used in a research framework and are awaiting clinical validation. Overall, the main biomarkers in AD can be broadly divided into cerebrospinal (CSF) and imaging biomarkers. Research is ongoing in the field of blood biomarkers, but large clinical studies are needed to assess their diagnostic potential [7]. In this review, we will focus on imaging biomarkers, both those currently available in clinical practice and those that are only part of the research framework [8]. Over CSF biomarkers that constitute an indirect measure of the ongoing pathological processes, imaging ones have the advantage of providing information on the in vivo pathological processes, giving a "window" to the changes happening in the brain at the different stages of the disease, and are less invasive and troublesome for the patients. We will focus on neurodegenerative imaging biomarkers (MRI and glucose metabolism), amyloid and tau imaging, and the newest in vivo biomarkers for neuroinflammation and BBB dysfunction.

2. Imaging of Neurodegeneration

2.1. Structural Magnetic Resonance Imaging (MRI)

Atrophy seems to be an unavoidable, inevitable progressive component of neurodegeneration. Brain tissue loss correlates well with cognitive deficits, both cross-sectionally and longitudinally in AD [9]. Structural brain changes are accurately consistent with upstream Braak stages of neurofibrillary tangle deposition [10,11] and downstream neuropsychological deficits [12]. Rates of change in several structural measures, including whole-brain [13], entorhinal cortex [14], hippocampus [15], and temporal lobe volumes [16], correlate closely with changes in cognitive performance, validating atrophy in these regions as markers of AD. For atrophy markers to be useful clinically, the subtleties should be known at the different stages of the disease, and their relationship with other imaging and biological markers should be understood. Atrophy measures change with disease progression over AD disease severity. Structural markers are more sensitive to change than are markers of Aβ deposition, both in MCI and in the moderate dementia stage of AD [17]. However, studies have shown that in the earliest forms of MCI, amyloid burden shows more abnormalities that are structural changes [18,19]. Atrophy is accompanied by microstructural changes, such as axonal loss and metabolite changes, all of which are measured with techniques other than MRI.

Structural MRI is still one of the most widely used neuroimaging techniques in the diagnosis of AD. T1-weighted scans are the most commonly used due to their ability to provide good contrast between grey and white matter and to detect subtle changes in grey matter. MRI gives the best spatial resolution of any clinical neuroimaging technique, so measures from an MRI include grey matter volume, cortical thickness, and volumetric measures of the hippocampus [20]. Measurements of grey matter are usually done visually, but recently there has been increased use of automated methods to calculate volume and cortical thickness [21,22] and subcortical segmentation of the hippocampus [23,24]. However, with these recent advances in more methodological techniques, visual reading is still the method most often used clinically to read an MRI. This shows the lack of a standardised protocol or method for the diagnosis of AD but is of high interest for researchers [24,25]. Clinicians also use structural MRI to determine whether cognitive impairment is due to reasons other than AD such as tumours or subdural hematomas [26].

Structural MRI studies have shown reduced hippocampal volume in individuals with amnestic MCI, and its reduction is thought to be one of the most predictive and sensitive measures of AD [27]; however, studies have shown other neuropsychiatric disorders such as schizophrenia [28] and depression [29] demonstrate a reduction in hippocampal volume as well. Figure 1, panel A, shows a coronal structural MRI session where hippocampal atrophy is shown (left larger than right). Therefore, the implementation of MRI-based biomarkers for clinical use requires validation across both clinical and analytical techniques. The diagnostic and prognostic accuracy of neuroimaging markers are dependent on both how the biomarker is measured (visual or quantitative) and which one is measured (MRI, Amyloid PET, fluorodeoxyglucose (FDG)-PET, etc.) [30]. Variation in methods and scanners can introduce noise and bias into the data which can impact the diagnostic accuracy.

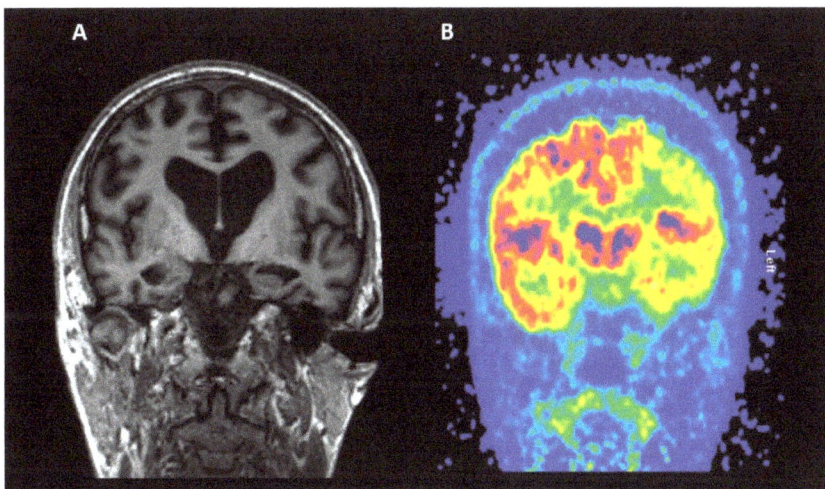

Figure 1. Imaging biomarkers of neurodegeneration. Coronal structural MRI section (panel **A**) and 18F-fluorodeoxyglucose (FDG) PET (panel **B**) from a patient with Alzheimer's disease (AD).

2.2. Fluorodeoxyglucose (FDG) PET

Glucose is the main source of energy used by the brain, which consumes around 25% of the amount circulating the whole body. The cerebral glucose metabolism is regulated by transport through the BBB, led by glucose transporters (GLUTs); GLUT1 is the main transporter on the BBB, while GLUT3 is the main transporter on neuron membranes, with a higher efficacy than GLUT1 [31]. GLUT1 is also present on astrocytes, which can uptake glucose in response to neuronal secretion of glutamate and produce lactate, another source of energy for neuronal activity [32]. The glucose consumption rate in the brain can be displayed in vivo using the PET tracer 18F-FDG, which reaches the neurons and enters the glycolytic process until the formation of FDG-6-phosphate, which will then stay trapped in the cells, at the same rate as the glucose [31]. The glucose consumption is not only an indicator of synaptic activity, whose loss is one of the main features of AD [33], but also reflects the excitatory glutamate release and recycling between astrocytes and neurons [34]. A reduction in the glucose metabolism is recognised as a biomarker of neurodegeneration, appearing years before the cognitive symptoms [33,35]. A pattern of reduced [18F]FDG uptake in posterior cingulate, hippocampi, and medial temporal structures is typical in AD and MCI, with subsequent spreading to the whole cortex as the disease progresses [33] (see Figure 1, panel B), while cerebellum, visual and primary motor cortices, and basal ganglia nuclei are less affected [36]. A different pattern of hypometabolism can be seen in other variants of AD, like posterior cortical atrophy and primary progressive aphasia [37]. It is interesting to note that the glucose hypometabolism is correlated with cognitive impairment and its severity, while the results of studies evaluating the same correlation between amyloid load and severity of cognitive impairment are less homogeneous [38,39]. The reduction in glucose metabolism in regions like the precuneus and posterior cingulate has been demonstrated to be associated with the severity of cognitive impairment [38]. A large study evaluated the baseline cerebral metabolic rate for glucose (CMRgl) in 298 subjects from the ADNI cohort (142 aMCI, 74 pAD, and 82 controls), correlating it with cognitive impairment severity; both the disease groups showed a reduction in the CMRgl in posterior cingulate, precuneus, and frontal and parietotemporal cortices compared with the cognitively intact subjects [40]. The CMRgl rate in the left frontal and temporal cortices was significantly correlated with low Mini-Mental State Examination (MMSE) scores when evaluating only the AD population [40]. In a different study, the pattern of regional hypometabolism appeared to be associated with specific cognitive domains, with visuospatial ability impairment correlated to a reduced metabolism in the posterior regions and impairment in language abilities with a left hemisphere reduction [38]. Interestingly, the impact of cognitive reserve in AD has also been studied with FDG-PET: Ewers et al. evaluated an ADNI cohort of cognitively normal subjects, classified as preclinical AD or healthy control based on the biomarkers profile, and they found that a higher level of education was associated with reduced FDG-PET in the amyloid-positive group [41]. This finding is in line with the literature, supporting the theory that high cognitive reserve can compensate the biological impairment, and highly educated subjects can show a degenerative profile worse than expected for the symptoms [41].

The accuracy of FDG-PET compared to serial clinical evaluation relative to post mortem pathological diagnosis was evaluated in a cohort of 44 subjects grouped as AD and not AD [42]. This study demonstrated that, in the diagnostic process, the FDG-PET is superior to clinical evaluation, which reached the same diagnostic power only later on in the follow-up [42].

Several studies demonstrated that FDG-PET is also a good predictor of disease progression from MCI to AD, according to a few longitudinal studies [43,44]. A longitudinal study aiming to establish the sensitivity and specificity of FDG-PET in patients evaluated and followed up for dementia proved an FDG-PET sensitivity of 93% in detecting progressive dementia and a specificity of 76%; it was also able to distinguish patients with AD from patients with other degenerative diseases with a sensitivity of 94% and a specificity of 73% for AD and 78% for other diseases [45]. It is also worth noticing that a negative scan at baseline indicates an unlikely progression across 3 years [45]. The use of FDG-PET in the clinical setting for the diagnostic process of MCI is more debated, with some studies showing hypometabolism in the cortex and others being inconclusive in the identification of MCIs [31]. In 2015, a Cochrane meta-analysis of 14 studies, for a total of 421 subjects, aimed to evaluate the effectiveness of FDG-PET in identifying MCI subjects converting to dementia at the follow-up [46]. According to the authors, the result of the meta-analysis did not support the use of FDG-PET in routine clinical use in MCI subjects. A limitation of this meta-analysis was the poor methodological quality of some of the studies, leading to risk of bias; therefore, more uniform protocols would be required to get to a satisfactory conclusion [46]. However, the use of FDG-PET is of high value in the diagnostic process, especially in the most difficult cases [37]. Few retrospective studies have actually demonstrated the usefulness of the FDG-PET in clarifying the diagnosis and increasing the cholinesterase inhibitor prescription; moreover, in atypical or uncertain cases, a repeated follow-up FDG-PET improved the diagnostic power and management [37]. FDG-PET is a widely used imaging technique, both in research and clinical settings, with a high predictive value and diagnostic power for Alzheimer's disease and different types of dementia. Together with the other biomarkers, such as cortical atrophy and amyloid and tau deposition, it is a fundamental tool for early diagnosis, selection criteria, and follow-up evaluation in clinical trials.

3. Amyloid Imaging

Accumulation of Aβ fibrils in the form of amyloid plaques is a neuropathological hallmark for autopsy-based diagnosis confirmation of dementia caused by AD [47]. Aβ deposition is thought to precede cognitive symptoms in AD and is therefore a potential preclinical marker of disease [48]. There have been different approaches to noninvasively visualise amyloid deposition in human brains with amyloid PET radiotracers. Typically, amyloid imaging agents bind to insoluble fibrillary forms of Aβ 40 and Aβ 42 deposits, which are major components of compact neuritic plaques and vascular deposits.

Clinical criteria for the suitable use of amyloid imaging in patients demonstrate the need to integrate scanning with detailed clinical and cognitive evaluations. These criteria state that amyloid imaging should only be used under certain circumstances such as in patients with persistent or progressive unexplained cognitive impairment or unclear clinical presentations [49]. Amyloid imaging, as stated by the clinical criteria, should not be used to determine severity of dementia or in patients with probable AD and of typical age, with a family history of dementia, and/or with the presence of the APOE4 allele [50,51]. 11C-Pittsburgh Compound B (PiB) was the first amyloid imaging PET agent used in human subjects in 2002 [52]. However, the PiB compound is labeled with 11C, with a short half-life of only 20 min, limiting its use. To overcome this problem, 18F-labeled Aβ tracers, with a longer half-life of 110 min, are used to show reliable assessment of brain amyloid in a single 15-minute scan. There are only three approved Aβ tracers for clinical use: 18F-Florbetapir [53], 18F-Florbetaben, and 18F-Flutemetamol.

18F-Florbetapir was the first tracer approved for the detection of in vivo amyloid and the first 18F-labelled tracer approved by the FDA since Fludeoxyglucose (FDG); subsequently, this has become the most widely used amyloid tracer. Multicentre studies showed that a high Aβ burden on 18F-Florbetapir PET was associated with poor memory performance in healthy participants [54]. It has also been shown that approximately 50% of MCI patients had a high Aβ burden on 18F-Florbetapir PET [55]. In phase III studies, 18F-Florbetapir demonstrated high sensitivity and specificity (92% and 100%, respectively) in detecting Aβ pathology with no tracer retention in control subjects [56,57]. 18F-Florbetaben reveals a high affinity for fibrillary Aβ in brain homogenates, selectively labelled Aβ plaques, and cerebral amyloid angiopathy in tissue sections from patients with AD [58]. 18F-Florbetaben PET can also detect Aβ pathology in a wide spectrum of neurodegenerative conditions such as frontotemporal lobar degeneration (FTLD). Cortical retention of 18F-Florbetaben was higher in patients with AD than in healthy controls or patients with frontotemporal dementia [59]. 18F-Flutemetamol, in phase I and II studies, was able to differentiate between patients with AD and healthy controls [60,61]. The prediction of progression to AD in patients with MCI was improved when combined with measures of brain atrophy [62]. The tracers discussed above have high affinity and selectivity for fibrillar Aβ in plaques and other Aβ-containing lesions [63,64]. When Aβ PET scans are visually read, cortical tracer retention is usually higher in patients with AD than in healthy controls, particularly in the frontal, cingulate, parietal, and lateral temporal cortices. Both visual and quantitative assessments of amyloid scans from different stages of disease progression reveal a consistent pattern of tracer retention that coincides with amyloid deposition found post mortem in patients with sporadic AD [65]. Longitudinal studies have shown that minute increases in Aβ deposition can be measured using PET; however, these changes can only be seen in those who have either have high or low burdens [66]. Acceptable Aβ loads in normal individuals have also been observed, and approximately 7% of these individuals have an increase of Aβ within 2.5 years above the threshold for "normal" levels [67].

The pivotal use of Aβ imaging is facilitating differential diagnosis in patients who present with atypical symptoms of dementia [68]. Clinical presentations of FTLD can be difficult to differentiate from early onset AD. FTLD does not have Aβ pathology, and these patients, for the most part, show no cortical retention of 11C-PiB—another amyloid tracer [69–71]. Therefore, using amyloid PET can help differentiate between FTLD and AD. The patterns of Aβ deposition can also help differential diagnosis. Patients with cognitively stable Parkinson's disease (PD) have no cortical Aβ deposition; however, Parkinson's disease dementia (PDD) shows signs of Aβ deposition [72,73].

4. Tau Imaging

Tau imaging is the latest innovation in the early detection of neurodegenerative proteinopathies. In the past few years, a number of first-generation tau-selective PET tracers have been developed. 18F-flortaucipir, 18F-THK5351, 18F-THK5317, and 11C-PBB3 have all been extensively used in research studies but have yet to be used clinically. Through imaging studies, tau tracer retention shows an affinity to not only known distributions of aggregated tau but also to mirror patterns of neuronal injury detected by FDG-PET [74,75]. FDG uptake and 18F-THK5317 retention show a negative correlation, primarily in frontal areas [76]. FDG also shows a mediating role in the association between tau pathology and cognitive decline in AD [77].

Tau imaging could be very useful to predict progression of AD due to the relationship between tau deposition, cognitive impairment, and neuronal injury. Tau imaging has the ability to assess the regional distribution and density of tau deposits in the brain which could also help with disease staging. While Aβ imaging studies indicate that total Aβ deposition in the brain is more important than regional differences in predicting cognitive decline, tau imaging data suggest that the topographical distribution of tau in the brain is more important than the total level of tau in the brain [78,79]. A combination of tau and Aβ imaging could be highly beneficial in predicting cognitive decline and neurodegeneration. Studies have demonstrated that high levels of cortical tau deposition in those with Aβ pathology showed increased cognitive impairment in several domains [80,81].

Most, if not all, applications of tau and amyloid imaging are used for the same purpose: accurate and early detection of AD pathology, disease staging, predicting disease progression, and use in disease-specific clinical treatment trials. However, several groups have suggested that tau imaging is better for disease staging and predicting progression than amyloid imaging [82,83]. These groups have compared patients with AD and non-AD tauopathies and have found significant differences in tracer retention between healthy controls, patients with AD, and patients presenting with atypical AD [84,85]. Interestingly, clinical presentations of patients with atypical AD significantly matched their tau deposits as assessed by 18F-flortaucipoir but not their Aβ burdens as assessed by 11C-PiB [86].

However, studies show that high levels of tau found in specific regions of interest (mesial and temporal lobes) are not found alongside a high level of Aβ. Conversely, high levels of tau are highly associated with high Aβ levels in the neocortex. This suggests that detectable levels of cortical Aβ deposits precede levels of cortical tau deposition. Post mortem studies have shown tau deposits in the mesial temporal cortex in elderly individuals, both healthy and with dementia [87]. These findings suggest that hippocampal tauopathy is age related, and not dependent on but magnified by Aβ pathology [74]; this is now known as primary age-related tauopathy (PART) [88].

The in vivo relationship between 18F-flortaucipir and grey matter intensity shows a negative correlation as measured by MRI in healthy controls. Moreover, a study by Wang et al. [89] showed that amyloid plaques affected the association between 18F-flortaucipir retention and cerebral atrophy. Amyloid-positive patients showed a significant association between tau imaging and volume loss, which suggests tau deposition and neuronal loss.

The best use of tau imaging would be a combination of amyloid imaging and selective tau imaging to explain whether Aβ accelerates or causes the spread of tau outside the mesial temporal cortex. This could also help elucidate whether this spreading into cortical areas corresponds clinically to the development of MCI [74,90].

Much like amyloid imaging, tau imaging can be used for differential diagnosis for neurodegenerative Aβ-related conditions such as Dementia Lewy Body (DLB) and other tauopathies such as progressive supranuclear palsy [91]. Also, approximately 40% of FTLD cases are caused by hyperphosphorylated tau, labelled FTLD-tau. As stated previously, Aβ deposition is not a pathological feature of FTLD; therefore, the tau imaging can help with correct diagnosis, especially for specific forms of the disease [92].

A low hippocampal signal has been observed in some tau tracers which is compounded by the unreliable and irregular tracer binding to the choroid plexus, which just lies above the hippocampus. Researchers have suggested that the tracers bind to the aggregated tau in the choroid plexus [93] despite the lack of in vitro autoradiographic studies showing a consistent failure of tracer binding [94]. Another theory suggests that the tracers actually bind to other β-sheet aggregated proteins, such as iron or transthyretin [95,96]. At the moment, no tau tracers have been validated for clinical use [97], and some researchers highlight the inconsistencies between the in vitro and in vivo binding profiles of the tracers [98].

Something that is even more alarming is the doubt over tau selectivity from some PET tracers. Studies show a there is "off-target" binding resulting from tracer binding to alternative targets. Selegiline, a selective and irreversible inhibitor of monoamine oxidase B, also known as MAO-B, can cause signal reductions in cortical and basal ganglia in 18F-THK5351 imaging. In fact, a single 5 mg dose of selegiline can cause signal reductions of up to 50%. This suggests that a certain percentage of tau binding seen in 18F-THK5351 is caused by MAO-B binding [99]. Newer second-generation tracers, such as 18F-RO69558948, have shown less off-target binding [100] with two other tracers (18F-MK6240 and 18F-PI2620) showing no off-target binding [101,102].

5. Imaging of Neuroinflammation

Neuroinflammation refers to the innate inflammatory response of the central nervous system (CNS) to any neuronal insult, such as infections, vascular lesions, trauma, and the presence of abnormal protein aggregates [103]. Data from studies conducted in the last decades indicate that in neurodegenerative diseases, and particularly in AD, neuroinflammation is not only an epiphenomenon secondary to Aβ and tau abnormalities, but it is an essential part of the disease pathophysiology. Results from genome-wide association studies indicate that many of the newly identified genetic risk variants associated with AD involve genes that play an important role in immune function [104]. The cellular players of inflammatory response in the brain are primarily microglia and astrocytes. Microglia activation and reactive astrocytosis can be evaluated in vivo by the use of PET imaging. Thus, in vivo detection of neuroinflammation could represent a useful tool to further clarify the role of immune response in AD pathology and to assess the effectiveness of novel treatments targeting neuroinflammation [105].

5.1. Imaging Microglia

Microglia are mononuclear resident phagocytes ubiquitously distributed in the brain, where they account for 10%–15% of non-neuronal cells [106]. Microglia are of myeloid lineage, originating from progenitors formed in the yolk sac, and their differentiation occurs in the CNS [107]. Under physiological conditions, microglial cells scan the brain parenchyma continuously in order to maintain the homeostasis and, in doing so, present in a ramified morphology. In this resting state they also provide supportive factors to tissue integrity and secrete trophic factors that help maintain neuronal plasticity [108]. Upon detection of any pathological triggers, mediated by membrane receptors, microglia become activated and migrate to the area of the lesion. They change their shape to an amoeboid one and start releasing proinflammatory cytokines, such as tumour necrosis factor-α and interleukin-1β, and free oxygen radicals, such as nitric oxide and superoxide [109]. Both post mortem and preclinical data indicate that in AD the accumulation of Aβ plaques is the main trigger for neuroinflammation. Activated microglia surround Aβ plaques in an attempt to phagocyte them or degrade them through the secretion of proteolytic enzymes [110,111]. Although the initial microglial activation aims at clearance of Aβ plaques and might exert a neuroprotective effect, its continuous triggering and the inefficacy in the clearing process might lead to a vicious cycle of sustained chronic inflammation, with an ultimately neurotoxic effect [112]. This dual function of microglia has been exemplified in the M1/M2 theory, which postulates that microglia switch from a M1 proinflammatory phenotype to a M2 anti-inflammatory one [113]. However, this theory seems to be over-simplistic, and it is likely that microglial phenotype switching and its dual function are a dynamic process.

Once activated, microglia express the Translocator Protein 18 kDa (TSPO), formerly known as peripheral benzodiazepine receptor (PBR). In physiologic conditions, TSPO expression is low within the CNS, primarily confined to endothelial cells, ependyma, choroid plexus, olfactory bulb, and glial cells. Following any brain injuries, TSPO expression on microglial outer mitochondrial membrane markedly increases, making it a suitable marker of glial activation [114]. Over the last decades, several TSPO radioligands have been developed, the most widely used being [11C]-PK11195. This tracer was initially used as a racemate, but the R-enantiomer has a greater affinity for TSPO than the S-enantiomer, and subsequent studies only used [11C]-(R)-PK11195 to investigate neuroinflammation in vivo [109,115]. Although [11C]-(R)-PK11195 has been widely used in several neurological diseases associated with neuroinflammation [116,117], this tracer suffers major limitations, such as a poor signal-to-noise ratio due to high nonspecific binding, high plasma protein binding, and the use of ^{11}C, which limits its use to PET research centers and hospitals with an on-site cyclotron. These difficulties led to the development of second-generation TSPO ligands, with higher TSPO affinity and better kinetics, such as [11C]-PBR28, [11C]-DAA1106, [18F]-DPA714, [18F]-FEPPA, and [18F]-GE180. However, the binding affinity of second generation TSPO tracers is affected by a single-nucleotide polymorphism (SNP) rs6971 in the TSPO gene, which causes an Alanine-to-Threonine substitution in the protein. Based on this, individuals are classified into high-affinity binders (HABs), mixed-affinity binders (MABs), and low-affinity binders (LABs), so that genotyping is essential for appropriate tracer quantification [118].

In AD, Cagnin et al. were the first to report an increase in [11C]-PK11195 binding in the temporal lobe [119], while other groups found no differences between AD patients and controls [120]. Using second-generation TSPO radioligands, other researchers have demonstrated a significant increase in AD subjects with [11C]-DAA1106 [121], [11C]-PBR28 [122], and [18F]-FEPPA [123]. The relationship between microglial activation and amyloid deposition in AD has also been evaluated, finding clusters of significant correlation in most cases [124,125]. Combined PET studies provided evidence for a significant inverse correlation between microglia activation and glucose metabolism in AD patients [126] as well as with hippocampal volume [127]. When looking at cognitive function, the results are varied: some authors have found a significant inverse correlation between TSPO binding and Mini-Mental State Examination (MMSE) scores [124,128], others found no correlation [125], and another group found a positive correlation between the global cortical index and MMSE score [122]. Using different cognitive measures, a negative correlation has been observed between [11C]-PBR28 binding in the inferior parietal lobule and performance on Block Design [123], as well as between [18F]-FEPPA binding in the parietal and prefrontal cortices and visuospatial tasks [129]. A PET multitracer study has recently demonstrated significant widespread correlation between levels of microglial activation and tau aggregation in both MCI and AD subjects, suggesting that these pathologies increase together as the disease progresses. Moreover, microglial activation and amyloid load were also correlated, with a different spatial distribution. The three processes seem to be often found in similar areas of the association cortex [130]. Results are more controversial in the MCI population: some studies have reported increased [11C]-PK11195 uptake in 38% of MCI subjects, while others have shown no differences compared to healthy controls. Similarly, using second-generation radioligands, Yasuno et al. showed significant increases in [11C]-DAA1106 binding in the cerebellum, medial prefrontal cortex, parietal cortex, lateral temporal cortex, anterior cingulate cortex, and striatum in MCI [129], while Kreisl et al. found no differences between MCI patients and controls using [11C]-PBR28 [122].

There are only few studies that have evaluated the longitudinal changes in microglial activation in the AD continuum. Fan et al. demonstrated that microglial activation detected by [11C]-PK11195 increases in AD as the disease progresses, while it is reduced in MCI [131]. A recent study on prodromal AD or MCI subjects using [11C]-PBR28 reported increased longitudinal binding in patients but not in controls, on average equal to 2.5%–7.5% per year [132]. In a study of 64 AD patients, significantly higher global cortical [18F]-DPA-714 binding has been demonstrated in slower decliners compared to fast decliners, further substantiating the concept that early microglial activation could be protective [125].

5.2. Imaging Astrocytes

Astrocytes are star-shaped glial cells, conventionally divided in two categories: protoplasmic astrocytes, located in the grey matter, and fibrous astrocytes, located in the white matter. Their main function is to provide nutritional support to neurons and insulate synaptic connections, regulating extracellular concentrations of ions and neurotransmitters. When activated, astrocytes increase the expression of the glial fibrillary acidic protein (GFAP), and the process of reactive astrogliosis aims at neuroprotection. In AD, it seems that astrocytes play an important role in the clearance of $A\beta$, and after exposure to $A\beta$ they can release cytokines, interleukins, and reactive oxygen species, contributing to the neuroinflammatory process [106].

During neuroinflammation, monoamine oxidase B (MAO B) is up-regulated in reactive astrocytes, and can be targeted in vivo using different PET tracers, such as [11C]-deuterium-L-deprenyl-[11C]-DED- and [11C]-deprenyl-D2 [105]. In a study on AD subjects and amyloid-positive MCI, increased [11C]-DED binding was observed in the frontal, parietal, and temporal cortices, and regional correlation between [11C]-DED uptake and amyloid burden was reported [133]. Results from a multitracer PET study using [11C]-DED, [11C]-PIB, and [18F]-FDG in genetic and sporadic AD patients showed divergent patterns of amyloid deposition and astrocytosis, with the latter process being elevated in the early presymptomatic stages of the disease, and the former increasing with disease progression [134]. Astrocytosis has also been imaged using ligands for the I2-imidazoline receptor, such as [11C]-BU99008. Studies with this PET tracer are underway in AD and MCI subjects.

Figure 2 shows the chemical structures of some of the most commonly used PET tracers mentioned so far [135].

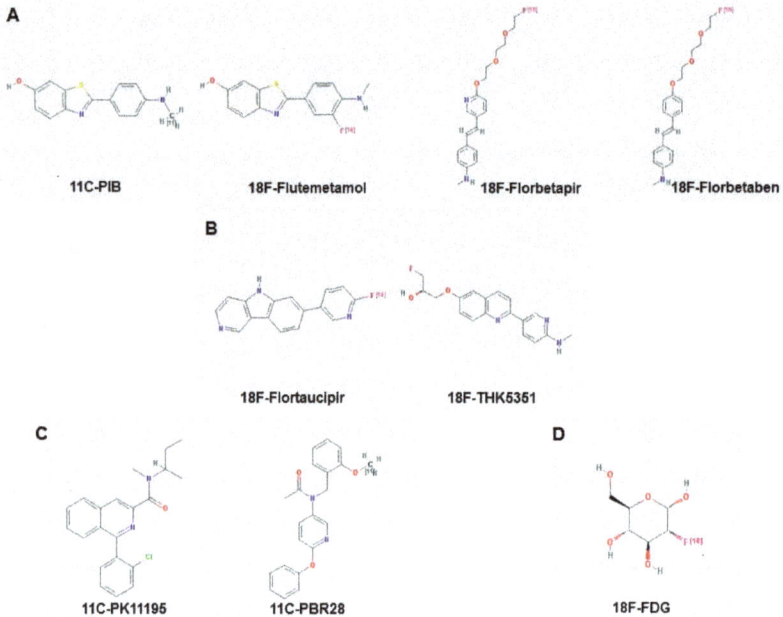

Figure 2. Chemical structure of some PET tracers. Panel **A** shows amyloid PET tracers, Panel **B** shows Tau tracers. Microglial tracers are shown in Panel **C**, and 18F-FDG is shown in Panel **D** (structures downloaded from [135]).

6. Imaging of Blood–Brain Barrier Dysfunction

The blood–brain barrier (BBB) is a highly functional, specialised barrier separating the intravascular system from the neurons, representing a fundamental interface between circulating cells in the bloodstream and the neuronal system. The BBB operates as part of the neurovascular unit (NVU), a multilayer barrier formed by endothelial cells expressing tight junction proteins, a basal lamina of extracellular proteins, astrocyte end-feet, and pericytes [136]. While low permeability is the usual state of the BBB, where protein and cell transport is led by the tight junction proteins (TJPs) and transporters [136], a breach in integrity and impaired function is a common finding in several diseases [137]. In AD, the deposition of Aβ fibrils in the vessel elicits the release of pro-inflammatory cytokines, contributing to BBB damage and an increase in its permeability [137]; moreover, cerebral amyloid angiopathy affects smooth muscle cells, pericytes, and endothelial cells, increasing the damage [137]. The timing of the BBB disruption and AD progression has been widely studied: an indirect measure of BBB breach is the CSF albumin index, demonstrating structural disruption in Alzheimer's and vascular dementia [138]. Post mortem studies reported BBB damage in subjects with AD, demonstrating the accumulation of several proteins in the hippocampus and the cortex and the degeneration of pericytes [139]. AD is also characterised by vascular changes in the endothelial and smooth muscle cells, partially secondary to amyloid toxicity; around amyloid deposits in the vessels, endothelial cells are less viable, and microvascular cerebral tissues showed reduced mitochondrial content and a higher concentration of pinocytotic vesicles [140].

While in case of tumour, strokes, or inflammatory diseases like multiple sclerosis the breach in the BBB permeability is to a major extent, in dementia it is more subtle and requires specific MRI imaging sequences [136] since other imaging techniques (PET and CT) failed to demonstrate any difference between patients with dementia and healthy controls [141,142]. A study conducted with a PET tracer [68Ga]ethylene-diamine-tetraacetic acid ([68Ga]EDTA) did not demonstrate a difference in

CNS permeability between a small group of AD subjects and healthy controls [142]. Similarly, a CT study with meglumine iothalamate failed to show any difference in BBB abnormality between AD and HC [141]. The measurement of BBB permeability with MRI is based on the use of paramagnetic contrast agent Gadolinium-based compounds and the measurement of its leakage from the intravascular space. The techniques used are either dynamic susceptibility contrast-enhanced MRI (DSC-MRI) or dynamic contrast-enhanced MRI (DCE-MRI) [136]. Very few studies have been conducted on small cohorts of AD or MCI patients. Degeneration of the BBB has been demonstrated in the hippocampus with the ageing process; however, that has been seen to appear earlier in subjects with Mild Cognitive Impairment when compared with cognitively intact subjects [139]. This evaluation was conducted using a DCE-MRI; with this technique, grey and white matter regions were simultaneously analysed. In a different MCI population compared to HC, DCE-MRI showed a lower contrast enhancement and slower contrast decay, respectively indicating lower vascular volume and higher BBB permeability in the hippocampi, suggesting impairment in the vasculature and possible BBB disruption [140]. Interestingly, the same difference was not seen in the cerebellum, but, considering that the cerebellum is spared by AD pathology, this is not surprising [140].

To investigate if the leakage could contribute to AD, a pilot study with a dynamic contrast-enhanced MRI was conducted on a population of MCI due to AD and early AD by Maastricht ad Leiden Universities [143]. The imaging protocol was designed with a resolution able to separate the vessel filling from the leakage. The authors also evaluated the relationship between BBB permeability and cognitive performance. The results of the study demonstrated a significantly higher BBB leakage rate in the patient group compared to the controls in the grey matter; the leakage volume was significantly higher in the grey matter, in the normal-appearing white matter, and in the cortex [143]. Considering all the subjects together, the leakage volume in the deep gray matter was higher when the MMSE was lower; a significantly higher leakage volume in the deep gray matter was found in the MCI group when compared to the controls. The overall results of the study supported the theory of BBB impairment as a contributing factor to the AD pathology, especially considering the association with the cognitive performance and the early phases of the subjects enrolled [143]. A case control MRI study was conducted on a cohort of 15 AD subjects and 15 healthy volunteers; for this dynamic contrast-enhanced MRI, regions of interest in the deep grey matter, cortical grey matter, white matter, CSF, and carotid and basilary arteries were selected [144]. In this study, the BBB permeability across the two groups did not differ significantly; however, a difference was seen in the temporal pattern after the injection, suggesting an early occurrence of the BBB permeability difference between healthy control and AD subjects [144]. Others have demonstrated that BBB permeability is increased in major dementia disorders but does not relate to amyloid pathology [145].

The evaluation of BBB damage and permeability is an interesting challenge, especially considering the complexity of the analysis required; certainly, more studies are needed to develop a reliable MRI protocol acquisition, and robust data results are necessary to be able to apply the technique on the larger scale of the clinical setting.

7. Limitations and Future Perspectives

Early diagnosis of sporadic neurodegenerative conditions can be very difficult, especially when patients present with nonspecific symptoms that can be attributable to any form of dementia or neurodegenerative disease. Recently, the NIA-AA research framework criteria for AD have developed the concept that diagnosis should be made based on the measurement of integrated biomarkers, moving towards a more biological definition of the disease. These biomarkers not only concern the presence of Aβ but also must include the tau status of the individual [146]. As the diagnostic criteria for AD continue to develop, the use of amyloid and tau PET imaging is likely to be at the forefront of use in clinical practice.

It is important to note, however, that PET imaging bears several methodological limitations, from the poor resolution of the PET itself to the presence of brain atrophy, which is certainly a crucial feature to be considered in AD imaging. Particularly regarding the latter, unfortunately, there is a lack of homogeneity in the approach to the atrophy [147]. Some studies' approach considered the partial volume effect, proportionate to the atrophy, applying partial volume correction; this has been done with different toolboxes or codes [148,149] and for different tracers [148–152]. Some other studies included the grey matter volume as a covariate in the analysis or excluded relative atrophy. The poor spatial resolution of the PET is also a limitation, especially when analysing small areas or areas where the cortical thickness and the voxel area have similar dimensions [147]. A possible solution is the use of combined PET and MR, which allows a better anatomical accuracy and a partial volume correction to the PET findings [153]. Another factor limiting broader and more routine use of PET imaging is the methodological quantification of the signal. Different techniques have been used in research, with various advantages and caveats to be considered. The standardised uptake value (SUV) technique, applicable to static images, is simple and practical to use; however, it is subjective to different variables, such as the tracer uptake time, dose measurement, and receiving body characteristics. More accurate estimates certainly come from kinetic parameters analysis; this is, however, less practical to use, requiring dynamic images and arterial input [154]. Despite the advances in imaging techniques, no single biomarker is likely to be able to provide the diagnostic certainty needed for early detection of neurodegenerative diseases. Identification or diagnosis requires a multimodal approach that combines biochemical and neuroimaging markers of pathology and neurodegeneration [155]. These biomarkers have now been incorporated into the new diagnostic criteria for the prodromal, preclinical, and overt stages of AD [156,157]. Furthermore, AD-specific interventional trials have been able to implement short-duration trials with smaller samples sizes due to the use of Aβ and/or tau biomarkers to confirm target and treatment efficacy [158]. Interpretation of amyloid positivity through PET is done either visually or quantitatively. Amyloid positivity is defined based on the presence of absence of tracer uptake in brain cortical regions compared to the cerebellum due to a lack of amyloid accumulation in this region. Visual analysis is usually performed using a binary scale while quantitative analysis involves receiver operating characteristic analysis without prespecified cut-off values. This causes data to over-fit which could result in sensitivity and specificity values that are overly optimistic [159]. Conversely, visual interpretation is dependent on the reader's experience, and while most scans are read by multiple readers to confirm positivity or negativity, this is against everyday clinical practice and will have an effect on diagnosis.

Multimodality imaging is the way forward in both research and clinical contexts in AD, suggesting that a combined use of MRI and PET may increase the accuracy of diagnosis due to the ability to detect pathological brain changes associated with AD in the earliest of stages (Tables 1 and 2). Moreover, in a research setting, and in particular in clinical trials with drugs targeting biomarkers, multimodal imaging also has the added value of allowing the monitoring of potential side effects of experimental drugs, which could be hindered by the cognitive impairment [160]. Even though in the diagnostic process of a neurodegenerative disease the results of imaging techniques have to be related to the clinical picture, there are some images with a very strong diagnostic power on their own, such as hippocampal atrophy for AD or DaTScan for Lewy Body Dementia [161]. However, studies have shown that high amyloid load or grey matter atrophy is not enough to give a clear predictive sign of AD, with many healthy individuals showing no signs of AD even with the hallmarks of the neuropathological changes [162,163]. The lack of actual multitracer studies, conducted longitudinally and exploring all the biomarkers at the same time-point, needs to be addressed, together with a strictly homogenous methodological protocol, to better facilitate a more detailed insight into disease pathology. Interest is now increasing in the use of plasma biomarkers for global organ diseases, which may be relevant in neurodegenerative disease, especially considering the link between nutrition, diet, and ageing. In particular, genomic, lipidomic, and proteomic biomarkers are increasingly interesting [164]. In particular, the study of genomics, i.e., the calorie-sensitive gene Sirt1, related to lipidomic and proteomic biomarkers, could be a sensitive tool in the assessment of a few chronic diseases which have showed association with AD (such as obesity and diabetes) [165]. Also, plasma biomarkers, due to their easy access, hold potential in terms of early diagnosis. Plasma Aβ levels seem to correlate with cognitive function and with CSF biomarkers [166], and the combination of clinical, imaging, and plasma markers can predict progression in MCI subjects [167]. This once again highlights the need for a clearer diagnostic route that does not rely solely on neuroimaging biomarkers.

Table 1. PET tracers in AD.

Target	Tracer	Clinical Correlates in AD	Ref.
Amyloid-β	18F-Florebetapir	Has demonstrated high sensitivity and specificity (92% and 100%, respectively) in detecting Aβ pathology	[13,14]
	18F-Florbetaben	High affinity for fibrillary Aβ, selectively labelled Aβ plaques, and cerebral amyloid angiopathy in tissue sections from patients with AD	[15]
	18F-Flutemetamol	In phase I and II studies, was able to differentiate between patients with AD and healthy controls	[17,18]
Tau protein	18F-flortaucipir, 18F-THK5351, 18F-THK5317, 11C-PBB3	Bind to neurofibrillary tangles with high selectivity and high signal-to-background ratio. Used for early detection of nerve fiber lesions in patients with AD	[39]
Microglial activation	11C-PK11195	Used to investigate neuroinflammation in vivo. There is an increase of binding in the temporal lobe of AD patients.	[73,74]
	11C-DAA1106, 11C-PBR28, 18F-FEPPA	Inverse correlation between microglia activation and glucose metabolism in AD patients as well as with hippocampal volume	[80,81]
	18F-DPA-714	Showed significantly higher global cortical binding in slower AD decliners compared to fast decliners	[79]
Astrocytes	[11C]-deuterium-L-deprenyl-[11C]-DED, [11C]-deprenyl-D2	In AD and amyloid-positive MCI, increased binding was observed in the frontal, parietal, and temporal cortices and regional correlation between 11C-DED uptake and amyloid burden	[59,86]
Glucose Metabolism	18F-FDG	Reduced uptake in posterior cingulate, hippocampi, and medial temporal structures is typical in AD and MCI, with a subsequent spreading to the whole cortex as the disease progresses. The reduction in glucose metabolism in regions like precuneus and posterior cingulate has been demonstrated to be associated with the severity of the cognitive impairment	[121, 126]

Table 2. MRI correlates in AD.

Target	Sequences	Clinical Correlates in AD	Ref.
Blood–brain barrier (BBB)	Dynamic susceptibility contrast-enhanced MRI (DSC-MRI)	Degeneration of the BBB has been demonstrated in the hippocampus with the ageing process; however, that has been seen to appear earlier in subjects with MCI when compared with cognitively intact subjects	[91]
	Dynamic contrast-enhanced MRI (DCE-MRI)	Significantly higher BBB leakage rate in AD compared to controls in the grey matter; the leakage volume was significantly higher in the grey matter, in the normal-appearing white matter, and in the cortex	[95]
Brain atrophy	Three-dimensional (3D) T1-weighted magnetisation-prepared rapid acquisition gradient-echo (T1-MPRAGE) sequence	Structural brain changes are accurately consistent with Braak stages of neurofibrillary tangle deposition and neuropsychological deficits. Rates of change in several structural measures, including whole-brain, entorhinal cortex, hippocampus, and temporal lobe volumes, correlate closely with changes in cognitive performance, validating atrophy in these regions as markers of AD.	[98–104]

Funding: This research received no external funding.

Conflicts of Interest: The authors declare no conflict of interest.

References

1. Prince, M.; Wimo, A.; Guerchet, M.; Ali, G.; Wu, Y.; Prina, M. *World Alzheimer Report 2015—The Global Impact of Dementia: An Analysis of Prevalence, Incidence, Cost and Trends*; Alzheimer's Disease International (ADI): London, UK, 2015.
2. Alzheimer's Association. 2018 Alzheimer's disease facts and figures. *Alzheimer's Dement.* **2018**, *14*, 367–429. [CrossRef]
3. Scheltens, P.; Blennow, K.; Breteler, M.M.; de Strooper, B.; Frisoni, G.B.; Salloway, S.; Van der Flier, W.M. Alzheimer's disease. *Lancet* **2016**, *388*, 505–517. [CrossRef]
4. McKhann, G.M.; Knopman, D.S.; Chertkow, H.; Hyman, B.T.; Jack, C.R., Jr.; Kawas, C.H.; Klunk, W.E.; Koroshetz, W.J.; Manly, J.J.; Mayeux, R.; et al. The diagnosis of dementia due to Alzheimer's disease: Recommendations from the National Institute on Aging-Alzheimer's Association workgroups on diagnostic guidelines for Alzheimer's disease. *Alzheimer's Dement.* **2011**, *7*, 263–269. [CrossRef] [PubMed]
5. Selkoe, D.J.; Hardy, J. The amyloid hypothesis of Alzheimer's disease at 25 years. *EMBO Mol. Med.* **2016**, *8*, 595–608. [CrossRef] [PubMed]
6. Small, S.A.; Duff, K. Linking Abeta and tau in late-onset Alzheimer's disease: A dual pathway hypothesis. *Neuron* **2008**, *60*, 534–542. [CrossRef] [PubMed]
7. Frisoni, G.B.; Boccardi, M.; Barkhof, F.; Blennow, K.; Cappa, S.; Chiotis, K.; Demonet, J.F.; Garibotto, V.; Giannakopoulos, P.; Gietl, A.; et al. Strategic roadmap for an early diagnosis of Alzheimer's disease based on biomarkers. *Lancet Neurol.* **2017**, *16*, 661–676. [CrossRef]
8. Kollack-Walker, S.; Liu, C.Y.; Fleisher, A.S. The Role of Neuroimaging in the Assessment of the Cognitively Impaired Elderly. *Neurol. Clin.* **2017**, *35*, 231–262. [CrossRef] [PubMed]
9. Frisoni, G.B.; Fox, N.C.; Jack, C.R., Jr.; Scheltens, P.; Thompson, P.M. The clinical use of structural MRI in Alzheimer disease. *Nat. Rev. Neurol.* **2010**, *6*, 67–77. [CrossRef] [PubMed]
10. Whitwell, J.L.; Josephs, K.A.; Murray, M.E.; Kantarci, K.; Przybelski, S.A.; Weigand, S.D.; Vemuri, P.; Senjem, M.L.; Parisi, J.E.; Knopman, D.S.; et al. MRI correlates of neurofibrillary tangle pathology at autopsy: A voxel-based morphometry study. *Neurology* **2008**, *71*, 743–749. [CrossRef] [PubMed]

11. Vemuri, P.; Wiste, H.J.; Weigand, S.D.; Shaw, L.M.; Trojanowski, J.Q.; Weiner, M.W.; Knopman, D.S.; Petersen, R.C.; Jack, C.R., Jr.; Alzheimer's Disease Neuroimaging Initiative. MRI and CSF biomarkers in normal, MCI, and AD subjects: Predicting future clinical change. *Neurology* **2009**, *73*, 294–301. [CrossRef] [PubMed]

12. Vemuri, P.; Whitwell, J.L.; Kantarci, K.; Josephs, K.A.; Parisi, J.E.; Shiung, M.S.; Knopman, D.S.; Boeve, B.F.; Petersen, R.C.; Dickson, D.W.; et al. Antemortem MRI based STructural Abnormality iNDex (STAND)-scores correlate with postmortem Braak neurofibrillary tangle stage. *Neuroimage* **2008**, *42*, 559–567. [CrossRef] [PubMed]

13. Sluimer, J.D.; van der Flier, W.M.; Karas, G.B.; Fox, N.C.; Scheltens, P.; Barkhof, F.; Vrenken, H. Whole-brain atrophy rate and cognitive decline: Longitudinal MR study of memory clinic patients. *Radiology* **2008**, *248*, 590–598. [CrossRef] [PubMed]

14. Cardenas, V.A.; Chao, L.L.; Studholme, C.; Yaffe, K.; Miller, B.L.; Madison, C.; Buckley, S.T.; Mungas, D.; Schuff, N.; Weiner, M.W. Brain atrophy associated with baseline and longitudinal measures of cognition. *Neurobiol. Aging* **2011**, *32*, 572–580. [CrossRef] [PubMed]

15. Jack, C.R., Jr.; Shiung, M.M.; Gunter, J.L.; O'Brien, P.C.; Weigand, S.D.; Knopman, D.S.; Boeve, B.F.; Ivnik, R.J.; Smith, G.E.; Cha, R.H.; et al. Comparison of different MRI brain atrophy rate measures with clinical disease progression in AD. *Neurology* **2004**, *62*, 591–600. [CrossRef] [PubMed]

16. Hua, X.; Lee, S.; Yanovsky, I.; Leow, A.D.; Chou, Y.Y.; Ho, A.J.; Gutman, B.; Toga, A.W.; Jack, C.R., Jr.; Bernstein, M.A.; et al. Optimizing power to track brain degeneration in Alzheimer's disease and mild cognitive impairment with tensor-based morphometry: An ADNI study of 515 subjects. *Neuroimage* **2009**, *48*, 668–681. [CrossRef] [PubMed]

17. Jack, C.R., Jr.; Lowe, V.J.; Weigand, S.D.; Wiste, H.J.; Senjem, M.L.; Knopman, D.S.; Shiung, M.M.; Gunter, J.L.; Boeve, B.F.; Kemp, B.J.; et al. Serial PIB and MRI in normal, mild cognitive impairment and Alzheimer's disease: Implications for sequence of pathological events in Alzheimer's disease. *Brain* **2009**, *132*, 1355–1365. [CrossRef] [PubMed]

18. Josephs, K.A.; Whitwell, J.L.; Ahmed, Z.; Shiung, M.M.; Weigand, S.D.; Knopman, D.S.; Boeve, B.F.; Parisi, J.E.; Petersen, R.C.; Dickson, D.W.; et al. Beta-amyloid burden is not associated with rates of brain atrophy. *Ann. Neurol.* **2008**, *63*, 204–212. [CrossRef] [PubMed]

19. Engler, H.; Forsberg, A.; Almkvist, O.; Blomquist, G.; Larsson, E.; Savitcheva, I.; Wall, A.; Ringheim, A.; Langstrom, B.; Nordberg, A. Two-year follow-up of amyloid deposition in patients with Alzheimer's disease. *Brain* **2006**, *129*, 2856–2866. [CrossRef] [PubMed]

20. Giorgio, A.; De Stefano, N. Clinical use of brain volumetry. *J. Magn. Reson. Imaging* **2013**, *37*, 1–14. [CrossRef] [PubMed]

21. Bozzali, M.; Filippi, M.; Magnani, G.; Cercignani, M.; Franceschi, M.; Schiatti, E.; Castiglioni, S.; Mossini, R.; Falautano, M.; Scotti, G.; et al. The contribution of voxel-based morphometry in staging patients with mild cognitive impairment. *Neurology* **2006**, *67*, 453–460. [CrossRef] [PubMed]

22. Dickerson, B.C.; Feczko, E.; Augustinack, J.C.; Pacheco, J.; Morris, J.C.; Fischl, B.; Buckner, R.L. Differential effects of aging and Alzheimer's disease on medial temporal lobe cortical thickness and surface area. *Neurobiol. Aging* **2009**, *30*, 432–440. [CrossRef] [PubMed]

23. Barnes, J.; Lewis, E.B.; Scahill, R.I.; Bartlett, J.W.; Frost, C.; Schott, J.M.; Rossor, M.N.; Fox, N.C. Automated measurement of hippocampal atrophy using fluid-registered serial MRI in AD and controls. *J. Comput. Assist. Tomogr.* **2007**, *31*, 581–587. [CrossRef] [PubMed]

24. Bishop, C.A.; Jenkinson, M.; Andersson, J.; Declerck, J.; Merhof, D. Novel Fast Marching for Automated Segmentation of the Hippocampus (FMASH): Method and validation on clinical data. *Neuroimage* **2011**, *55*, 1009–1019. [CrossRef] [PubMed]

25. Jack, C.R., Jr.; Barkhof, F.; Bernstein, M.A.; Cantillon, M.; Cole, P.E.; Decarli, C.; Dubois, B.; Duchesne, S.; Fox, N.C.; Frisoni, G.B.; et al. Steps to standardization and validation of hippocampal volumetry as a biomarker in clinical trials and diagnostic criterion for Alzheimer's disease. *Alzheimer's Dement.* **2011**, *7*, 474–485.e4. [CrossRef] [PubMed]

26. Jack, C.R., Jr.; Bernstein, M.A.; Borowski, B.J.; Gunter, J.L.; Fox, N.C.; Thompson, P.M.; Schuff, N.; Krueger, G.; Killiany, R.J.; Decarli, C.S.; et al. Update on the magnetic resonance imaging core of the Alzheimer's disease neuroimaging initiative. *Alzheimer's Dement.* **2010**, *6*, 212–220. [CrossRef] [PubMed]

27. Kehoe, E.G.; McNulty, J.P.; Mullins, P.G.; Bokde, A.L. Advances in MRI biomarkers for the diagnosis of Alzheimer's disease. *Biomark. Med.* **2014**, *8*, 1151–1169. [CrossRef] [PubMed]

28. Steen, R.G.; Mull, C.; McClure, R.; Hamer, R.M.; Lieberman, J.A. Brain volume in first-episode schizophrenia: Systematic review and meta-analysis of magnetic resonance imaging studies. *Br. J. Psychiatry* **2006**, *188*, 510–518. [CrossRef] [PubMed]

29. Arnone, D.; McIntosh, A.M.; Ebmeier, K.P.; Munafo, M.R.; Anderson, I.M. Magnetic resonance imaging studies in unipolar depression: Systematic review and meta-regression analyses. *Eur. Neuropsychopharmacol.* **2012**, *22*, 1–16. [CrossRef] [PubMed]

30. Frisoni, G.B.; Jack, C.R. Harmonization of magnetic resonance-based manual hippocampal segmentation: A mandatory step for wide clinical use. *Alzheimer's Dement.* **2011**, *7*, 171–174. [CrossRef] [PubMed]

31. Calsolaro, V.; Edison, P. Alterations in Glucose Metabolism in Alzheimer's Disease. *Recent Pat. Endocr. Metab. Immune Drug Discov.* **2016**, *10*, 31–39. [CrossRef] [PubMed]

32. Shah, K.; Desilva, S.; Abbruscato, T. The role of glucose transporters in brain disease: Diabetes and Alzheimer's Disease. *Int. J. Mol. Sci.* **2012**, *13*, 12629–12655. [CrossRef] [PubMed]

33. Femminella, G.D.; Edison, P. Evaluation of neuroprotective effect of glucagon-like peptide 1 analogs using neuroimaging. *Alzheimer's Dement.* **2014**, *10*, S55–S61. [CrossRef] [PubMed]

34. Herholz, K. Use of FDG PET as an imaging biomarker in clinical trials of Alzheimer's disease. *Biomark. Med.* **2012**, *6*, 431–439. [CrossRef] [PubMed]

35. Jack, C.R., Jr.; Knopman, D.S.; Jagust, W.J.; Shaw, L.M.; Aisen, P.S.; Weiner, M.W.; Petersen, R.C.; Trojanowski, J.Q. Hypothetical model of dynamic biomarkers of the Alzheimer's pathological cascade. *Lancet Neurol.* **2010**, *9*, 119–128. [CrossRef]

36. Chen, Z.C.; Zhong, C.J. Decoding Alzheimer's disease from perturbed cerebral glucose metabolism: Implications for diagnostic and therapeutic strategies. *Prog. Neurobiol.* **2013**, *108*, 21–43. [CrossRef] [PubMed]

37. Laforce, R., Jr.; Soucy, J.P.; Sellami, L.; Dallaire-Theroux, C.; Brunet, F.; Bergeron, D.; Miller, B.L.; Ossenkoppele, R. Molecular imaging in dementia: Past, present, and future. *Alzheimer's Dement.* **2018**, *14*, 1522–1552. [CrossRef] [PubMed]

38. Furst, A.J.; Rabinovici, G.D.; Rostomian, A.H.; Steed, T.; Alkalay, A.; Racine, C.; Miller, B.L.; Jagust, W.J. Cognition, glucose metabolism and amyloid burden in Alzheimer's disease. *Neurobiol. Aging* **2012**, *33*, 215–225. [CrossRef] [PubMed]

39. Edison, P.; Archer, H.A.; Hinz, R.; Hammers, A.; Pavese, N.; Tai, Y.F.; Hotton, G.; Cutler, D.; Fox, N.; Kennedy, A.; et al. Amyloid, hypometabolism, and cognition in Alzheimer disease: An [^{11}C]PIB and [^{18}F]FDG PET study. *Neurology* **2007**, *68*, 501–508. [CrossRef] [PubMed]

40. Langbaum, J.B.S.; Chen, K.; Lee, W.; Reschke, C.; Bandy, D.; Fleisher, A.S.; Alexander, G.E.; Foster, N.L.; Weiner, M.W.; Koeppe, R.A.; et al. Categorical and correlational analyses of baseline fluorodeoxyglucose positron emission tomography images from the Alzheimer's Disease Neuroimaging Initiative (ADNI). *Neuroimage* **2009**, *45*, 1107–1116. [CrossRef] [PubMed]

41. Ewers, M.; Insel, P.S.; Stern, Y.; Weiner, M.W.; Alzheimer's Disease Neuroimaging Initiative (ADNI). Cognitive reserve associated with FDG-PET in preclinical Alzheimer disease. *Neurology* **2013**, *80*, 1194–1201. [CrossRef] [PubMed]

42. Jagust, W.; Reed, B.; Mungas, D.; Ellis, W.; DeCarli, C. What does fluorodeoxyglucose PET imaging add to a clinical diagnosis of dementia? *Neurology* **2007**, *69*, 871–877. [CrossRef] [PubMed]

43. Drzezga, A.; Lautenschlager, N.; Siebner, H.; Riemenschneider, M.; Willoch, F.; Minoshima, S.; Schwaiger, M.; Kurz, A. Cerebral metabolic changes accompanying conversion of mild cognitive impairment into Alzheimer's disease: A PET follow-up study. *Eur. J. Nucl. Med. Mol. Imaging* **2003**, *30*, 1104–1113. [PubMed]

44. Anchisi, D.; Borroni, B.; Franceschi, M.; Kerrouche, N.; Kalbe, E.; Beuthien-Beumann, B.; Cappa, S.; Lenz, O.; Ludecke, S.; Marcone, A.; et al. Heterogeneity of brain glucose metabolism in mild cognitive impairment and clinical progression to Alzheimer disease. *Arch. Neurol.* **2005**, *62*, 1728–1733. [CrossRef] [PubMed]

45. Silverman, D.H.; Small, G.W.; Chang, C.Y.; Lu, C.S.; Kung De Aburto, M.A.; Chen, W.; Czernin, J.; Rapoport, S.I.; Pietrini, P.; Alexander, G.E.; et al. Positron emission tomography in evaluation of dementia: Regional brain metabolism and long-term outcome. *JAMA* **2001**, *286*, 2120–2127. [CrossRef] [PubMed]

46. Smailagic, N.; Vacante, M.; Hyde, C.; Martin, S.; Ukoumunne, O.; Sachpekidis, C. [18F]-FDG PET for the early diagnosis of Alzheimer's disease dementia and other dementias in people with mild cognitive impairment (MCI). *Cochrane Database Syst. Rev.* **2015**, *1*, CD010632. [CrossRef] [PubMed]

47. Pike, K.E.; Savage, G.; Villemagne, V.L.; Ng, S.; Moss, S.A.; Maruff, P.; Mathis, C.A.; Klunk, W.E.; Masters, C.L.; Rowe, C.C. Beta-amyloid imaging and memory in non-demented individuals: Evidence for preclinical Alzheimer's disease. *Brain* **2007**, *130*, 2837–2844. [CrossRef] [PubMed]

48. Johnson, K.A.; Minoshima, S.; Bohnen, N.I.; Donohoe, K.J.; Foster, N.L.; Herscovitch, P.; Karlawish, J.H.; Rowe, C.C.; Hedrick, S.; Pappas, V.; et al. Update on appropriate use criteria for amyloid PET imaging: Dementia experts, mild cognitive impairment, and education. Amyloid Imaging Task Force of the Alzheimer's Association and Society for Nuclear Medicine and Molecular Imaging. *Alzheimer's Dement.* **2013**, *9*, e106–e109. [CrossRef] [PubMed]

49. Villemagne, V.L.; Dore, V.; Burnham, S.C.; Masters, C.L.; Rowe, C.C. Imaging tau and amyloid-beta proteinopathies in Alzheimer disease and other conditions. *Nat. Rev. Neurol.* **2018**, *14*, 225–236. [CrossRef] [PubMed]

50. Johnson, K.A.; Minoshima, S.; Bohnen, N.I.; Donohoe, K.J.; Foster, N.L.; Herscovitch, P.; Karlawish, J.H.; Rowe, C.C.; Carrillo, M.C.; Hartley, D.M.; et al. Appropriate use criteria for amyloid PET: A report of the Amyloid Imaging Task Force, the Society of Nuclear Medicine and Molecular Imaging, and the Alzheimer's Association. *Alzheimer's Dement.* **2013**, *9*, E1–E16. [CrossRef] [PubMed]

51. Apostolova, L.G.; Haider, J.M.; Goukasian, N.; Rabinovici, G.D.; Chetelat, G.; Ringman, J.M.; Kremen, S.; Grill, J.D.; Restrepo, L.; Mendez, M.F.; et al. Critical review of the Appropriate Use Criteria for amyloid imaging: Effect on diagnosis and patient care. *Alzheimer's Dement.* **2016**, *5*, 15–22. [CrossRef] [PubMed]

52. Mathis, C.A.; Bacskai, B.J.; Kajdasz, S.T.; McLellan, M.E.; Frosch, M.P.; Hyman, B.T.; Holt, D.P.; Wang, Y.; Huang, G.F.; Debnath, M.L.; et al. A lipophilic thioflavin-T derivative for positron emission tomography (PET) imaging of amyloid in brain. *Bioorg. Med. Chem. Lett.* **2002**, *12*, 295–298. [CrossRef]

53. Lister-James, J.; Pontecorvo, M.J.; Clark, C.; Joshi, A.D.; Mintun, M.A.; Zhang, W.; Lim, N.; Zhuang, Z.; Golding, G.; Choi, S.R.; et al. Florbetapir f-18: A histopathologically validated Beta-amyloid positron emission tomography imaging agent. *Semin. Nucl. Med.* **2011**, *41*, 300–304. [CrossRef] [PubMed]

54. Sperling, R.A.; Johnson, K.A.; Doraiswamy, P.M.; Reiman, E.M.; Fleisher, A.S.; Sabbagh, M.N.; Sadowsky, C.H.; Carpenter, A.; Davis, M.D.; Lu, M.; et al. Amyloid deposition detected with florbetapir F 18 ((18)F-AV-45) is related to lower episodic memory performance in clinically normal older individuals. *Neurobiol. Aging* **2013**, *34*, 822–831. [CrossRef] [PubMed]

55. Fleisher, A.S.; Chen, K.; Liu, X.; Roontiva, A.; Thiyyagura, P.; Ayutyanont, N.; Joshi, A.D.; Clark, C.M.; Mintun, M.A.; Pontecorvo, M.J.; et al. Using positron emission tomography and florbetapir F18 to image cortical amyloid in patients with mild cognitive impairment or dementia due to Alzheimer disease. *Arch. Neurol.* **2011**, *68*, 1404–1411. [CrossRef] [PubMed]

56. Clark, C.M.; Schneider, J.A.; Bedell, B.J.; Beach, T.G.; Bilker, W.B.; Mintun, M.A.; Pontecorvo, M.J.; Hefti, F.; Carpenter, A.P.; Flitter, M.L.; et al. Use of florbetapir-PET for imaging beta-amyloid pathology. *JAMA* **2011**, *305*, 275–283. [CrossRef] [PubMed]

57. Clark, C.M.; Pontecorvo, M.J.; Beach, T.G.; Bedell, B.J.; Coleman, R.E.; Doraiswamy, P.M.; Fleisher, A.S.; Reiman, E.M.; Sabbagh, M.N.; Sadowsky, C.H.; et al. Cerebral PET with florbetapir compared with neuropathology at autopsy for detection of neuritic amyloid-beta plaques: A prospective cohort study. *Lancet Neurol.* **2012**, *11*, 669–678. [CrossRef]

58. Zhang, W.; Oya, S.; Kung, M.P.; Hou, C.; Maier, D.L.; Kung, H.F. F-18 stilbenes as PET imaging agents for detecting beta-amyloid plaques in the brain. *J. Med. Chem.* **2005**, *48*, 5980–5988. [CrossRef] [PubMed]

59. Rowe, C.C.; Ackerman, U.; Browne, W.; Mulligan, R.; Pike, K.L.; O'Keefe, G.; Tochon-Danguy, H.; Chan, G.; Berlangieri, S.U.; Jones, G.; et al. Imaging of amyloid beta in Alzheimer's disease with 18F-BAY94-9172, a novel PET tracer: Proof of mechanism. *Lancet Neurol.* **2008**, *7*, 129–135. [CrossRef]

60. Vandenberghe, R.; Van Laere, K.; Ivanoiu, A.; Salmon, E.; Bastin, C.; Triau, E.; Hasselbalch, S.; Law, I.; Andersen, A.; Korner, A.; et al. 18F-flutemetamol amyloid imaging in Alzheimer disease and mild cognitive impairment: A phase 2 trial. *Ann. Neurol.* **2010**, *68*, 319–329. [CrossRef] [PubMed]

61. Nelissen, N.; Van Laere, K.; Thurfjell, L.; Owenius, R.; Vandenbulcke, M.; Koole, M.; Bormans, G.; Brooks, D.J.; Vandenberghe, R. Phase 1 study of the Pittsburgh compound B derivative 18F-flutemetamol in healthy volunteers and patients with probable Alzheimer disease. *J. Nucl. Med.* **2009**, *50*, 1251–1259. [CrossRef] [PubMed]

62. Thurfjell, L.; Lotjonen, J.; Lundqvist, R.; Koikkalainen, J.; Soininen, H.; Waldemar, G.; Brooks, D.J.; Vandenberghe, R. Combination of biomarkers: PET [^{18}F]flutemetamol imaging and structural MRI in dementia and mild cognitive impairment. *Neurodegener. Dis.* **2012**, *10*, 246–249. [CrossRef] [PubMed]

63. Ye, L.; Morgenstern, J.L.; Gee, A.D.; Hong, G.; Brown, J.; Lockhart, A. Delineation of positron emission tomography imaging agent binding sites on beta-amyloid peptide fibrils. *J. Biol. Chem.* **2005**, *280*, 23599–23604. [CrossRef] [PubMed]

64. Cohen, A.D.; Rabinovici, G.D.; Mathis, C.A.; Jagust, W.J.; Klunk, W.E.; Ikonomovic, M.D. Using Pittsburgh Compound B for in vivo PET imaging of fibrillar amyloid-beta. *Adv. Pharmacol.* **2012**, *64*, 27–81. [PubMed]

65. Braak, H.; Braak, E. Frequency of stages of Alzheimer-related lesions in different age categories. *Neurobiol. Aging* **1997**, *18*, 351–357. [CrossRef]

66. Villain, N.; Chetelat, G.; Grassiot, B.; Bourgeat, P.; Jones, G.; Ellis, K.A.; Ames, D.; Martins, R.N.; Eustache, F.; Salvado, O.; et al. Regional dynamics of amyloid-beta deposition in healthy elderly, mild cognitive impairment and Alzheimer's disease: A voxelwise PiB-PET longitudinal study. *Brain* **2012**, *135*, 2126–2139. [CrossRef] [PubMed]

67. Vlassenko, A.G.; Mintun, M.A.; Xiong, C.; Sheline, Y.I.; Goate, A.M.; Benzinger, T.L.; Morris, J.C. Amyloid-beta plaque growth in cognitively normal adults: Longitudinal [^{11}C]Pittsburgh compound B data. *Ann. Neurol.* **2011**, *70*, 857–861. [CrossRef] [PubMed]

68. Ng, S.Y.; Villemagne, V.L.; Masters, C.L.; Rowe, C.C. Evaluating atypical dementia syndromes using positron emission tomography with carbon 11 labeled Pittsburgh Compound B. *Arch. Neurol.* **2007**, *64*, 1140–1144. [CrossRef] [PubMed]

69. Rabinovici, G.D.; Jagust, W.J.; Furst, A.J.; Ogar, J.M.; Racine, C.A.; Mormino, E.C.; O'Neil, J.P.; Lal, R.A.; Dronkers, N.F.; Miller, B.L.; et al. Abeta amyloid and glucose metabolism in three variants of primary progressive aphasia. *Ann. Neurol.* **2008**, *64*, 388–401. [CrossRef] [PubMed]

70. Drzezga, A.; Grimmer, T.; Henriksen, G.; Stangier, I.; Perneczky, R.; Diehl-Schmid, J.; Mathis, C.A.; Klunk, W.E.; Price, J.; DeKosky, S.; et al. Imaging of amyloid plaques and cerebral glucose metabolism in semantic dementia and Alzheimer's disease. *Neuroimage* **2008**, *39*, 619–633. [CrossRef] [PubMed]

71. Engler, H.; Santillo, A.F.; Wang, S.X.; Lindau, M.; Savitcheva, I.; Nordberg, A.; Lannfelt, L.; Langstrom, B.; Kilander, L. In vivo amyloid imaging with PET in frontotemporal dementia. *Eur. J. Nucl. Med. Mol. Imaging* **2008**, *35*, 100–106. [CrossRef] [PubMed]

72. Edison, P.; Rowe, C.C.; Rinne, J.O.; Ng, S.; Ahmed, I.; Kemppainen, N.; Villemagne, V.L.; O'Keefe, G.; Nagren, K.; Chaudhury, K.R.; et al. Amyloid load in Parkinson's disease dementia and Lewy body dementia measured with [^{11}C]PIB positron emission tomography. *J. Neurol. Neurosurg. Psychiatry* **2008**, *79*, 1331–1338. [CrossRef] [PubMed]

73. Kalaitzakis, M.E.; Walls, A.J.; Pearce, R.K.; Gentleman, S.M. Striatal Abeta peptide deposition mirrors dementia and differentiates DLB and PDD from other parkinsonian syndromes. *Neurobiol. Dis.* **2011**, *41*, 377–384. [CrossRef] [PubMed]

74. Delacourte, A.; Sergeant, N.; Wattez, A.; Maurage, C.A.; Lebert, F.; Pasquier, F.; David, J.P. Tau aggregation in the hippocampal formation: An ageing or a pathological process? *Exp. Gerontol.* **2002**, *37*, 1291–1296. [CrossRef]

75. Chiotis, K.; Saint-Aubert, L.; Rodriguez-Vieitez, E.; Leuzy, A.; Almkvist, O.; Savitcheva, I.; Jonasson, M.; Lubberink, M.; Wall, A.; Antoni, G.; et al. Longitudinal changes of tau PET imaging in relation to hypometabolism in prodromal and Alzheimer's disease dementia. *Mol. Psychiatry* **2018**, *23*, 1666–1673. [CrossRef] [PubMed]

76. Chiotis, K.; Saint-Aubert, L.; Savitcheva, I.; Jelic, V.; Andersen, P.; Jonasson, M.; Eriksson, J.; Lubberink, M.; Almkvist, O.; Wall, A.; et al. Imaging in-vivo tau pathology in Alzheimer's disease with THK5317 PET in a multimodal paradigm. *Eur. J. Nucl. Med. Mol. Imaging* **2016**, *43*, 1686–1699. [CrossRef] [PubMed]

77. Saint-Aubert, L.; Almkvist, O.; Chiotis, K.; Almeida, R.; Wall, A.; Nordberg, A. Regional tau deposition measured by [^{18}F]THK5317 positron emission tomography is associated to cognition via glucose metabolism in Alzheimer's disease. *Alzheimer's Res. Ther.* **2016**, *8*, 38. [CrossRef] [PubMed]

78. Royall, D.R. Location, location, location! *Neurobiol. Aging* **2007**, *28*, 1481–1482, discussion 1483. [CrossRef] [PubMed]

79. Delacourte, A.; David, J.P.; Sergeant, N.; Buee, L.; Wattez, A.; Vermersch, P.; Ghozali, F.; Fallet-Bianco, C.; Pasquier, F.; Lebert, F.; et al. The biochemical pathway of neurofibrillary degeneration in aging and Alzheimer's disease. *Neurology* **1999**, *52*, 1158–1165. [CrossRef] [PubMed]

80. Scholl, M.; Lockhart, S.N.; Schonhaut, D.R.; O'Neil, J.P.; Janabi, M.; Ossenkoppele, R.; Baker, S.L.; Vogel, J.W.; Faria, J.; Schwimmer, H.D.; et al. PET Imaging of Tau Deposition in the Aging Human Brain. *Neuron* **2016**, *89*, 971–982. [CrossRef] [PubMed]

81. Pontecorvo, M.J.; Devous, M.D., Sr.; Navitsky, M.; Lu, M.; Salloway, S.; Schaerf, F.W.; Jennings, D.; Arora, A.K.; McGeehan, A.; Lim, N.C.; et al. Relationships between flortaucipir PET tau binding and amyloid burden, clinical diagnosis, age and cognition. *Brain* **2017**, *140*, 748–763. [CrossRef] [PubMed]

82. Johnson, K.A.; Schultz, A.; Betensky, R.A.; Becker, J.A.; Sepulcre, J.; Rentz, D.; Mormino, E.; Chhatwal, J.; Amariglio, R.; Papp, K.; et al. Tau positron emission tomographic imaging in aging and early Alzheimer disease. *Ann. Neurol.* **2016**, *79*, 110–119. [CrossRef] [PubMed]

83. Lockhart, S.N.; Baker, S.L.; Okamura, N.; Furukawa, K.; Ishiki, A.; Furumoto, S.; Tashiro, M.; Yanai, K.; Arai, H.; Kudo, Y.; et al. Dynamic PET Measures of Tau Accumulation in Cognitively Normal Older Adults and Alzheimer's Disease Patients Measured Using [^{18}F] THK-5351. *PLoS ONE* **2016**, *11*, e0158460. [CrossRef] [PubMed]

84. Harada, R.; Okamura, N.; Furumoto, S.; Furukawa, K.; Ishiki, A.; Tomita, N.; Tago, T.; Hiraoka, K.; Watanuki, S.; Shidahara, M.; et al. 18F-THK5351: A Novel PET Radiotracer for Imaging Neurofibrillary Pathology in Alzheimer Disease. *J. Nucl. Med.* **2016**, *57*, 208–214. [CrossRef] [PubMed]

85. Cho, H.; Choi, J.Y.; Hwang, M.S.; Lee, J.H.; Kim, Y.J.; Lee, H.M.; Lyoo, C.H.; Ryu, Y.H.; Lee, M.S. Tau PET in Alzheimer disease and mild cognitive impairment. *Neurology* **2016**, *87*, 375–383. [CrossRef] [PubMed]

86. Ossenkoppele, R.; Cohn-Sheehy, B.I.; La Joie, R.; Vogel, J.W.; Moller, C.; Lehmann, M.; van Berckel, B.N.; Seeley, W.W.; Pijnenburg, Y.A.; Gorno-Tempini, M.L.; et al. Atrophy patterns in early clinical stages across distinct phenotypes of Alzheimer's disease. *Hum. Brain Mapp.* **2015**, *36*, 4421–4437. [CrossRef] [PubMed]

87. Tomlinson, B.E.; Blessed, G.; Roth, M. Observations on the brains of demented old people. *J. Neurol. Sci.* **1970**, *11*, 205–242. [CrossRef]

88. Jellinger, K.A.; Alafuzoff, I.; Attems, J.; Beach, T.G.; Cairns, N.J.; Crary, J.F.; Dickson, D.W.; Hof, P.R.; Hyman, B.T.; Jack, C.R., Jr.; et al. PART, a distinct tauopathy, different from classical sporadic Alzheimer disease. *Acta Neuropathol.* **2015**, *129*, 757–762. [CrossRef] [PubMed]

89. Wang, L.; Benzinger, T.L.; Su, Y.; Christensen, J.; Friedrichsen, K.; Aldea, P.; McConathy, J.; Cairns, N.J.; Fagan, A.M.; Morris, J.C.; et al. Evaluation of Tau Imaging in Staging Alzheimer Disease and Revealing Interactions Between beta-Amyloid and Tauopathy. *JAMA Neurol.* **2016**, *73*, 1070–1077. [CrossRef] [PubMed]

90. Price, J.L.; Morris, J.C. Tangles and plaques in nondemented aging and "preclinical" Alzheimer's disease. *Ann. Neurol.* **1999**, *45*, 358–368. [CrossRef]

91. Ishiki, A.; Harada, R.; Okamura, N.; Tomita, N.; Rowe, C.C.; Villemagne, V.L.; Yanai, K.; Kudo, Y.; Arai, H.; Furumoto, S.; et al. Tau imaging with [^{18}F]THK-5351 in progressive supranuclear palsy. *Eur. J. Neurol.* **2017**, *24*, 130–136. [CrossRef] [PubMed]

92. Josephs, K.A.; Holton, J.L.; Rossor, M.N.; Godbolt, A.K.; Ozawa, T.; Strand, K.; Khan, N.; Al-Sarraj, S.; Revesz, T. Frontotemporal lobar degeneration and ubiquitin immunohistochemistry. *Neuropathol. Appl. Neurobiol.* **2004**, *30*, 369–373. [CrossRef] [PubMed]

93. Ikonomovic, M.D.; Abrahamson, E.E.; Price, J.C.; Mathis, C.A.; Klunk, W.E. [F-18]AV-1451 positron emission tomography retention in choroid plexus: More than "off-target" binding. *Ann. Neurol.* **2016**, *80*, 307–308. [CrossRef] [PubMed]

94. Marquie, M.; Normandin, M.D.; Meltzer, A.C.; Siao Tick Chong, M.; Andrea, N.V.; Anton-Fernandez, A.; Klunk, W.E.; Mathis, C.A.; Ikonomovic, M.D.; Debnath, M.; et al. Pathological correlations of [F-18]-AV-1451 imaging in non-alzheimer tauopathies. *Ann. Neurol.* **2017**, *81*, 117–128. [CrossRef] [PubMed]

95. Chen, R.; Chen, C.P.; Preston, J.E. Effects of transthyretin on thyroxine and beta-amyloid removal from cerebrospinal fluid in mice. *Clin. Exp. Pharmacol. Physiol.* **2016**, *43*, 844–850. [CrossRef] [PubMed]

96. Lowe, V.J.; Curran, G.; Fang, P.; Liesinger, A.M.; Josephs, K.A.; Parisi, J.E.; Kantarci, K.; Boeve, B.F.; Pandey, M.K.; Bruinsma, T.; et al. An autoradiographic evaluation of AV-1451 Tau PET in dementia. *Acta Neuropathol. Commun.* **2016**, *4*, 58. [CrossRef] [PubMed]

97. Beach, T.G.; Monsell, S.E.; Phillips, L.E.; Kukull, W. Accuracy of the clinical diagnosis of Alzheimer disease at National Institute on Aging Alzheimer Disease Centers, 2005–2010. *J. Neuropathol. Exp. Neurol.* **2012**, *71*, 266–273. [CrossRef] [PubMed]

98. Marquie, M.; Normandin, M.D.; Vanderburg, C.R.; Costantino, I.M.; Bien, E.A.; Rycyna, L.G.; Klunk, W.E.; Mathis, C.A.; Ikonomovic, M.D.; Debnath, M.L.; et al. Validating novel tau positron emission tomography tracer [F-18]-AV-1451 (T807) on postmortem brain tissue. *Ann. Neurol.* **2015**, *78*, 787–800. [CrossRef] [PubMed]

99. Ng, K.P.; Pascoal, T.A.; Mathotaarachchi, S.; Therriault, J.; Kang, M.S.; Shin, M.; Guiot, M.C.; Guo, Q.; Harada, R.; Comley, R.A.; et al. Monoamine oxidase B inhibitor, selegiline, reduces ^{18}F-THK5351 uptake in the human brain. *Alzheimer's Res. Ther.* **2017**, *9*, 25. [CrossRef] [PubMed]

100. Gobbi, L.C.; Knust, H.; Korner, M.; Honer, M.; Czech, C.; Belli, S.; Muri, D.; Edelmann, M.R.; Hartung, T.; Erbsmehl, I.; et al. Identification of Three Novel Radiotracers for Imaging Aggregated Tau in Alzheimer's Disease with Positron Emission Tomography. *J. Med. Chem.* **2017**, *60*, 7350–7370. [CrossRef] [PubMed]

101. Walji, A.M.; Hostetler, E.D.; Selnick, H.; Zeng, Z.; Miller, P.; Bennacef, I.; Salinas, C.; Connolly, B.; Gantert, L.; Holahan, M.; et al. Discovery of 6-(Fluoro-(18)F)-3-(1H-pyrrolo[2,3-c]pyridin-1-yl)isoquinolin-5-amine ([(18)F]-MK-6240): A Positron Emission Tomography (PET) Imaging Agent for Quantification of Neurofibrillary Tangles (NFTs). *J. Med. Chem.* **2016**, *59*, 4778–4789. [CrossRef] [PubMed]

102. Stephens, A.; Kroth, H.; Berndt, M.; Capotosti, F.; Mueller, A. Characterization of novel PET tracers for the assessment of tau pathology in Alzheimer's disease and other tauopathies. *Neurodegener. Dis.* **2017**, *17*.

103. Calsolaro, V.; Edison, P. Neuroinflammation in Alzheimer's disease: Current evidence and future directions. *Alzheimer's Dement.* **2016**, *12*, 719–732. [CrossRef] [PubMed]

104. Cuyvers, E.; Sleegers, K. Genetic variations underlying Alzheimer's disease: Evidence from genome-wide association studies and beyond. *Lancet Neurol.* **2016**, *15*, 857–868. [CrossRef]

105. Cerami, C.; Iaccarino, L.; Perani, D. Molecular Imaging of Neuroinflammation in Neurodegenerative Dementias: The Role of In Vivo PET Imaging. *Int. J. Mol. Sci.* **2017**, *18*, 993. [CrossRef] [PubMed]

106. Edison, P.; Brooks, D.J. Role of Neuroinflammation in the Trajectory of Alzheimer's Disease and in vivo Quantification Using PET. *J. Alzheimer's Dis.* **2018**, *64*, S339–S351. [CrossRef] [PubMed]

107. Heppner, F.L.; Ransohoff, R.M.; Becher, B. Immune attack: The role of inflammation in Alzheimer disease. *Nat. Rev. Neurosci.* **2015**, *16*, 358–372. [CrossRef] [PubMed]

108. Heneka, M.T.; Carson, M.J.; El Khoury, J.; Landreth, G.E.; Brosseron, F.; Feinstein, D.L.; Jacobs, A.H.; Wyss-Coray, T.; Vitorica, J.; Ransohoff, R.M.; et al. Neuroinflammation in Alzheimer's disease. *Lancet Neurol.* **2015**, *14*, 388–405. [CrossRef]

109. Knezevic, D.; Mizrahi, R. Molecular imaging of neuroinflammation in Alzheimer's disease and mild cognitive impairment. *Prog. Neuropsychopharmacol. Biol. Psychiatry* **2018**, *80*, 123–131. [CrossRef] [PubMed]

110. Schwab, C.; McGeer, P.L. Inflammatory aspects of Alzheimer disease and other neurodegenerative disorders. *J. Alzheimer's Dis.* **2008**, *13*, 359–369. [CrossRef]

111. Rogers, J.; Luber-Narod, J.; Styren, S.D.; Civin, W.H. Expression of immune system-associated antigens by cells of the human central nervous system: Relationship to the pathology of Alzheimer's disease. *Neurobiol. Aging* **1988**, *9*, 339–349. [CrossRef]

112. Heneka, M.T.; Kummer, M.P.; Latz, E. Innate immune activation in neurodegenerative disease. *Nat. Rev. Immunol.* **2014**, *14*, 463–477. [CrossRef] [PubMed]

113. Ransohoff, R.M. A polarizing question: Do M1 and M2 microglia exist? *Nat. Neurosci.* **2016**, *19*, 987–991. [CrossRef] [PubMed]

114. Lagarde, J.; Sarazin, M.; Bottlaender, M. In vivo PET imaging of neuroinflammation in Alzheimer's disease. *J. Neural Transm.* **2018**, *125*, 847–867. [CrossRef] [PubMed]

115. Vivash, L.; O'Brien, T.J. Imaging Microglial Activation with TSPO PET: Lighting Up Neurologic Diseases? *J. Nucl. Med.* **2016**, *57*, 165–168. [CrossRef] [PubMed]

116. Politis, M.; Giannetti, P.; Su, P.; Turkheimer, F.; Keihaninejad, S.; Wu, K.; Waldman, A.; Malik, O.; Matthews, P.M.; Reynolds, R.; et al. Increased PK11195 PET binding in the cortex of patients with MS correlates with disability. *Neurology* **2012**, *79*, 523–530. [CrossRef] [PubMed]

117. Gerhard, A.; Pavese, N.; Hotton, G.; Turkheimer, F.; Es, M.; Hammers, A.; Eggert, K.; Oertel, W.; Banati, R.B.; Brooks, D.J. In vivo imaging of microglial activation with [^{11}C](R)-PK11195 PET in idiopathic Parkinson's disease. *Neurobiol. Dis.* **2006**, *21*, 404–412. [CrossRef] [PubMed]

118. Owen, D.R.; Yeo, A.J.; Gunn, R.N.; Song, K.; Wadsworth, G.; Lewis, A.; Rhodes, C.; Pulford, D.J.; Bennacef, I.; Parker, C.A.; et al. An 18-kDa translocator protein (TSPO) polymorphism explains differences in binding affinity of the PET radioligand PBR28. *J. Cereb. Blood Flow Metab.* **2012**, *32*, 1–5. [CrossRef] [PubMed]

119. Cagnin, A.; Brooks, D.J.; Kennedy, A.M.; Gunn, R.N.; Myers, R.; Turkheimer, F.E.; Jones, T.; Banati, R.B. In-vivo measurement of activated microglia in dementia. *Lancet* **2001**, *358*, 461–467. [CrossRef]

120. Wiley, C.A.; Lopresti, B.J.; Venneti, S.; Price, J.; Klunk, W.E.; DeKosky, S.T.; Mathis, C.A. Carbon 11-labeled Pittsburgh Compound B and carbon 11-labeled (R)-PK11195 positron emission tomographic imaging in Alzheimer disease. *Arch. Neurol.* **2009**, *66*, 60–67. [CrossRef] [PubMed]

121. Yasuno, F.; Ota, M.; Kosaka, J.; Ito, H.; Higuchi, M.; Doronbekov, T.K.; Nozaki, S.; Fujimura, Y.; Koeda, M.; Asada, T.; et al. Increased binding of peripheral benzodiazepine receptor in Alzheimer's disease measured by positron emission tomography with [11C]DAA1106. *Biol. Psychiatry* **2008**, *64*, 835–841. [CrossRef] [PubMed]

122. Kreisl, W.C.; Lyoo, C.H.; McGwier, M.; Snow, J.; Jenko, K.J.; Kimura, N.; Corona, W.; Morse, C.L.; Zoghbi, S.S.; Pike, V.W.; et al. In vivo radioligand binding to translocator protein correlates with severity of Alzheimer's disease. *Brain* **2013**, *136*, 2228–2238. [CrossRef] [PubMed]

123. Suridjan, I.; Pollock, B.G.; Verhoeff, N.P.; Voineskos, A.N.; Chow, T.; Rusjan, P.M.; Lobaugh, N.J.; Houle, S.; Mulsant, B.H.; Mizrahi, R. In-vivo imaging of grey and white matter neuroinflammation in Alzheimer's disease: A positron emission tomography study with a novel radioligand, [18F]-FEPPA. *Mol. Psychiatry* **2015**, *20*, 1579–1587. [CrossRef] [PubMed]

124. Edison, P.; Archer, H.A.; Gerhard, A.; Hinz, R.; Pavese, N.; Turkheimer, F.E.; Hammers, A.; Tai, Y.F.; Fox, N.; Kennedy, A.; et al. Microglia, amyloid, and cognition in Alzheimer's disease: An [11C](R)PK11195-PET and [11C]PIB-PET study. *Neurobiol. Dis.* **2008**, *32*, 412–419. [CrossRef] [PubMed]

125. Hamelin, L.; Lagarde, J.; Dorothee, G.; Leroy, C.; Labit, M.; Comley, R.A.; de Souza, L.C.; Corne, H.; Dauphinot, L.; Bertoux, M.; et al. Early and protective microglial activation in Alzheimer's disease: A prospective study using 18F-DPA-714 PET imaging. *Brain* **2016**, *139*, 1252–1264. [CrossRef] [PubMed]

126. Fan, Z.; Aman, Y.; Ahmed, I.; Chetelat, G.; Landeau, B.; Ray Chaudhuri, K.; Brooks, D.J.; Edison, P. Influence of microglial activation on neuronal function in Alzheimer's and Parkinson's disease dementia. *Alzheimer's Dement.* **2015**, *11*, 608–621.e7. [CrossRef] [PubMed]

127. Femminella, G.D.; Ninan, S.; Atkinson, R.; Fan, Z.; Brooks, D.J.; Edison, P. Does Microglial Activation Influence Hippocampal Volume and Neuronal Function in Alzheimer's Disease and Parkinson's Disease Dementia? *J. Alzheimer's Dis.* **2016**, *51*, 1275–1289. [CrossRef] [PubMed]

128. Yokokura, M.; Mori, N.; Yagi, S.; Yoshikawa, E.; Kikuchi, M.; Yoshihara, Y.; Wakuda, T.; Sugihara, G.; Takebayashi, K.; Suda, S.; et al. In vivo changes in microglial activation and amyloid deposits in brain regions with hypometabolism in Alzheimer's disease. *Eur. J. Nucl. Med. Mol. Imaging* **2011**, *38*, 343–351. [CrossRef] [PubMed]

129. Yasuno, F.; Kosaka, J.; Ota, M.; Higuchi, M.; Ito, H.; Fujimura, Y.; Nozaki, S.; Takahashi, S.; Mizukami, K.; Asada, T.; et al. Increased binding of peripheral benzodiazepine receptor in mild cognitive impairment-dementia converters measured by positron emission tomography with [^{11}C]DAA1106. *Psychiatry Res.* **2012**, *203*, 67–74. [CrossRef] [PubMed]

130. Dani, M.; Wood, M.; Mizoguchi, R.; Fan, Z.; Walker, Z.; Morgan, R.; Hinz, R.; Biju, M.; Kuruvilla, T.; Brooks, D.J.; et al. Microglial activation correlates in vivo with both tau and amyloid in Alzheimer's disease. *Brain* **2018**, *141*, 2740–2754. [CrossRef] [PubMed]

131. Fan, Z.; Brooks, D.J.; Okello, A.; Edison, P. An early and late peak in microglial activation in Alzheimer's disease trajectory. *Brain* **2017**, *140*, 792–803. [CrossRef] [PubMed]

132. Kreisl, W.C.; Lyoo, C.H.; Liow, J.S.; Wei, M.; Snow, J.; Page, E.; Jenko, K.J.; Morse, C.L.; Zoghbi, S.S.; Pike, V.W.; et al. ^{11}C-PBR28 binding to translocator protein increases with progression of Alzheimer's disease. *Neurobiol. Aging* **2016**, *44*, 53–61. [CrossRef] [PubMed]

133. Santillo, A.F.; Gambini, J.P.; Lannfelt, L.; Langstrom, B.; Ulla-Marja, L.; Kilander, L.; Engler, H. In vivo imaging of astrocytosis in Alzheimer's disease: An ^{11}C-L-deuteriodeprenyl and PIB PET study. *Eur. J. Nucl. Med. Mol. Imaging* **2011**, *38*, 2202–2208. [CrossRef] [PubMed]

134. Rodriguez-Vieitez, E.; Saint-Aubert, L.; Carter, S.F.; Almkvist, O.; Farid, K.; Scholl, M.; Chiotis, K.; Thordardottir, S.; Graff, C.; Wall, A.; et al. Diverging longitudinal changes in astrocytosis and amyloid PET in autosomal dominant Alzheimer's disease. *Brain* **2016**, *139*, 922–936. [CrossRef] [PubMed]

135. Kim, S.; Thiessen, P.A.; Bolton, E.E.; Chen, J.; Fu, G.; Gindulyte, A.; Han, L.; He, J.; He, S.; Shoemaker, B.A.; et al. PubChem Substance and Compound databases. *Nucleic Acids Res.* **2016**, *44*, D1202–D1213. [CrossRef] [PubMed]

136. Raja, R.; Rosenberg, G.A.; Caprihan, A. MRI measurements of Blood-Brain Barrier function in dementia: A review of recent studies. *Neuropharmacology* **2018**, *134*, 259–271. [CrossRef] [PubMed]

137. Zenaro, E.; Piacentino, G.; Constantin, G. The blood-brain barrier in Alzheimer's disease. *Neurobiol. Dis.* **2017**, *107*, 41–56. [CrossRef] [PubMed]

138. Bowman, G.L.; Kaye, J.A.; Quinn, J.F. Dyslipidemia and blood-brain barrier integrity in Alzheimer's disease. *Curr. Gerontol. Geriatr. Res.* **2012**, *2012*, 184042. [CrossRef] [PubMed]

139. Montagne, A.; Barnes, S.R.; Sweeney, M.D.; Halliday, M.R.; Sagare, A.P.; Zhao, Z.; Toga, A.W.; Jacobs, R.E.; Liu, C.Y.; Amezcua, L.; et al. Blood-brain barrier breakdown in the aging human hippocampus. *Neuron* **2015**, *85*, 296–302. [CrossRef] [PubMed]

140. Wang, H.; Golob, E.J.; Su, M.Y. Vascular volume and blood-brain barrier permeability measured by dynamic contrast enhanced MRI in hippocampus and cerebellum of patients with MCI and normal controls. *J. Magn. Reson. Imaging* **2006**, *24*, 695–700. [CrossRef] [PubMed]

141. Caserta, M.T.; Caccioppo, D.; Lapin, G.D.; Ragin, A.; Groothuis, D.R. Blood-brain barrier integrity in Alzheimer's disease patients and elderly control subjects. *J. Neuropsychiatry Clin. Neurosci.* **1998**, *10*, 78–84. [CrossRef] [PubMed]

142. Schlageter, N.L.; Carson, R.E.; Rapoport, S.I. Examination of blood-brain barrier permeability in dementia of the Alzheimer type with [68Ga]EDTA and positron emission tomography. *J. Cereb. Blood Flow Metab.* **1987**, *7*, 1–8. [CrossRef] [PubMed]

143. Van de Haar, H.J.; Burgmans, S.; Jansen, J.F.; van Osch, M.J.; van Buchem, M.A.; Muller, M.; Hofman, P.A.; Verhey, F.R.; Backes, W.H. Blood-Brain Barrier Leakage in Patients with Early Alzheimer Disease. *Radiology* **2016**, *281*, 527–535. [CrossRef] [PubMed]

144. Starr, J.M.; Farrall, A.J.; Armitage, P.; McGurn, B.; Wardlaw, J. Blood-brain barrier permeability in Alzheimer's disease: A case-control MRI study. *Psychiatry Res.* **2009**, *171*, 232–241. [CrossRef] [PubMed]

145. Janelidze, S.; Hertze, J.; Nagga, K.; Nilsson, K.; Nilsson, C.; Swedish Bio, F.S.G.; Wennstrom, M.; van Westen, D.; Blennow, K.; Zetterberg, H.; et al. Increased blood-brain barrier permeability is associated with dementia and diabetes but not amyloid pathology or APOE genotype. *Neurobiol. Aging* **2017**, *51*, 104–112. [CrossRef] [PubMed]

146. Jack, C.R., Jr.; Bennett, D.A.; Blennow, K.; Carrillo, M.C.; Dunn, B.; Haeberlein, S.B.; Holtzman, D.M.; Jagust, W.; Jessen, F.; Karlawish, J.; et al. NIA-AA Research Framework: Toward a biological definition of Alzheimer's disease. *Alzheimer's Dement.* **2018**, *14*, 535–562. [CrossRef] [PubMed]

147. Habib, M.; Mak, E.; Gabel, S.; Su, L.; Williams, G.; Waldman, A.; Wells, K.; Ritchie, K.; Ritchie, C.; O'Brien, J.T. Functional neuroimaging findings in healthy middle-aged adults at risk of Alzheimer's disease. *Ageing Res. Rev.* **2017**, *36*, 88–104. [CrossRef] [PubMed]

148. Baker, S.L.; Maass, A.; Jagust, W.J. Considerations and code for partial volume correcting [^{18}F]-AV-1451 tau PET data. *Data Brief* **2017**, *15*, 648–657. [CrossRef] [PubMed]

149. Gonzalez-Escamilla, G.; Lange, C.; Teipel, S.; Buchert, R.; Grothe, M.J.; Alzheimer's Disease Neuroimaging Initiative. PETPVE12: An SPM toolbox for Partial Volume Effects correction in brain PET—Application to amyloid imaging with AV45-PET. *Neuroimage* **2017**, *147*, 669–677. [CrossRef] [PubMed]

150. Minhas, D.S.; Price, J.C.; Laymon, C.M.; Becker, C.R.; Klunk, W.E.; Tudorascu, D.L.; Abrahamson, E.E.; Hamilton, R.L.; Kofler, J.K.; Mathis, C.A.; et al. Impact of partial volume correction on the regional correspondence between in vivo [C-11]PiB PET and postmortem measures of A beta load. *Neuroimage-Clin.* **2018**, *19*, 182–189. [CrossRef] [PubMed]

151. Rullmann, M.; Dukart, J.; Hoffmann, K.T.; Luthardt, J.; Tiepolt, S.; Patt, M.; Gertz, H.J.; Schroeter, M.L.; Seiby, J.; Schulz-Schaeffer, W.J.; et al. Partial-Volume Effect Correction Improves Quantitative Analysis of F-18-Florbetaben beta-Amyloid PET Scans. *J. Nucl. Med.* **2016**, *57*, 198–203. [CrossRef] [PubMed]

152. Shidahara, M.; Thomas, B.A.; Okamura, N.; Ibaraki, M.; Matsubara, K.; Oyama, S.; Ishikawa, Y.; Watanuki, S.; Iwata, R.; Furumoto, S.; et al. A comparison of five partial volume correction methods for Tau and Amyloid PET imaging with [F-18]THK5351 and [C-11]PIB. *Ann. Nucl. Med.* **2017**, *31*, 563–569. [CrossRef] [PubMed]

153. Jena, A.; Renjen, P.N.; Taneja, S.; Gambhir, A.; Negi, P. Integrated (18)F-fluorodeoxyglucose positron emission tomography magnetic resonance imaging ((18)F-FDG PET/MRI), a multimodality approach for comprehensive evaluation of dementia patients: A pictorial essay. *Indian J. Radiol. Imaging* **2015**, *25*, 342–352. [PubMed]

154. Wangerin, K.A.; Muzi, M.; Peterson, L.M.; Linden, H.M.; Novakova, A.; Mankoff, D.A.; Kinahan, P.E. A virtual clinical trial comparing static versus dynamic PET imaging in measuring response to breast cancer therapy. *Phys. Med. Biol.* **2017**, *62*, 3639–3655. [CrossRef] [PubMed]

155. Shaw, L.M.; Korecka, M.; Clark, C.M.; Lee, V.M.; Trojanowski, J.Q. Biomarkers of neurodegeneration for diagnosis and monitoring therapeutics. *Nat. Rev. Drug Discov.* **2007**, *6*, 295–303. [CrossRef] [PubMed]

156. Dubois, B.; Hampel, H.; Feldman, H.H.; Scheltens, P.; Aisen, P.; Andrieu, S.; Bakardjian, H.; Benali, H.; Bertram, L.; Blennow, K.; et al. Preclinical Alzheimer's disease: Definition, natural history, and diagnostic criteria. *Alzheimer's Dement.* **2016**, *12*, 292–323. [CrossRef] [PubMed]

157. Jack, C.R., Jr.; Bennett, D.A.; Blennow, K.; Carrillo, M.C.; Feldman, H.H.; Frisoni, G.B.; Hampel, H.; Jagust, W.J.; Johnson, K.A.; Knopman, D.S.; et al. A/T/N: An unbiased descriptive classification scheme for Alzheimer disease biomarkers. *Neurology* **2016**, *87*, 539–547. [CrossRef] [PubMed]

158. Ostrowitzki, S.; Deptula, D.; Thurfjell, L.; Barkhof, F.; Bohrmann, B.; Brooks, D.J.; Klunk, W.E.; Ashford, E.; Yoo, K.; Xu, Z.X.; et al. Mechanism of amyloid removal in patients with Alzheimer disease treated with gantenerumab. *Arch. Neurol.* **2012**, *69*, 198–207. [CrossRef] [PubMed]

159. Altman, D.G.; Lausen, B.; Sauerbrei, W.; Schumacher, M. Dangers of using "optimal" cutpoints in the evaluation of prognostic factors. *J. Natl. Cancer Inst.* **1994**, *86*, 829–835. [CrossRef] [PubMed]

160. Johnson, K.A.; Fox, N.C.; Sperling, R.A.; Klunk, W.E. Brain Imaging in Alzheimer Disease. *CSH Perspect. Med.* **2012**, *2*, a006213. [CrossRef] [PubMed]

161. Scheltens, P. Imaging in Alzheimer's disease. *Dialogues Clin. Neurosci.* **2009**, *11*, 191–199. [PubMed]

162. Wolk, D.A.; Price, J.C.; Saxton, J.A.; Snitz, B.E.; James, J.A.; Lopez, O.L.; Aizenstein, H.J.; Cohen, A.D.; Weissfeld, L.A.; Mathis, C.A.; et al. Amyloid imaging in mild cognitive impairment subtypes. *Ann. Neurol.* **2009**, *65*, 557–568. [CrossRef] [PubMed]

163. Bourgeat, P.; Chetelat, G.; Villemagne, V.L.; Fripp, J.; Raniga, P.; Pike, K.; Acosta, O.; Szoeke, C.; Ourselin, S.; Ames, D.; et al. Beta-amyloid burden in the temporal neocortex is related to hippocampal atrophy in elderly subjects without dementia. *Neurology* **2010**, *74*, 121–127. [CrossRef] [PubMed]

164. Martins, I.J. The future of biomarkers tests and genomic medicine in global organ disease. *Int. J. Microbiol. Infect. Dis.* **2017**, *1*, 1–6.

165. Martins, I.J. The Role of Clinical Proteomics, Lipidomics, and Genomics in the Diagnosis of Alzheimer's Disease. *Proteomes* **2016**, *4*, 14. [CrossRef] [PubMed]

166. Hanon, O.; Vidal, J.S.; Lehmann, S.; Bombois, S.; Allinquant, B.; Treluyer, J.M.; Gele, P.; Delmaire, C.; Blanc, F.; Mangin, J.F.; et al. Plasma amyloid levels within the Alzheimer's process and correlations with central biomarkers. *Alzheimer's Dement.* **2018**, *14*, 858–868. [CrossRef] [PubMed]

167. Korolev, I.O.; Symonds, L.L.; Bozoki, A.C.; Alzheimer's Disease Neuroimaging Initiative. Predicting Progression from Mild Cognitive Impairment to Alzheimer's Dementia Using Clinical, MRI, and Plasma Biomarkers via Probabilistic Pattern Classification. *PLoS ONE* **2016**, *11*, e0138866. [CrossRef] [PubMed]

International Journal of
Molecular Sciences

MDPI

Review

Sleep Disturbance as a Potential Modifiable Risk Factor for Alzheimer's Disease

Eiko N. Minakawa [1,*], Keiji Wada [1] and Yoshitaka Nagai [1,2,*]

[1] Department of Degenerative Neurological Diseases, National Institute of Neuroscience, National Center of Neurology and Psychiatry, Kodaira 187-8502, Japan; wada@ncnp.go.jp
[2] Department of Neurotherapeutics, Osaka University Graduate School of Medicine, Osaka 565-0871, Japan
* Correspondence: minakawa@ncnp.go.jp (E.N.M.); nagai@neurother.med.osaka-u.ac.jp (Y.N.);
    Tel.: +81-42-341-1715 (E.N.M.); +81-6-6879-3563 (Y.N.)

Received: 6 January 2019; Accepted: 3 February 2019; Published: 13 February 2019

Abstract: Sleep disturbance is a common symptom in patients with various neurodegenerative diseases, including Alzheimer's disease (AD), and it can manifest in the early stages of the disease. Impaired sleep in patients with AD has been attributed to AD pathology that affects brain regions regulating the sleep–wake or circadian rhythm. However, recent epidemiological and experimental studies have demonstrated an association between impaired sleep and an increased risk of AD. These studies have led to the idea of a bidirectional relationship between AD and impaired sleep; in addition to the conventional concept that impaired sleep is a consequence of AD pathology, various evidence strongly suggests that impaired sleep is a risk factor for the initiation and progression of AD. Despite this recent progress, much remains to be elucidated in order to establish the benefit of therapeutic interventions against impaired sleep to prevent or alleviate the disease course of AD. In this review, we provide an overview of previous studies that have linked AD and sleep. We then highlight the studies that have tested the causal relationship between impaired sleep and AD and will discuss the molecular and cellular mechanisms underlying this link. We also propose future works that will aid the development of a novel disease-modifying therapy and prevention of AD via targeting impaired sleep through non-pharmacological and pharmacological interventions.

Keywords: Alzheimer's disease; sleep disturbance; sleep fragmentation; slow-wave sleep; amyloid beta; tau; proteostasis; default-mode network; cognitive behavioral therapy for insomnia

---

1. Introduction

Sleep disturbance is a common symptom associated with Alzheimer's disease (AD), which is the leading cause of dementia worldwide [1]. More than 60% of patients with AD develop sleep disturbance, which often occurs at the early stages of the disease or even before the onset of major cognitive decline [2]. Impaired sleep in these patients has been attributed to the progression of AD pathology to brain regions that regulate the sleep–wake or circadian rhythm (Figure 1) [3]. However, various epidemiological studies have demonstrated the association between impaired sleep and an increased risk of AD or AD-related pathology [3]. Multiple studies using animal models of AD have also indicated that impaired sleep exacerbates memory decline and AD-related pathology (Figure 1) [4]. These recent findings suggest that sleep disturbance is a potential modifiable risk factor for AD and could be a novel target for disease-modifying therapies to prevent the development of AD and/or ameliorate the cognitive decline in patients with AD [3]. In this review, we will first provide an overview on the epidemiological and experimental studies that have linked AD and sleep. We will then describe experimental studies that have examined the causal relationship between impaired sleep and AD, and will discuss the molecular and cellular mechanisms that might underlie this link. Finally, we will propose future research directions, including the establishment of a novel disease-modifying therapy and the prevention of AD via targeting impaired sleep in patients with AD and cognitively normal people.

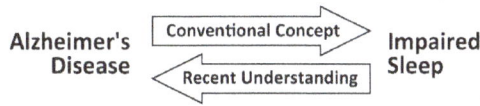

Figure 1. Bidirectional relationship between Alzheimer's disease (AD) and impaired sleep. Impaired sleep is prevalent in patients with AD. Both epidemiological and experimental studies have led to the recent concept of a bidirectional relationship between AD and impaired sleep. In addition to the conventional concept that impaired sleep is a consequence of AD pathology affecting brain regions regulating the sleep–wake or circadian rhythm, impaired sleep has been suggested as a risk factor for the initiation and progression of AD.

2. Age-Related Sleep Alterations

Physiological sleep in mammals, including humans, is composed of rapid eye movement (REM) sleep and non-REM (NREM) sleep. Human NREM sleep can be classified into three stages according to its depth, namely stage N1, N2, and N3, using electroencephalogram (EEG) findings that are characteristic of each stage [5]. Stage N3, the deepest NREM sleep, is characterized by a dominant EEG activity that consists of high-voltage slow waves with a frequency range of 1–4 Hz and is thus referred to as slow-wave sleep (SWS) [6].

The age-associated alterations in sleep architecture have been well characterized. The most prominent changes are increased sleep fragmentation by intermittent nocturnal arousals and a reduced amount of SWS, which is associated with shorter overall sleep duration and increased N1 and N2 duration [7]. Compared with these changes in NREM sleep, REM sleep is relatively spared except for a decreased REM latency with age [7] until around 80 years old, after which its duration is also reduced [8].

3. Sleep Disturbance in Alzheimer's Disease (AD)

3.1. Sleep Abnormalities in Patients with Alzheimer's Disease (AD)

Patients with AD often experience difficulty falling asleep, repeated nocturnal arousals, early arousals in the morning, and excessive sleepiness during daytime [9]. One or more sleep disorders, including insomnia, circadian rhythm sleep–wake disorders, sleep-related breathing disorders (SRBD), and sleep-related movement disorders, underlie these symptoms [10].

The most consistently reported changes in sleep architecture in patients with mild to moderate AD are sleep fragmentation, which is due to an increased number and duration of intermittent nocturnal arousals, a reduced amount of SWS, a resulting decrease in overall sleep duration, and an increase in N1 [11]. These AD-associated changes in NREM sleep seem to be an exaggeration of sleep alterations that are associated with normal aging, which become more pronounced with an increase in the severity of AD [12]. In addition, sleep spindles and K complexes, which are the EEG markers of stage N2, exhibit poorer formation, lower amplitude, shorter duration and smaller number [11]. These changes are mostly similar to the age-associated change in these stage N2 markers, except for the spindle formation, whose age-related change is still controversial [13]. Meanwhile, the total duration of REM sleep, which is relatively spared in normal aging, is reduced in patients with AD due to a reduced duration of each REM episode [14]. Other REM sleep variables, such as the number of REM episodes and REM latency, are usually spared in AD [14].

The alterations in the diurnal rhythm of activity and sleep due to circadian rhythm dysregulation are also present in patients with preclinical AD and symptomatic AD [15]. A disrupted circadian rhythm can cause sundowning syndrome or nocturnal delirium [16], in which patients often become agitated, restless or anxious in the early evening. These symptoms usually resolve during the daytime, but greatly impair the quality of life of patients, families and caregivers [16].

3.2. Sleep Disturbance as a Consequence of AD Pathology

Sleep alterations in patients with AD have been interpreted as a consequence of the progression of AD pathology to brain regions that are involved in the regulation of the sleep–wake or circadian rhythm [3]. AD pathology affects galaninergic neurons in the intermediate nucleus of the hypothalamus [17]. This area is a homolog of the ventrolateral preoptic nucleus of rodents, which is selectively active during sleep [18] and sends inhibitory projections to wake-promoting areas [19]. The number of remaining galaninergic neurons in the intermediate nucleus of autopsied AD brains has been found to be negatively correlated with the severity of ante-mortem sleep fragmentation evaluated within one year of death by actigraphy [17], which suggests that galaninergic neuronal loss due to AD pathology leads to sleep fragmentation in patients with AD.

AD pathology also affects the cholinergic neuronal network [14], which comprises the brainstem, thalamus, basal forebrain and cerebral cortex. This network regulates the initiation and maintenance of REM sleep [14]. The primary circadian pacemaker in the mammalian brain is the hypothalamic suprachiasmatic nucleus (SCN), which is also affected by AD pathology. AD is associated with a significant loss of vasopressin- and vasoactive intestinal peptide-expressing neurons, which are involved in the maintenance of circadian function in the SCN [20–22].

Various transgenic or knock-in mouse models of AD also develop sleep abnormalities, such as increased wakefulness [23–25], a decrease in NREM sleep [23,25] and REM sleep [23], circadian rhythm delay [24], or a reduced amplitude of the circadian rhythm [25]. The sleep–wake patterns of APPswe/PS1dE9 transgenic mice [26] are normal before Aβ deposition, but a tendency of increased wakefulness and decreased sleep starts at the age when Aβ deposition is initially observed. Furthermore, these sleep abnormalities exacerbte with age and increased Aβ deposition [23]. Furthermore, APPswe/PS1dE9 mice that are actively immunized with Aβ, which decreases Aβ deposition in the brain, showed a normal sleep–wake pattern [23]. Taken together, these findings suggest that sleep disturbance is caused not only by a neuronal loss in the brain regions regulating sleep or circadian rhythm, but also by Aβ accumulation in the brain.

4. Sleep Disturbance as a Risk Factor of AD

4.1. Epidemiological Studies

Contrary to the conventional understanding that impaired sleep in patients with AD is a consequence of AD-related pathology, multiple recent epidemiological studies have suggested that sleep disturbance could be a risk factor for cognitive decline and AD. According to a recent meta-analysis, sleep disturbance or sleep disorders, including short or long sleep duration, poor sleep quality (difficulty in falling asleep or increased intermittent nocturnal arousal), circadian rhythm abnormality, insomnia or SRBD, were associated with a significant increase in the risk ratio (RR) for cognitive impairment (RR: 1.64, 95% CI: 1.45–1.87), preclinical AD (RR: 3.78, 95% CI: 2.27–6.30) and AD diagnoses based on the ICD-9 (International Classification of Diseases, Ninth edition) or DSM-IV (Diagnostic Statistical Manual, Fourth edition) (RR: 1.55, 95% CI: 1.25–1.93) [27].

In a prospective study that used actigraphy to quantitatively assess the sleep of 737 community-dwelling older adults without dementia, a higher level of sleep fragmentation due to increased intermittent nocturnal arousal was associated with an increased risk of AD (hazard ratio = 1.22, 95% CI: 1.03–1.44) [28]. Individuals with high sleep fragmentation (in the 90th percentile) at baseline had a 1.5-fold higher risk of developing AD compared to those with low sleep fragmentation (in the 10th percentile) during the 6-year follow-up period (mean = 3.3 years) [28]. In addition, in a positron emission tomography (PET) study that examined the association between sleep variables and amyloid beta (Aβ) deposition in older people without dementia, a self-reported shorter sleep duration and poorer sleep quality were associated with significantly greater in vivo Aβ deposition in the precuneus [29], which is affected by Aβ pathology in preclinical AD [30].

Although these studies indicate an association between impaired sleep and AD, epidemiological observational studies conducted so far are limited in discerning the causal relationship between impaired sleep and AD, especially considering the relatively short follow-up periods compared to the long disease course of AD [3]. For example, a subgroup meta-analysis for the effect of sleep disturbance demonstrated that both short and long sleep duration were associated with a higher risk of cognitive decline or AD [27]. Additional studies are needed to determine whether both short and long sleep duration do indeed affect the disease course of AD, or whether either of these is a prodromal symptom of AD or reflecting the comorbidities of AD, such as depression.

4.2. The Causal Relationship between Sleep Disturbance and AD Pathology

Various animal models of AD and sleep disturbance have been used to assess the causal relationship between sleep disturbance and AD and the molecular or cellular mechanisms potentially underlying this link. Kang et al. (2009) were the first to report that chronic sleep restriction accelerates Aβ deposition in the brain using two transgenic AD mouse models (APPswe and APPswe/PS1dE9 mice) [31]. Other studies have also demonstrated that sleep deprivation or restriction in various AD models exacerbates AD-related biochemical or pathological changes in mice brains, such as an increase in Aβ or phosphorylated tau [32,33], an increase in insoluble phosphorylated tau and glial fibrillary acidic protein levels [34], and an increase in $A\beta_{40}$, $A\beta_{42}$ and β-site amyloid-precursor-protein-cleaving enzyme 1 (BACE1), which produce toxic Aβ species [35].

In these studies, sleep disturbance was induced in the mice either by intermittent gentle tactile stimuli, resulting in total deprivation of sleep [31], by the platform-over-water technique, resulting in elimination of REM sleep and a decrease in SWS [31–33,35], or by alteration of the light–dark cycle, resulting in a disrupted circadian rhythm [34]. The limitation of these studies is that the resultant sleep–wake patterns using the above methods are different from those observed in patients with AD or in normal aging. In addition, these methods induce relatively high levels of stress in mice. These acute or chronic behavioral stresses could aggravate the AD pathology via an increase in Aβ [36] and might therefore be a confounding factor. In a recent study, we took advantage of a novel device that induces impaired sleep closely resembling that of patients with AD (i.e., an increase in sleep fragmentation and a decrease in the amount of SWS) without severe stress [37] and found that chronic sleep fragmentation indeed aggravates Aβ deposition in the AD mice brain [38]. Notably, the severity of Aβ deposition showed a significant positive correlation with the severity of sleep fragmentation [38]. Since all mice were subjected to sleep impairment by a unified protocol, our results strongly suggest that the aggravation of Aβ pathology is more directly related to sleep impairment than the behavioral stress, if any, that was induced by the device we used to induce sleep impairment. Considering this point, our results are consistent with a previous epidemiological study that demonstrated an association between sleep fragmentation and an increased risk of AD [28]. Thus, our evidence supports the view that sleep disturbance in older people and patients with AD affects the disease course of AD.

5. Molecular/Cellular Mechanisms that Link AD and Sleep

5.1. Impaired Sleep Alters the Dynamics of Aβ and Tau in the Brain

The two major pathological hallmarks of AD are senile plaques, which are the extracellular deposits that are mainly composed of insoluble Aβ, and neurofibrillary tangles (NFT), which are the intracytoplasmic deposits that are mainly composed of hyperphosphorylated insoluble tau [39]. The dynamics of extracellular Aβ in relation to neuronal activity and the sleep–wake cycle have been extensively studied using various in vitro and in vivo animal and humans. Several recent studies have also examined the relationship between the dynamics of tau and sleep.

Extracellular Aβ in the central nervous system can be detected as a soluble form in cerebrospinal fluid (CSF) in humans as well as in CSF or interstitial fluid (ISF) in mice. The soluble Aβ shows diurnal fluctuation in both healthy young humans and mice, with an increase during wakefulness and a decrease during sleep [31,40]. The amplitude of this diurnal fluctuation is decreased in the CSF of older people without Aβ deposition and disappears in older people with Aβ deposition [40].

In APPswe/PS1dE9 mice, the diurnal Aβ fluctuation in ISF disappears when mice develop Aβ deposition [23]. These studies suggest that the dynamics of extracellular soluble Aβ is one of the potential mechanisms linking sleep and an increased risk of AD. Therefore, the mechanisms that affect the dynamics of soluble Aβ in ISF or CSF have been extensively studied.

Various studies have confirmed that Aβ production is regulated by neuronal action potential firing. An increase in neuronal firing leads to an increase in the extracellular secretion of soluble Aβ in an activity-dependent manner [41,42]. In vivo experiments have also demonstrated a direct relationship between increased neuronal activity and increased production of extracellular soluble Aβ in the brain, which was detected in the interstitial fluid (ISF) [43]. Furthermore, a sustained increase in the neuronal activity by optogenetic stimulation induces an increase in soluble Aβ in ISF followed by insoluble Aβ deposition in the projection area of the stimulated neurons [44]. Consistent with these studies, extended wakefulness by total sleep deprivation results in an increased level of soluble Aβ in the ISF or CSF [31,45]. Interestingly, specific disruption of SWS, but not sleep duration or sleep efficiency, induces an increase in CSF Aβ [46]. This suggests that each sleep component may influence the dynamics of extracellular Aβ in different ways.

The mechanism underlying the decrease in soluble Aβ during sleep is still controversial. The interchanging convective flow of ISF and CSF in the interstitial space of the brain has been reported to play a crucial role in the removal of the extracellular metabolites, including Aβ, in the brain [47]. Furthermore, Xie et al. reported that natural sleep is associated with a 60% increase in the interstitial space in the brain, which results in an increase in the clearance efficiency of interstitial metabolites, including Aβ, by the increased convective flow of CSF and ISF [48]. The removal of interstitial Aβ by this clearance system, named the glymphatic system, may be one of the mechanisms underlying the decrease in CSF and ISF Aβ during sleep. Meanwhile, recent human studies have analyzed Aβ turnover in CSF by radioactive labeling of Aβ [49,50]. These studies concluded that a decreased production of Aβ due to reduced neuronal activity rather than the increased clearance of Aβ is a necessary and critical factor for the decrease in CSF Aβ during sleep [49,50].

Extracellular soluble tau is another important component in ISF and CSF that is related to AD pathology, while intracellular aggregated tau is a pathological hallmark of AD. Recent studies have indicated that the total tau and phosphorylated tau in CSF are biomarkers that differentiate patients with AD from healthy controls as well as those with mild cognitive impairments due to preclinical AD from those due to other conditions [51,52].

Similarly to Aβ, neuronal activity has been found to induce the extracellular release of tau in an in vitro model [53]. Neuronal activity also induces the propagation of aggregated tau pathology in vivo via the extracellular release of tau and uptake of released tau by nearby neurons [54]. Extracellularly released tau is indeed detectable in the ISF of tau transgenic mouse models [55–58]. Multiple recent studies have examined the in vivo dynamics of the extracellular tau in ISF and CSF in relation to neuronal activity and the sleep–wake cycle. In tau transgenic mice with regulatable expression, the half-life of extracellular soluble ISF tau was revealed to be 17.3 days [56]. This is remarkably longer than that of Aβ, which shows diurnal fluctuation. Consistent with this finding, poorer sleep quality, which was measured for six consecutive nights before CSF collection, was found to have a significant negative correlation with an increase in CSF tau, while acute deprivation of SWS did not lead to CSF tau elevation [46]. Meanwhile, a very recent study demonstrated that acute sleep deprivation leads to a remarkable increase of tau in both mice ISF and human CSF [59]. Importantly, another recent study that used a combination of sleep monitoring by single-channel EEG with PET imaging and CSF analysis of both Aβ and tau revealed that a decrease in SWS, especially at the lowest frequencies of 1–2 Hz, was more associated with the accumulation of tau than that of Aβ [60].

Together, these studies suggest that impaired sleep affects the dynamics of both Aβ and tau, which may lead to the exacerbation of AD-related pathology. Further studies are awaited to determine whether the dynamics of Aβ and tau are regulated via same mechanisms of production and clearance and via similar components of sleep.

5.2. Prolonged Wakefulness Induces Impaired Proteostasis, a Common Pathomechanism Underlying Neurodegenerative Diseases

Proteins with proper functions are indispensable for living organisms. Intracellular and in vivo protein quality is maintained in a homeostatic manner through the coordination of multiple intra- and extracellular systems that regulate protein synthesis, folding, disaggregation, and degradation [61]. The resultant homeostasis of protein quality (Figure 2; left), which is called proteostasis, is of general importance for maintaining human health.

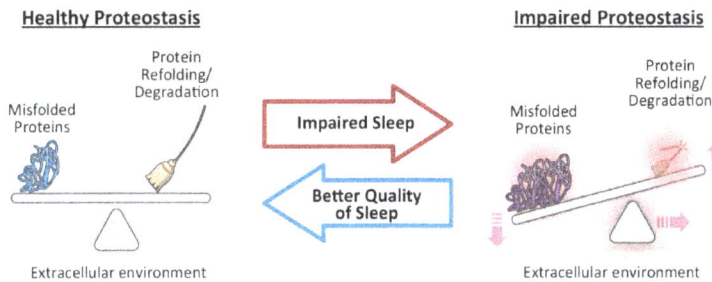

Figure 2. Impaired sleep as a potential therapeutic target to restore proteostasis. Healthy proteostasis is maintained through the coordination of various intra- and extracellular systems that regulate protein synthesis, folding, disaggregation, and degradation (left). Increased synthesis of misfolded proteins, dysfunction of protein refolding or degradation systems, or changes in extracellular environment can lead to impaired proteostasis and result in the accumulation of misfolded and aggregation-prone toxic proteins (right), which is a common pathomechanism underlying neurodegenerative diseases. Based on recent studies that have indicated that impaired sleep leads to impaired proteostasis (middle; red arrow), future studies that better examine the relationship between sleep and proteostasis could lead to the development of novel therapeutics that restore healthy proteostasis via better quality of sleep (middle; blue arrow).

Impaired proteostasis (Figure 2; right) is a common pathomechanism underlying neurodegenerative diseases, such as AD, Parkinson's disease (PD), dementia with Lewy bodies (DLB), amyotrophic lateral sclerosis and Huntington's disease [39]. Neurodegenerative diseases are characterized by selective and progressive neuronal degeneration, which is accompanied by abnormal protein aggregates in the regions of the central nervous system (CNS) that are characteristic of each disease [39]. Patients exhibit slowly progressive neurological or psychiatric symptoms of various types, such as cognitive or motor impairment, or involuntary movements, depending on the affected regions specific to each disease. Recent studies have reported that impaired proteostasis and the resultant accumulation of misfolded and aggregation-prone proteins (Figure 2; right) exhibit neurotoxicity and lead to neuronal dysfunction followed by neurodegeneration [61].

Sleep affects proteostasis in the brain. A detailed transcriptomic study revealed that the most abundant categories of genes that are upregulated in the mice brain during sleep are those involved in macromolecule biosynthesis, such as structural components of ribosomes, translation initiation and elongation factors and tRNA activators [62]. In addition, genes involved in intracellular transport, such as vesicle-mediated protein trafficking, are also upregulated during sleep [62].

Among the multiple molecules/pathways involved in the refolding or degradation of misfolded proteins to maintain proteostasis, such as chaperones, the ubiquitin–proteasome system and autophagy, the relationship between sleep and the unfolded protein response (UPR) pathway has been studied in detail. Prolonged wakefulness by sleep deprivation activates the UPR pathway, which is one of the major mechanisms that prevent the accumulation of misfolded proteins and maintains proteostasis [63]. When the endoplasmic reticulum (ER), a major site of protein folding and post-translational modification, is overloaded and stressed by the accumulation of misfolded and potentially toxic proteins, the UPR is activated and triggers different levels of downstream pathways according to the duration and the severity of the ER stress [64]. Mild or transient ER stress induces adaptive or protective pathways, such as increased transcription of chaperones for proper protein refolding, attenuation of general protein translation and removal of misfolded proteins for degradation at the proteasome. When ER stress is not alleviated by these pathways, the pro-apoptotic signaling pathway is activated, which leads to cellular injury or cell death [64]. Various studies have shown that prolonged wakefulness by sleep deprivation for six hours or longer leads to the upregulation of protective or adaptive pathways downstream of UPR activation in the rodent brain, such as the increased production of BiP/GRP78, a major ER chaperone and a marker of UPR activation [63,65–67]. However, in the aged mice brain, six hours of sleep deprivation failed to induce the protective pathways downstream of UPR activation, such as the upregulation of BiP/GRP78 or inhibition of general protein translation. On the contrary, six hours of sleep deprivation did activate pro-apoptotic signaling pathways [68].

These studies suggest that prolonged wakefulness by acute sleep deprivation is sufficient to at least transiently impair proteostasis in the brain (Figure 2, middle; red arrow), and that aging impairs the protective responses against impaired sleep, which could in turn lead to neurodegeneration. Further studies on the role of sleep in the maintenance of proteostasis via the UPR and other pathways could aid the development of novel therapeutics that can restore healthy proteostasis via better quality of sleep and could represent disease modification strategies for neurodegenerative diseases (Figure 2, middle; blue arrow).

5.3. Impaired Sleep May Aggravate the Propagation of AD-Related Pathology via Impaired Functional Connectivity in the Brain

Functional connectivity in the brain is defined as inter-regional correlations in the neuronal activation patterns of anatomically separate brain regions [69]. Functional connectivity reflects the integrity of communication between two functionally related brain regions [70]. Independent component analysis of the functional connectivity at the resting state, when individuals are awake but not focused on their external environment, has identified several functional resting-state networks (RSNs) in the cerebral cortex that exhibit increased activity, specifically at resting state [70]. The default mode network (DMN) is one of the major RSNs and underlies most of the baseline brain activity at rest [71]. The core regions of the DMN include the medial prefrontal cortex, posterior cingulate cortex, precuneus and parietal cortex, all of which have structural interconnections and functional connectivity [72].

Intriguingly, a recent study demonstrated that functional connectivity in the brain shows a diurnal patternand that nocturnal sleep restores morning-to-evening connectivity changes [73]. A lack of sleep has been associated with a deficit in the recovery of functional connectivity on the following morning within various networks, including the DMN [73]. Another study demonstrated that the significant functional correlations between frontal and posterior areas of the DMN become non-significant during SWS, which suggests that the integrity of the DMN is decreased during deep sleep [74]. These studies indicate the potential importance of sleep on the maintenance of the DMN during arousal.

It has also been well established that all regions of the DMN are vulnerable to AD-related pathology [75]. Indeed, DMN impairment is present in early symptomatic AD and progresses with the disease course [76]. DMN impairment is even observed in preclinical AD, when AD-related histopathology accumulates before overt clinical symptoms appear [76]. Furthermore, the carriers of the *ApoE ε4* allele, which is the most potent risk factor for AD, also show DMN impairment similar to that of preclinical AD, even in the absence of Aβ deposition in the brain [77].

The precise mechanism underlying the relationship between the DMN and the progression of AD pathology has yet to be fully elucidated. However, recent studies have strongly suggested that misfolded neurotoxic proteins, such as Aβ and tau in the case of AD, are transmitted along interconnected neural networks [78,79]. Consistent with this, it is plausible that misfolded toxic proteins can be propagated from the brain regions that are initially affected by AD pathology to adjacent healthy brain regions [76]. This protein propagation via interconnected brain regions could eventually lead to the gradual deterioration of the entire brain network from a semi-functional state to a dysfunctional state as misfolded proteins accumulate over the years [76]. From this point of view, the alterations in functional connectivity due to impaired sleep, especially at the preclinical or early stages of AD, might be an additional pathomechanism underlying the progression of AD-related pathology that results from impaired sleep.

5.4. Other Mechanisms that May Link Impaired Sleep and AD-Related Pathology

Inflammatory immune responses, blood–brain barrier (BBB) disruption, and oxidative stress are known to affect AD-related pathology, which can also be induced by impaired sleep [3,80].

Acute and chronic sleep loss in humans result in the induction of both cellular and humoral immunological responses. An increase in the number of circulating leukocytes (mainly monocytes and neutrophils) and increased levels of proinflammatory cytokines, such as interleukin-1β (IL-1β) and IL-6 and tumor necrosis factor-α (TNF-α), are observed after acute sleep deprivation or subacute sleep restriction [81]. The resulting low-grade systemic inflammation could facilitate neuroinflammation when sleep impairment is sustained, which could aggravate AD-related brain pathology [82]. Indeed, chronic sleep loss in rodents has been associated with microglial activation and astrocytic phagocytosis in the brain [83]. In addition, chronic low-grade inflammation has been proposed to underlie the BBB breakdown following sleep loss observed in rodent models [84], which could also worsen AD-related pathology [85]. Furthermore, sleep deprivation promotes oxidative stress in the rodent brain [86]. A recent prospective epidemiological study indicated that obstructive sleep apnea (OSA) in cognitively normal older people is associated with increased Aβ deposition [87]. Besides the sleep fragmentation itself due to OSA, which could affect Aβ dynamics (as discussed in Section 5.1), a combination of hypoxemia, neuroinflammation, and oxidative stress could be additional mechanisms underlying the exacerbation of AD pathology in patients with OSA.

6. Conclusions and Future Directions

Impaired sleep is prevalent in patients with AD, which often occurs in the early or even preclinical stages of AD. Both epidemiological and experimental studies have led to the recent concept of a bidirectional relationship between AD and impaired sleep (Figure 1). In addition to the conventional concept that impaired sleep is a consequence of AD-related pathology, impaired sleep has been suggested to be a risk factor for the initiation and progression of AD, at least in cognitively normal older people and in patients with AD. Despite this recent progress, much remains to be elucidated in future works that will aid the development of therapeutic interventions against impaired sleep to prevent or alleviate the disease course of AD.

First, the essential components of "better sleep" that reduce the risk for AD need to be determined. A recent study demonstrated that an acute inhibition of SWS is sufficient to affect Aβ dynamics in humans [46]. While the importance of REM sleep in regulating NREM sleep has been established [88], additional studies are crucial in obtaining a more comprehensive understanding of the roles and interactions between the different components of sleep, including REM sleep, light NREM sleep and SWS. Furthermore, the molecular and cellular mechanisms underlying the link between AD and these different components of sleep remain to be determined. It would also be necessary to determine the contribution of other sleep-related factors to AD-related pathology, such as the optimal duration of sleep that reduces the risk of AD.

Second, potential therapeutic methods to achieve "better sleep" need to be investigated. Recent meta-analyses have demonstrated the effect of non-pharmacological treatment by cognitive behavioral therapy for insomnia (CBT-I) on primary chronic insomnia [89,90]. In addition, several randomized control studies have shown that CBT-I is more effective than pharmacotherapy using conventional hypnotics that target γ-aminobutyric acid (GABA)$_A$ receptor-mediated systems [91]. CBT-I provided via cost-effective and accessible ways, such as computerized and online platforms or video conferencing, has also shown therapeutic benefits [91]. While these non-pharmacologic methods are recommended as first-line treatments for primary chronic insomnia [92], the recent development of novel hypnotics with different mechanisms of action and potentially better safety, especially in elderly patients, might provide better therapeutic opportunities compared to traditional hypnotics [93]. Whether these non-pharmacological and pharmacological treatments can also achieve "better sleep" that reduces the risk for AD development and progression remains to be determined.

Furthermore, chronic short sleep is highly prevalent in both healthy young adults and adolescents, especially in developed countries [94]. These people generally have insufficient sleep during weekdays and use weekends to catch up on sleep, which leads to the subjective normalization of sleepiness. However, several studies have demonstrated that weekend sleep is not sufficient to fully recover the cognitive performance deficit induced by sleep insufficiency during weekdays [95–97]. Whether the accumulation of sleep insufficiency that begins from adolescence or young adulthood affects the molecular or cellular links between sleep and AD and whether this could lead to an increased risk of AD development would be particularly important for the primary prevention of AD.

Last but not least, impaired sleep mainly due to sleep fragmentation and a decrease in SWS is also prevalent in patients with various neurodegenerative diseases other than AD. Considering that neurodegenerative diseases, including AD, share a common pathomechanism of misfolded protein accumulation and impaired proteostasis, "better sleep" that reduces the risk for AD might also alleviate the disease course of other neurodegenerative diseases. Elucidating the link between impaired sleep and the dynamics of misfolded proteins that accumulate in each disease, such as α-synuclein in PD and DLB as well as Aβ and tau in AD, could lead to the development of a novel disease-modifying therapy that has far-reaching implications for neurodegenerative diseases in general.

Funding: This work was funded in part by Grants-in-Aid for Young Scientists (B) (26860681 to E.N.M.), Young Scientists (18K15474 to E.N.M.) and Scientific Research (B) (18H02585 to E.N.M.) from the Japan Society for the Promotion of Science, Japan, Research Grant from Japan Foundation for Neuroscience and Mental Health (to E.N.M.), grants for Practical Research Project for Rare/Intractable Diseases (JP16ek0109018, JP18ek0109222 to Y.N.) from the Japan Agency for Medical Research and Development and Intramural Research Grants for Neurological and Psychiatric Disorders (27-9, 30-3 to K.W. and Y.N.) from NCNP, Japan.

Acknowledgments: We thank Nia Cason from Edanz Group (www.edanzediting.com/ac) for editing a draft of this manuscript.

Conflicts of Interest: The authors declare no conflict of interest.

Abbreviations

AD	Alzheimer's disease
REM	Rapid Eye Movement
NREM	Non-Rapid Eye Movement
EEG	Electroencephalogram
SWS	Slow-wave Sleep
SRBD	Sleep-related breathing disorders
SCN	Suprachiasmatic Nucleus
RR	Risk Ratio
ICD-9	International Classification of Diseases, Ninth Edition

DSM-IV	Diagnostic Statistical Manual, Fourth Edition
Aβ	Amyloid β
BACE1	β-site amyloid precursor protein cleaving enzyme 1
NFT	Neurofibrillary Tangles
CSF	Cerebrospinal Fluid
ISF	Interstitial Fluid
PD	Parkinson's Disease
DLB	Dementia with Lewy Bodies
CNS	Central Nervous System
UPR	Unfolded Protein Response
ER	Endoplasmic Reticulum
RSN	Resting State Network
DMN	Default Mode Network
TNF-α	Tumor Necrosis Factor-α
OSA	Obstructive Sleep Apnea
CBT-I	Cognitive Behavioral Therapy for Insomnia
GABA	γ-aminobutyric Acid

References

1. Brzecka, A.; Leszek, J.; Ashraf, G.M.; Ejma, M.; Ávila-Rodriguez, M.F.; Yarla, N.S.; Tarasov, V.V.; Chubarev, V.N.; Samsonova, A.N.; Barreto, G.E.; et al. Sleep disorders associated with Alzheimer's disease: A perspective. *Front. Neurosci.* **2018**, *12*, 330. [CrossRef] [PubMed]
2. Guarnieri, B.; Adorni, F.; Musicco, M.; Appollonio, I.; Bonanni, E.; Caffarra, P.; Caltagirone, C.; Cerroni, G.; Concari, L.; Cosentino, F.I.I.; et al. Prevalence of sleep disturbances in mild cognitive impairment and dementing disorders: A multicenter italian clinical cross-sectional study on 431 patients. *Dement. Geriatr. Cogn. Disord.* **2012**, *33*, 50–58. [CrossRef] [PubMed]
3. Ju, Y.-E.S.; Lucey, B.P.; Holtzman, D.M. Sleep and Alzheimer disease pathology—A bidirectional relationship. *Nat. Rev. Neurol.* **2014**, *10*, 115–119. [CrossRef]
4. Dufort-Gervais, J.; Mongrain, V.; Brouillette, J. Bidirectional relationships between sleep and amyloid-beta in the hippocampus. *Neurobiol. Learn. Mem.* **2018**, (in press). [CrossRef] [PubMed]
5. Moser, D.; Anderer, P.; Gruber, G.; Parapatics, S.; Loretz, E.; Boeck, M.; Kloesch, G.; Heller, E.; Schmidt, A.; Danker-Hopfe, H.; et al. Sleep classification according to AASM and Rechtschaffen & Kales: Effects on sleep scoring parameters. *Sleep* **2009**, *32*, 139–149. [PubMed]
6. Léger, D.; Debellemaniere, E.; Rabat, A.; Bayon, V.; Benchenane, K.; Chennaoui, M. Slow-wave sleep: From the cell to the clinic. *Sleep Med. Rev.* **2018**, *41*, 113–132. [CrossRef] [PubMed]
7. Ohayon, M.M.; Carskadon, M.A.; Guilleminault, C.; Vitiello, M.V. Meta-analysis of quantitative sleep parameters from childhood to old age in healthy individuals: Developing normative sleep values across the human lifespan. *Sleep* **2004**, *27*, 1255–1273. [CrossRef]
8. Mander, B.A.; Winer, J.R.; Walker, M.P. Sleep and human aging. *Neuron* **2017**, *94*, 19–36. [CrossRef]
9. Peter-Derex, L.; Yammine, P.; Bastuji, H.; Croisile, B. Sleep and Alzheimer's disease. *Sleep Med. Rev.* **2015**, *19*, 29–38. [CrossRef]
10. Yaffe, K.; Falvey, C.M.; Hoang, T. Connections between sleep and cognition in older adults. *Lancet. Neurol.* **2014**, *13*, 1017–1028. [CrossRef]
11. Petit, D.; Gagnon, J.-F.; Fantini, M.L.; Ferini-Strambi, L.; Montplaisir, J. Sleep and quantitative EEG in neurodegenerative disorders. *J. Psychosom. Res.* **2004**, *56*, 487–496. [CrossRef] [PubMed]
12. Prinz, P.N.; Peskind, E.R.; Vitaliano, P.P.; Raskind, M.A.; Eisdorfer, C.; Zemcuznikov, H.N.; Gerber, C.J. Changes in the sleep and waking EEGs of nondemented and demented elderly subjects. *J. Am. Geriatr. Soc.* **1982**, *30*, 86–92. [CrossRef] [PubMed]
13. Crowley, K.; Trinder, J.; Kim, Y.; Carrington, M.; Colrain, I.M. The effects of normal aging on sleep spindle and K-complex production. *Clin. Neurophysiol.* **2002**, *113*, 1615–1622. [CrossRef]

14. Montplaisir, J.; Petit, D.; Lorrain, D.; Gauthier, S.; Nielsen, T. Sleep in Alzheimer's disease: Further considerations on the role of brainstem and forebrain cholinergic populations in sleep–wake mechanisms. *Sleep* **1995**, *18*, 145–148. [CrossRef] [PubMed]

15. Musiek, E.S.; Bhimasani, M.; Zangrilli, M.A.; Morris, J.C.; Holtzman, D.M.; Ju, Y.-E.S. Circadian rest-activity pattern changes in aging and preclinical alzheimer disease. *JAMA Neurol.* **2018**, *75*, 582–590. [CrossRef] [PubMed]

16. Bedrosian, T.A.; Nelson, R.J. Sundowning syndrome in aging and dementia: Research in mouse models. *Exp. Neurol.* **2013**, *243*, 67–73. [CrossRef] [PubMed]

17. Lim, A.S.P.; Ellison, B.A.; Wang, J.L.; Yu, L.; Schneider, J.A.; Buchman, A.S.; Bennett, D.A.; Saper, C.B. Sleep is related to neuron numbers in the ventrolateral preoptic/intermediate nucleus in older adults with and without Alzheimer's disease. *Brain* **2014**, *137*, 2847–2861. [CrossRef]

18. Sherin, J.E.; Shiromani, P.J.; McCarley, R.W.; Saper, C.B. Activation of ventrolateral preoptic neurons during sleep. *Science* **1996**, *271*, 216–219. [CrossRef]

19. Sherin, J.E.; Elmquist, J.K.; Torrealba, F.; Saper, C.B. Innervation of histaminergic tuberomammillary neurons by GABAergic and galaninergic neurons in the ventrolateral preoptic nucleus of the rat. *J. Neurosci.* **1998**, *18*, 4705–4721. [CrossRef]

20. Swaab, D.F.; Fliers, E.; Partiman, T.S. The suprachiasmatic nucleus of the human brain in relation to sex, age and senile dementia. *Brain Res.* **1985**, *342*, 37–44. [CrossRef]

21. Zhou, J.N.; Hofman, M.A.; Swaab, D.F. VIP neurons in the human SCN in relation to sex, age and Alzheimer's disease. *Neurobiol. Aging* **1995**, *16*, 571–576. [CrossRef]

22. Harper, D.G.; Stopa, E.G.; Kuo-Leblanc, V.; McKee, A.C.; Asayama, K.; Volicer, L.; Kowall, N.; Satlin, A. Dorsomedial SCN neuronal subpopulations subserve different functions in human dementia. *Brain* **2008**, *131*, 1609–1617. [CrossRef] [PubMed]

23. Roh, J.H.; Huang, Y.; Bero, A.W.; Kasten, T.; Stewart, F.R.; Bateman, R.J.; Holtzman, D.M. Disruption of the sleep–wake cycle and diurnal fluctuation of β-amyloid in mice with Alzheimer's disease pathology. *Sci. Transl. Med.* **2012**, *4*, 150ra122. [CrossRef] [PubMed]

24. Duncan, M.J.; Smith, J.T.; Franklin, K.M.; Beckett, T.L.; Murphy, M.P.; St Clair, D.K.; Donohue, K.D.; Striz, M.; O'Hara, B.F. Effects of aging and genotype on circadian rhythms, sleep and clock gene expression in APPxPS1 knock-in mice, a model for Alzheimer's disease. *Exp. Neurol.* **2012**, *236*, 249–258. [CrossRef] [PubMed]

25. Platt, B.; Drever, B.; Koss, D.; Stoppelkamp, S.; Jyoti, A.; Plano, A.; Utan, A.; Merrick, G.; Ryan, D.; Melis, V.; et al. Abnormal cognition, sleep, EEG and brain metabolism in a novel knock-in Alzheimer mouse, PLB1. *PLoS ONE* **2011**, *6*, e27068. [CrossRef]

26. Savonenko, A.; Xu, G.M.; Melnikova, T.; Morton, J.L.; Gonzales, V.; Wong, M.P.F.; Price, D.L.; Tang, F.; Markowska, A.L.; Borchelt, D.R. Episodic-like memory deficits in the APPswe/PS1dE9 mouse model of Alzheimer's disease: Relationships to β-amyloid deposition and neurotransmitter abnormalities. *Neurobiol. Dis.* **2005**, *18*, 602–617. [CrossRef] [PubMed]

27. Bubu, O.M.; Brannick, M.; Mortimer, J.; Umasabor-Bubu, O.; Sebastião, Y.V.; Wen, Y.; Schwartz, S.; Borenstein, A.R.; Wu, Y.; Morgan, D.; et al. Sleep, Cognitive impairment and Alzheimer's disease: A Systematic Review and Meta-Analysis. *Sleep* **2017**, *40*. [CrossRef]

28. Lim, A.S.P.; Kowgier, M.; Yu, L.; Buchman, A.S.; Bennett, D.A. Sleep fragmentation and the risk of incident Alzheimer's disease and cognitive decline in older persons. *Sleep* **2013**, *36*, 1027–1032. [CrossRef]

29. Spira, A.P.; Gamaldo, A.A.; An, Y.; Wu, M.N.; Simonsick, E.M.; Bilgel, M.; Zhou, Y.; Wong, D.F.; Ferrucci, L.; Resnick, S.M. Self-reported sleep and β-amyloid deposition in community-dwelling older adults. *JAMA Neurol.* **2013**, *70*, 1537–1543. [CrossRef]

30. Palmqvist, S.; Schöll, M.; Strandberg, O.; Mattsson, N.; Stomrud, E.; Zetterberg, H.; Blennow, K.; Landau, S.; Jagust, W.; Hansson, O. Earliest accumulation of β-amyloid occurs within the default-mode network and concurrently affects brain connectivity. *Nat. Commun.* **2017**, *8*, 1214. [CrossRef]

31. Kang, J.-E.; Lim, M.M.; Bateman, R.J.; Lee, J.J.; Smyth, L.P.; Cirrito, J.R.; Fujiki, N.; Nishino, S.; Holtzman, D.M. Amyloid-beta dynamics are regulated by orexin and the sleep–wake cycle. *Science* **2009**, *326*, 1005–1007. [CrossRef] [PubMed]

32. Rothman, S.M.; Herdener, N.; Frankola, K.A.; Mughal, M.R.; Mattson, M.P. Chronic mild sleep restriction accentuates contextual memory impairments and accumulations of cortical Aβ and pTau in a mouse model of Alzheimer's disease. *Brain Res.* **2013**, *1529*, 200–208. [CrossRef] [PubMed]

33. Qiu, H.; Zhong, R.; Liu, H.; Zhang, F.; Li, S.; Le, W. Chronic Sleep Deprivation Exacerbates Learning-Memory Disability and Alzheimer's Disease-Like Pathologies in AβPP(swe)/PS1(ΔE9) Mice. *J. Alzheimer's Dis.* **2016**, *50*, 669–685. [CrossRef] [PubMed]

34. Di Meco, A.; Joshi, Y.B.; Praticò, D. Sleep deprivation impairs memory, tau metabolism and synaptic integrity of a mouse model of Alzheimer's disease with plaques and tangles. *Neurobiol. Aging* **2014**, *35*, 1813–1820. [CrossRef] [PubMed]

35. Chen, L.; Huang, J.; Yang, L.; Zeng, X.-A.; Zhang, Y.; Wang, X.; Chen, M.; Li, X.; Zhang, Y.; Zhang, M. Sleep deprivation accelerates the progression of Alzheimer's disease by influencing Aβ-related metabolism. *Neurosci. Lett.* **2017**, *650*, 146–152. [CrossRef] [PubMed]

36. Kang, J.E.; Cirrito, J.R.; Dong, H.; Csernansky, J.G.; Holtzman, D.M. Acute stress increases interstitial fluid amyloid-beta via corticotropin-releasing factor and neuronal activity. *Proc. Natl. Acad. Sci. USA* **2007**, *104*, 10673–10678. [CrossRef] [PubMed]

37. Miyazaki, K.; Itoh, N.; Ohyama, S.; Kadota, K.; Oishi, K. Continuous exposure to a novel stressor based on water aversion induces abnormal circadian locomotor rhythms and sleep–wake cycles in mice. *PLoS ONE* **2013**, *8*, e55452. [CrossRef] [PubMed]

38. Minakawa, E.N.; Miyazaki, K.; Maruo, K.; Yagihara, H.; Fujita, H.; Wada, K.; Nagai, Y. Chronic sleep fragmentation exacerbates amyloid β deposition in Alzheimer's disease model mice. *Neurosci. Lett.* **2017**, *653*, 362–369. [CrossRef] [PubMed]

39. Nagai, Y.; Minakawa, E.N. Drug Development for Neurodegenerative Diseases. In *Neurodegenerative Disorders as Systemic Diseases*, 1st ed.; Wada, K., Ed.; Springer: Tokyo, Japan, 2015; ISBN 978-4-431-54541-5.

40. Huang, Y.; Potter, R.; Sigurdson, W.; Santacruz, A.; Shih, S.; Ju, Y.-E.; Kasten, T.; Morris, J.C.; Mintun, M.; Duntley, S.; et al. Effects of age and amyloid deposition on Aβ dynamics in the human central nervous system. *Arch. Neurol.* **2012**, *69*, 51–58. [CrossRef] [PubMed]

41. Nitsch, R.M.; Farber, S.A.; Growdon, J.H.; Wurtman, R.J. Release of amyloid beta-protein precursor derivatives by electrical depolarization of rat hippocampal slices. *Proc. Natl. Acad. Sci. USA* **1993**, *90*, 5191–5193. [CrossRef] [PubMed]

42. Kamenetz, F.; Tomita, T.; Hsieh, H.; Seabrook, G.; Borchelt, D.; Iwatsubo, T.; Sisodia, S.; Malinow, R. APP Processing and Synaptic Function. *Neuron* **2003**, *37*, 925–937. [CrossRef]

43. Cirrito, J.R.; Yamada, K.A.; Finn, M.B.; Sloviter, R.S.; Bales, K.R.; May, P.C.; Schoepp, D.D.; Paul, S.M.; Mennerick, S.; Holtzman, D.M. Synaptic activity regulates interstitial fluid amyloid-β levels in vivo. *Neuron* **2005**, *48*, 913–922. [CrossRef] [PubMed]

44. Yamamoto, K.; Tanei, Z.; Hashimoto, T.; Wakabayashi, T.; Okuno, H.; Naka, Y.; Yizhar, O.; Fenno, L.E.; Fukayama, M.; Bito, H.; et al. Chronic optogenetic activation augments aβ pathology in a mouse model of Alzheimer disease. *Cell Rep.* **2015**, *11*, 859–865. [CrossRef] [PubMed]

45. Ooms, S.; Overeem, S.; Besse, K.; Rikkert, M.O.; Verbeek, M.; Claassen, J.A.H.R. Effect of 1 night of total sleep deprivation on cerebrospinal fluid β-amyloid 42 in healthy middle-aged men: A randomized clinical trial. *JAMA Neurol.* **2014**, *71*, 971–977. [CrossRef] [PubMed]

46. Ju, Y.-E.S.; Ooms, S.J.; Sutphen, C.; Macauley, S.L.; Zangrilli, M.A.; Jerome, G.; Fagan, A.M.; Mignot, E.; Zempel, J.M.; Claassen, J.A.H.R.; et al. Slow wave sleep disruption increases cerebrospinal fluid amyloid-β levels. *Brain* **2017**, *140*, 2104–2111. [CrossRef] [PubMed]

47. Iliff, J.J.; Wang, M.; Liao, Y.; Plogg, B.A.; Peng, W.; Gundersen, G.A.; Benveniste, H.; Vates, G.E.; Deane, R.; Goldman, S.A.; et al. A paravascular pathway facilitates CSF flow through the brain parenchyma and the clearance of interstitial solutes, including amyloid β. *Sci. Transl. Med.* **2012**, *4*, 147ra111. [CrossRef]

48. Xie, L.; Kang, H.; Xu, Q.; Chen, M.J.; Liao, Y.; Thiyagarajan, M.; O'Donnell, J.; Christensen, D.J.; Nicholson, C.; Iliff, J.J.; et al. Sleep drives metabolite clearance from the adult brain. *Science* **2013**, *342*, 373–377. [CrossRef]

49. Lucey, B.P.; Mawuenyega, K.G.; Patterson, B.W.; Elbert, D.L.; Ovod, V.; Kasten, T.; Morris, J.C.; Bateman, R.J. Associations between β-amyloid kinetics and the β-amyloid diurnal pattern in the central nervous system. *JAMA Neurol.* **2017**, *74*, 207–215. [CrossRef] [PubMed]

50. Lucey, B.P.; Hicks, T.J.; McLeland, J.S.; Toedebusch, C.D.; Boyd, J.; Elbert, D.L.; Patterson, B.W.; Baty, J.; Morris, J.C.; Ovod, V.; et al. Effect of sleep on overnight cerebrospinal fluid amyloid β kinetics. *Ann. Neurol.* **2018**, *83*, 197–204. [CrossRef] [PubMed]

51. Polanco, J.C.; Li, C.; Bodea, L.-G.; Martinez-Marmol, R.; Meunier, F.A.; Götz, J. Amyloid-β and tau complexity—towards improved biomarkers and targeted therapies. *Nat. Rev. Neurol.* **2018**, *14*, 22–39. [CrossRef] [PubMed]

52. Olsson, B.; Lautner, R.; Andreasson, U.; Öhrfelt, A.; Portelius, E.; Bjerke, M.; Hölttä, M.; Rosén, C.; Olsson, C.; Strobel, G.; et al. CSF and blood biomarkers for the diagnosis of Alzheimer's disease: A systematic review and meta-analysis. *Lancet Neurol.* **2016**, *15*, 673–684. [CrossRef]

53. Pooler, A.M.; Phillips, E.C.; Lau, D.H.W.; Noble, W.; Hanger, D.P. Physiological release of endogenous tau is stimulated by neuronal activity. *EMBO Rep.* **2013**, *14*, 389–394. [CrossRef]

54. Wu, J.W.; Hussaini, S.A.; Bastille, I.M.; Rodriguez, G.A.; Mrejeru, A.; Rilett, K.; Sanders, D.W.; Cook, C.; Fu, H.; Boonen, R.A.C.M.; et al. Neuronal activity enhances tau propagation and tau pathology in vivo. *Nat. Neurosci.* **2016**, *19*, 1085–1092. [CrossRef] [PubMed]

55. Yamada, K.; Cirrito, J.R.; Stewart, F.R.; Jiang, H.; Finn, M.B.; Holmes, B.B.; Binder, L.I.; Mandelkow, E.-M.; Diamond, M.I.; Lee, V.M.-Y.; et al. In vivo microdialysis reveals age-dependent decrease of brain interstitial fluid tau levels in P301S human tau transgenic mice. *J. Neurosci.* **2011**, *31*, 13110–13117. [CrossRef] [PubMed]

56. Yamada, K.; Patel, T.K.; Hochgräfe, K.; Mahan, T.E.; Jiang, H.; Stewart, F.R.; Mandelkow, E.-M.; Holtzman, D.M. Analysis of in vivo turnover of tau in a mouse model of tauopathy. *Mol. Neurodegen.* **2015**, *10*, 55. [CrossRef] [PubMed]

57. Barten, D.M.; Cadelina, G.W.; Hoque, N.; DeCarr, L.B.; Guss, V.L.; Yang, L.; Sankaranarayanan, S.; Wes, P.D.; Flynn, M.E.; Meredith, J.E.; et al. Tau transgenic mice as models for cerebrospinal fluid tau biomarkers. *J. Alzheimer's Dis.* **2011**, *24*, 127–141. [CrossRef] [PubMed]

58. Takeda, S.; Wegmann, S.; Cho, H.; DeVos, S.L.; Commins, C.; Roe, A.D.; Nicholls, S.B.; Carlson, G.A.; Pitstick, R.; Nobuhara, C.K.; et al. Neuronal uptake and propagation of a rare phosphorylated high-molecular-weight tau derived from Alzheimer's disease brain. *Nat. Commun.* **2015**, *6*, 8490. [CrossRef] [PubMed]

59. Holth, J.K.; Fritschi, S.K.; Wang, C.; Pedersen, N.P.; Cirrito, J.R.; Mahan, T.E.; Finn, M.B.; Manis, M.; Geerling, J.C.; Fuller, P.M.; et al. The sleep–wake cycle regulates brain interstitial fluid tau in mice and CSF tau in humans. *Science* **2019**, in press. [CrossRef]

60. Lucey, B.P.; McCullough, A.; Landsness, E.C.; Toedebusch, C.D.; McLeland, J.S.; Zaza, A.M.; Fagan, A.M.; McCue, L.; Xiong, C.; Morris, J.C.; et al. Reduced non-rapid eye movement sleep is associated with tau pathology in early Alzheimer's disease. *Sci. Transl. Med.* **2019**, *11*. [CrossRef]

61. Labbadia, J.; Morimoto, R.I. The biology of proteostasis in aging and disease. *Annu. Rev. Biochem.* **2015**, *84*, 435–464. [CrossRef] [PubMed]

62. Mackiewicz, M.; Shockley, K.R.; Romer, M.A.; Galante, R.J.; Zimmerman, J.E.; Naidoo, N.; Baldwin, D.A.; Jensen, S.T.; Churchill, G.A.; Pack, A.I. Macromolecule biosynthesis: A key function of sleep. *Physiol. Genom.* **2007**, *31*, 441–457. [CrossRef] [PubMed]

63. Elliott, A.S.; Huber, J.D.; O'Callaghan, J.P.; Rosen, C.L.; Miller, D.B. A review of sleep deprivation studies evaluating the brain transcriptome. *SpringerPlus* **2014**, *3*, 728. [CrossRef] [PubMed]

64. Naidoo, N. Cellular stress/the unfolded protein response: Relevance to sleep and sleep disorders. *Sleep Med. Rev.* **2009**, *13*, 195–204. [CrossRef]

65. Naidoo, N.; Giang, W.; Galante, R.J.; Pack, A.I. Sleep deprivation induces the unfolded protein response in mouse cerebral cortex. *J. Neurochem.* **2005**, *92*, 1150–1157. [CrossRef] [PubMed]

66. Cirelli, C.; Gutierrez, C.M.; Tononi, G. Extensive and Divergent Effects of Sleep and Wakefulness on Brain Gene Expression. *Neuron* **2004**, *41*, 35–43. [CrossRef]

67. Cirelli, C.; Faraguna, U.; Tononi, G. Changes in brain gene expression after long-term sleep deprivation. *J. Neurochem.* **2006**, *98*, 1632–1645. [CrossRef]

68. Naidoo, N.; Ferber, M.; Master, M.; Zhu, Y.; Pack, A.I. Aging impairs the unfolded protein response to sleep deprivation and leads to proapoptotic signaling. *J. Neurosci.* **2008**, *28*, 6539–6548. [CrossRef]

69. Fox, M.D.; Raichle, M.E. Spontaneous fluctuations in brain activity observed with functional magnetic resonance imaging. *Nat. Rev. Neurosci.* **2007**, *8*, 700–711. [CrossRef]

70. Van den Heuvel, M.P.; Pol, H.E.H. Exploring the brain network: A review on resting-state fMRI functional connectivity. *Eur. Neuropsychopharmacol.* **2010**, *20*, 519–534. [CrossRef]

71. Raichle, M.E.; MacLeod, A.M.; Snyder, A.Z.; Powers, W.J.; Gusnard, D.A.; Shulman, G.L. A default mode of brain function. *Proc. Natl. Acad. Sci. USA* **2001**, *98*, 676–682. [CrossRef]

72. Buckner, R.L.; Andrews-Hanna, J.R.; Schacter, D.L. The brain's default network. *Ann. N. Y. Acad. Sci.* **2008**, *1124*, 1–38. [CrossRef] [PubMed]

73. Kaufmann, T.; Elvsåshagen, T.; Alnæs, D.; Zak, N.; Pedersen, P.Ø.; Norbom, L.B.; Quraishi, S.H.; Tagliazucchi, E.; Laufs, H.; Bjørnerud, A.; et al. The brain functional connectome is robustly altered by lack of sleep. *NeuroImage* **2016**, *127*, 324–332. [CrossRef] [PubMed]

74. Horovitz, S.G.; Braun, A.R.; Carr, W.S.; Picchioni, D.; Balkin, T.J.; Fukunaga, M.; Duyn, J.H. Decoupling of the brain's default mode network during deep sleep. *Proc. Natl. Acad. Sci. USA* **2009**, *106*, 11376–11381. [CrossRef] [PubMed]

75. Fjell, A.M.; McEvoy, L.; Holland, D.; Dale, A.M.; Walhovd, K.B. What is normal in normal aging? Effects of aging, amyloid and Alzheimer's disease on the cerebral cortex and the hippocampus. *Prog. Neurobiol.* **2014**, *117*, 20–40. [CrossRef] [PubMed]

76. Brier, M.R.; Thomas, J.B.; Ances, B.M. Network dysfunction in Alzheimer's disease: Refining the disconnection hypothesis. *Brain Connect.* **2014**, *4*, 299–311. [CrossRef]

77. Sheline, Y.I.; Morris, J.C.; Snyder, A.Z.; Price, J.L.; Yan, Z.; D'Angelo, G.; Liu, C.; Dixit, S.; Benzinger, T.; Fagan, A.; et al. APOE4 allele disrupts resting state fMRI connectivity in the absence of amyloid plaques or decreased CSF Aβ42. *J. Neurosci.* **2010**, *30*, 17035–17040. [CrossRef] [PubMed]

78. Walker, L.C.; Diamond, M.I.; Duff, K.E.; Hyman, B.T. Mechanisms of protein seeding in neurodegenerative diseases. *JAMA Neurol.* **2013**, *70*, 304–310. [CrossRef] [PubMed]

79. Zhou, J.; Gennatas, E.D.; Kramer, J.H.; Miller, B.L.; Seeley, W.W. Predicting regional neurodegeneration from the healthy brain functional connectome. *Neuron* **2012**, *73*, 1216–1227. [CrossRef]

80. Ingiosi, A.M.; Opp, M.R.; Krueger, J.M. Sleep and immune function: Glial contributions and consequences of aging. *Curr. Opin. Neurobiol.* **2013**, *23*, 806–811. [CrossRef]

81. Hurtado-Alvarado, G.; Pavón, L.; Castillo-García, S.A.; Hernández, M.E.; Domínguez-Salazar, E.; Velázquez-Moctezuma, J.; Gómez-González, B. Sleep loss as a factor to induce cellular and molecular inflammatory variations. *Clin. Dev. Immunol.* **2013**, *2013*, 801341. [CrossRef]

82. Heneka, M.T.; Carson, M.J.; El Khoury, J.; Landreth, G.E.; Brosseron, F.; Feinstein, D.L.; Jacobs, A.H.; Wyss-Coray, T.; Vitorica, J.; Ransohoff, R.M.; et al. Neuroinflammation in Alzheimer's disease. *Lancet Neurol.* **2015**, *14*, 388–405. [CrossRef]

83. Bellesi, M.; de Vivo, L.; Chini, M.; Gilli, F.; Tononi, G.; Cirelli, C. Sleep loss promotes astrocytic phagocytosis and microglial activation in mouse cerebral cortex. *J. Neurosci.* **2017**, *37*, 5263–5273. [CrossRef] [PubMed]

84. Hurtado-Alvarado, G.; Domínguez-Salazar, E.; Pavon, L.; Velazquez-Moctezuma, J.; Gomez-Gonzalez, B. Blood-brain barrier disruption induced by chronic sleep loss: Low-grade inflammation may be the link. *J. Immunol. Res.* **2016**, *2016*, e4576012. [CrossRef]

85. Sweeney, M.D.; Sagare, A.P.; Zlokovic, B.V. Blood–brain barrier breakdown in Alzheimer's disease and other neurodegenerative disorders. *Nat. Rev. Neurol.* **2018**, *14*, 133–150. [CrossRef] [PubMed]

86. Villafuerte, G.; Miguel-Puga, A.; Murillo Rodríguez, E.; Machado, S.; Manjarrez, E.; Arias-Carrión, O. Sleep deprivation and oxidative stress in animal models: A systematic review. *Oxid. Med. Cell. Longev.* **2015**, *2015*, 234952. [CrossRef]

87. Sharma, R.A.; Varga, A.W.; Bubu, O.M.; Pirraglia, E.; Kam, K.; Parekh, A.; Wohlleber, M.; Miller, M.D.; Andrade, A.; Lewis, C.; et al. Obstructive sleep apnea severity affects amyloid burden in cognitively normal elderly. A longitudinal study. *Am. J. Respir. Crit. Care. Med.* **2017**, *197*, 933–943. [CrossRef] [PubMed]

88. Hayashi, Y.; Kashiwagi, M.; Yasuda, K.; Ando, R.; Kanuka, M.; Sakai, K.; Itohara, S. Cells of a common developmental origin regulate REM/non-REM sleep and wakefulness in mice. *Science* **2015**, aad1023. [CrossRef]

89. Trauer, J.M.; Qian, M.Y.; Doyle, J.S.; Rajaratnam, S.M.W.; Cunnington, D. Cognitive behavioral therapy for chronic insomnia: A systematic review and meta-analysis. *Ann. Intern. Med.* **2015**, *163*, 191–204. [CrossRef]

90. Van Straten, A.; van der Zweerde, T.; Kleiboer, A.; Cuijpers, P.; Morin, C.M.; Lancee, J. Cognitive and behavioral therapies in the treatment of insomnia: A meta-analysis. *Sleep. Med. Rev.* **2018**, *38*, 3–16. [CrossRef]

91. Kay-Stacey, M.; Attarian, H. Advances in the management of chronic insomnia. *BMJ* **2016**, *354*, i2123. [CrossRef]

92. Wilson, S.; Nutt, D.; Alford, C.; Argyropoulos, S.; Baldwin, D.; Bateson, A.; Britton, T.; Crowe, C.; Dijk, D.-J.; Espie, C.; et al. British Association for Psychopharmacology consensus statement on evidence-based treatment of insomnia, parasomnias and circadian rhythm disorders. *J. Psychopharmacol.* **2010**, *24*, 1577–1601. [CrossRef] [PubMed]

93. Abad, V.C.; Guilleminault, C. Insomnia in elderly patients: Recommendations for pharmacological management. *Drugs Aging.* **2018**, *35*, 791–817. [CrossRef] [PubMed]

94. Zhao, Z.; Zhao, X.; Veasey, S.C. Neural consequences of chronic short sleep: Reversible or lasting? *Front. Neurol.* **2017**, *8*, 235. [CrossRef] [PubMed]

95. Van Dongen, H.P.A.; Maislin, G.; Mullington, J.M.; Dinges, D.F. The cumulative cost of additional wakefulness: Dose-response effects on neurobehavioral functions and sleep physiology from chronic sleep restriction and total sleep deprivation. *Sleep* **2003**, *26*, 117–126. [CrossRef]

96. Belenky, G.; Wesensten, N.J.; Thorne, D.R.; Thomas, M.L.; Sing, H.C.; Redmond, D.P.; Russo, M.B.; Balkin, T.J. Patterns of performance degradation and restoration during sleep restriction and subsequent recovery: A sleep dose-response study. *J. Sleep Res.* **2003**, *12*, 1–12. [CrossRef] [PubMed]

97. Pejovic, S.; Basta, M.; Vgontzas, A.N.; Kritikou, I.; Shaffer, M.L.; Tsaoussoglou, M.; Stiffler, D.; Stefanakis, Z.; Bixler, E.O.; Chrousos, G.P. Effects of recovery sleep after one work week of mild sleep restriction on interleukin-6 and cortisol secretion and daytime sleepiness and performance. *Am. J. Physiol. Endocrinol. Metab.* **2013**, *305*, E890–E896. [CrossRef] [PubMed]

International Journal of
Molecular Sciences

MDPI

Article

Differential Methylation in *APOE* (Chr19; Exon Four; from 44,909,188 to 44,909,373/hg38) and Increased Apolipoprotein E Plasma Levels in Subjects with Mild Cognitive Impairment

Oscar Mancera-Páez [1,2,3,4,*], Kelly Estrada-Orozco [2,5,6], María Fernanda Mahecha [3], Francy Cruz [2,3,7], Kely Bonilla-Vargas [2,3], Nicolás Sandoval [3], Esneyder Guerrero [3], David Salcedo-Tacuma [3], Jesús D. Melgarejo [3,8], Edwin Vega [2], Jenny Ortega-Rojas [3], Gustavo C. Román [9,10], Rodrigo Pardo-Turriago [1,2,3,11] and Humberto Arboleda [2,3]

[1] Department of Neurology, Faculty of Medicine, Universidad Nacional de Colombia, Bogotá ZC 57, Colombia; rpardot@unal.edu.co
[2] Neurosciences Research Group, Faculty of Medicine, Universidad Nacional de Colombia, Bogotá ZC 57, Colombia; kpestradao@unal.edu.co (K.E.-O.); Fjcruzs@unal.edu.co (F.C.); kjbonillav@unal.edu.co (K.B.-V.); edwin.alberto.vega@gmail.com (E.V.); Harboledag@unal.edu.co (H.A.)
[3] Genetic Institute, Universidad Nacional de Colombia, Bogotá ZC 57, Colombia; mfmahechaa@unal.edu.co (M.F.M.); nicocardavid@hotmail.com (N.S.); emguerrerog@unal.edu.co (E.G.); drsalcedot@unal.edu.co (D.S.-T.); jesus.melgarejo1024@gmail.com (J.D.M.); jcortegar@unal.edu.co (J.O.-R.)
[4] David Cabello International Alzheimer Disease Scholarship Fund, Houston Methodist Hospital, Houston, TX 77030, USA
[5] Center for Evidence to Implementation, Bogotá ZC 57, Colombia
[6] Health Technologies and Politics Assessment Group, Clinical Research Institute, Faculty of Medicine, Universidad Nacional de Colombia, Bogotá ZC 57, Colombia
[7] PhD Program in Clinical and Translational Science, Department of Translational Research and of New Surgical and Medical Technologies, University of Pisa, 56128 Pisa, Italy
[8] Laboratory of Neuroscience, University of Zulia, Maracaibo 4001, Venezuela
[9] Department of Neurology, Methodist Neurological Institute and the Institute for Academic Medicine Houston Methodist Research Institute, Houston Methodist Hospital, Houston, TX 77030, USA; Gcroman@houstonmethodist.org
[10] Weill Cornell Medical College, Department of Neurology, Cornell University, New York, NY 10065, USA
[11] Hospital Universitario Nacional de Colombia, Bogotá ZC 57, Colombia
*   Correspondence: ogmancerap@unal.edu.co; Tel.: +57-3175717209

Received: 30 January 2019; Accepted: 12 March 2019; Published: 20 March 2019

Abstract: Background: Biomarkers are essential for identification of individuals at high risk of mild cognitive impairment (MCI) for potential prevention of dementia. We investigated DNA methylation in the *APOE* gene and apolipoprotein E (ApoE) plasma levels as MCI biomarkers in Colombian subjects with MCI and controls. Methods: In total, 100 participants were included (71% women; average age, 70 years; range, 43–91 years). MCI was diagnosed by neuropsychological testing, medical and social history, activities of daily living, cognitive symptoms and neuroimaging. Using multivariate logistic regression models adjusted by age and gender, we examined the risk association of MCI with plasma ApoE and *APOE* methylation. Results: MCI was diagnosed in 41 subjects (average age, 66.5 ± 9.6 years) and compared with 59 controls. Elevated plasma ApoE and *APOE* methylation of CpGs 165, 190, and 198 were risk factors for MCI ($p < 0.05$). Higher CpG-227 methylation correlated with lower risk for MCI ($p = 0.002$). Only CpG-227 was significantly correlated with plasma ApoE levels (correlation coefficient = -0.665; $p = 0.008$). Conclusion: Differential *APOE* methylation and increased plasma ApoE levels were correlated with MCI. These epigenetic patterns require confirmation in larger samples but could potentially be used as biomarkers to identify early stages of MCI.

Keywords: *APOE* gene; apolipoprotein E; DNA methylation; mild cognitive impairment; Hispanics

---

1. Introduction

Mild Cognitive Impairment (MCI) affects 3–20% of individuals older than 65 years, with prevalence rates varying according to geographic regions [1–4]. Approximately 20% of elderly individuals with diagnosed MCI would develop dementia [5,6]. Interestingly, although Hispanics from Latin America (LA) have almost two-fold higher risk of developing Late-Onset Alzheimer's disease (LOAD) than Caucasian North Americans [7,8], the rates of MCI reported in individuals from the United States (US) [9,10] are notably higher than among Hispanics in LA (20% vs. <10%) [11]. This could be attributed to underdiagnosis of MCI in many regions of LA and failure to identify individuals at high risk for dementia. Additionally, numerous vascular risk factors associated with both MCI and dementia occur in Hispanics at higher rates and often with insufficient treatment, including hypertension, diabetes mellitus, smoking, sedentary lifestyle, hyperhomocysteinemia, obesity and dyslipidemia [12–16]. Thus, from the public health perspective, it is critical to implement new strategies to identify subjects at high risk for MCI to prevent and/or delay the development of dementia in this highly susceptible population.

The study of genetic traits is important to investigate the early stages of complex diseases such as AD. In fact, the ApoE-ε4 variant has been demonstrated to be the major genetic risk factor for AD in the general population [17]. The apolipoprotein E (ApoE) has three isoforms, ApoE-ε2, ApoE-ε3, and ApoE-ε4, with direct genetic correspondence to the ε2, ε3, and ε4 alleles. Besides the allelic variant, increased plasmatic apolipoprotein E levels have been examined in relation to AD risk [18]. However, reduced plasma apolipoprotein E levels have been considered a marker of progression of cognitive impairment independently of the *APOE* genotype [19,20]. Moreover, subjects with different dementia types and with one or two copies of the ε4 allele of the *APOE* gene exhibit decreased expression levels of serum apolipoprotein E with regard to both earlier onset of symptoms and deposits of beta-amyloid plaques [21–23].

Epigenetic modifications such as DNA methylation at CpG sites within the genome influence protein expression levels [24]. Hypermethylated promoters are primarily associated with gene expression inhibition [25]; however, in some instances, hypermethylation has been associated with enhanced expression of some genes such as TREM2 in LOAD [26]. The *APOE* gene has a bimodal methylation structure, with a hypomethylated CpG promoter and with comparatively hypermethylated CpG sites located in the *APOE* exon 4 to 3′ UTR region. In AD brains, the *APOE* CpG sites are differentially methylated in both a tissue-specific and an *APOE* genotype-specific manner [27].

Although the expression of *APOE* and its differential methylation levels in LOAD have been explored, there are no studies in subjects with MCI in LA describing the relationships between *APOE* methylation levels and apolipoprotein E differential expression. Therefore, we conducted this research study to estimate the DNA methylation levels for the *APOE* gene (Chr19; exon four; from 44,909,188 to 44,909,373) (Figure 1) and plasma levels of apolipoprotein E (ApoE) in a sample of LA subjects from Colombia with MCI; furthermore, we explored the relationships among *APOE* genotype, DNA methylation of the *APOE* gene and the risk of MCI.

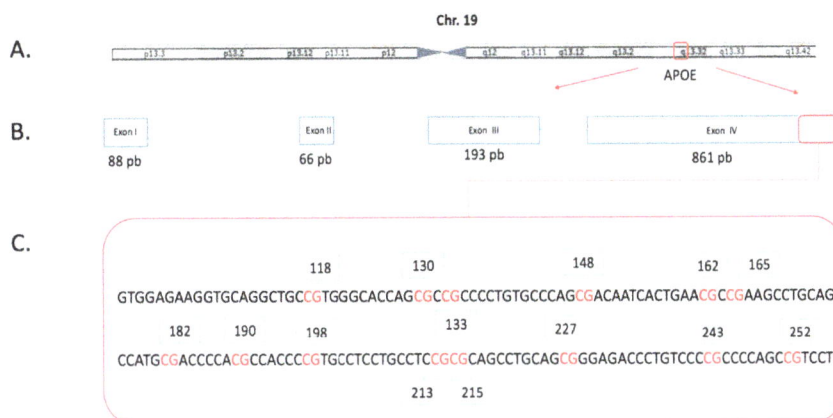

Figure 1. Schematic map of the human APOE gene. (**A**) The *APOE* gene contains four exons, is on the long arm of chromosome 19. (**B**) APOE gene structure (Red arrow) and the region evaluated in this study (Red box), (**C**). DNA sequence of the region evaluated by the BSP methodology, CpGs dinucleotides in red text and identification of each one in blue boxes. The genomic positions are described in Table 2.

2. Results

2.1. Baseline Characteristics

Of the total 100 participants evaluated, 41 had MCI and 59 were controls (Table 1). The mean age of the whole selected sample was 68.9 ± 9.5 years, and 71% ($n = 71$) were women with an age range of 43–91 years old. There was no statistically significant difference in the ApoE-ε4 distribution between MCI and controls.

Table 1. Baseline characteristics of the selected sample according to individuals with mild cognitive impairment and normal controls.

Baseline Characteristics	Whole Sample ($n = 100$)	MCI ($n = 41$)	Control ($n = 59$)	*p* Value *
Demographic data				
Age, years				
Average	68.9 ± 9.5	66.5 ± 9.6	70.5 ± 9.1	0.029
Range	43–91	43–91	50–88	
Gender				0.008
Women, *n* (%)	71 (71.0)	35 (85.4)	36 (61.0)	
Men, *n* (%)	29 (29.0)	6 (14.6)	23 (39.0)	
Genetic traits				
APOE-ε4	25 (25.0)	10 (24.4)	15 (25.4)	0.999

MCI, mild cognitive impairment. * *p* value of comparison between controls and individuals with MCI. The Student *t*-test was used to calculate differences for average age and the Chi-square test for gender and *APOE*-ε4.

2.2. Plasma ApoE Levels and APOE Methylation

Table 2 shows the genomic position of each CpG sites and compares the genotype traits between individuals with MCI and the normal control group. The plasma ApoE levels were higher among those with MCI ($p < 0.001$). *APOE* methylation of CpGs 118 ($p = 0.009$), 165 ($p = 0.040$), 190 ($p = 0.045$), 198 ($p = 0.010$) and 227 ($p < 0.001$) were lower in MCI participants (CpGs = 118, 165, 190, and 198) and only one was reversed (CpG-227). Comparisons between non-*APOE*-ε4 carriers and *APOE*-ε4 carriers (Table 3) showed that only CpG-148 was differently distributed ($p = 0.003$), being higher among *APOE*-ε4 carriers; the remaining CpG sites were similarly distributed ($p > 0.05$).

Table 2. Results of ApoE plasma levels and *APOE* methylation of CpGs sites and comparison between individuals with mild cognitive impairment and normal controls.

Variables	Genomic Position (hg38)	Whole Sample (n = 100)	MCI (n = 41)	Control (n = 59)	p Value *
ApoE plasmatic Levels, mcg/mL	-	103.2 ± 26.5	113.8 ± 26.4	86.0 ± 15.7	<0.0001 [†]
Global methylation	-	91.9 ± 3.0	92.8 ± 2.6	91.6 ± 3.1	0.154 [†]
Methylation by CpG sites					
CpG118	chr19:44,909,208	85.6 ± 5.1	89.6 ± 4.1	84.7 ± 4.9	0.009 [†]
CpG130	chr19:44,909,220	88.9 ± 7.1	89.0 ± 3.9	88.9 ± 7.6	0.484
CpG133	chr19:44,909,223	87.9 ± 10.1	85.7 ± 13.3	88.4 ± 9.3	0.620
CpG148	chr19:44,909,238	94.2 ± 4.2	94.9 ± 3.4	93.9 ± 4.4	0.255
CpG162	chr19:44,909,252	89.7 ± 6.1	90.9 ± 3.9	89.3 ± 6.7	0.361
CpG165	chr19:44,909,254	92 ± 5.2	94.5 ± 2.3	91.2 ± 5.6	0.040 [†]
CpG182	chr19:44,909,272	95.2 ± 8.2	93.3 ± 11.7	95.9 ± 6.6	0.324
CpG190	chr19:44,909,280	93.8 ± 5.8	96.4 ± 2.3	93.0 ± 6.3	0.045 [†]
CpG198	chr19:44,909,288	90.8 ± 7.9	96.2 ± 3.1	89.3 ± 8.2	0.01 [†]
CpG213	chr19:44,909,303	95.1 ± 6.7	96.6 ± 6.8	94.7 ± 6.7	0.212
CpG215	chr19:44,909,305	90.7 ± 11.5	91.3 ± 12.9	90.5 ± 11.2	0.272
CpG227	chr19:44,909,317	97.7 ± 2.3	95.6 ± 2.5	98.4 ± 1.8	<0.0001
CpG243	chr19:44,909,333	91.9 ± 10.4	92.5 ± 11.9	91.6 ± 9.9	0.157
CpG252	chr19:44,909,342	90.4 ± 6.2	90.0 ± 6.1	90.6 ± 6.4	0.833

MCI, mild cognitive impairment. * *p* value of comparison between controls and MCI. [†] Student *t*-test was used to calculate differences; U-Mann–Whitney for variables with non-parametric distribution.

Table 3. Values of ApoE plasma levels and *APOE* methylation of CpGs sites according to *APOE* genotype.

Variables	Non-*APOE*-ε4 Carriers (n = 70)	*APOE*-ε4 Carriers (n = 25)	p Value *
ApoE plasma levels, mcg/mL	106.0 ± 31.3	103.0 ± 26.2	0.738 [†]
Global methylation	92.2 ± 2.2	91.6 ± 3.3	0.557 [†]
Methylation by CpG Sites			
CpG118	84.5 ± 6.0	85.8 ± 4.5	0.428 [†]
CpG130	90.9 ± 6.5	88.0 ± 7.3	0.079
CpG133	84.8 ± 11.9	89.9 ± 7.0	0.087
CpG148	96.2 ± 2.9	93.3 ± 4.4	0.003
CpG162	90.8 ± 3.1	89.4 ± 7.0	0.447
CpG165	89.7 ± 4.3	92.6 ± 5.4	0.076 [†]
CpG182	94.2 ± 10.1	95.4 ± 7.7	0.474
CpG190	95.0 ± 2.0	94.1 ± 3.1	0.316 [†]
CpG198	91.1 ± 4.6	91.4 ± 5.6	0.860 [†]
CpG213	95.6 ± 5.2	94.8 ± 7.4	0.872
CpG215	93.0 ± 8.1	89.6 ± 12.7	0.650
CpG227	97.5 ± 2.4	98.0 ± 2.1	0.482
CpG243	95.0 ± 3.6	90.0 ± 12.3	0.363
CpG252	89.9 ± 3.5	90.5 ± 7.3	0.200

* *p* Value of comparison between controls and individuals with MCI. [†] The Student *t*-test was implemented for calculating differences. The remaining quantitative variables were analyzed by using U-Mann–Whitney as they followed a non-parametric distribution.

2.3. Plasma ApoE Levels and APOE Methylation Levels as Risk Factors for MCI

Logistic regression models adjusted by age and sex (Table 4) demonstrated that the increment on plasma ApoE levels (OR = 1.07; 95% CI = 1.02–1.13; p = 0.003), CpG-165 (OR = 1.20; 95% CI = 1.01–1.43; p = 0.045), and CpG-190 (OR = 1.52; 95% CI = 1.06–2.19; p = 0.042) can be considered risk factors for MCI. Higher CpG-227 methylation (OR = 0.49; 95% CI = 0.31–0.78; p = 0.002) correlated with lower risk for MCI.

Table 4. Adjusted logistic regression models to examine the association of plasma ApoE levels and *APOE* methylation with mild cognitive impairment.

Variables	Risk for Mild Cognitive Impairment		
	Odds Ratios	95% Confident Interval	p Value
Plasma ApoE levels	1.07	1.02–1.13	0.003
APOE methylation		-	
CpG 118	1.25	0.96–1.62	0.092
CpG 165	1.20	1.01–1.43	0.045
CpG 190	1.52	1.06–2.19	0.023
CpG 198	1.30	1.01–1.67	0.042
CpG 227 *	10.05	1.50–67.30	0.017

Models were adjusted by age and sex. * As CpG227 followed a non-parametric distribution, we divided it into four quartiles and <25th percentile was considered as the risk reference.

2.4. Correlation between Plasma ApoE Levels and APOE Methylation Levels

The direct comparisons of plasma ApoE levels and *APOE* methylation are shown in Figure 2. We observed a trend for CpG-165 and CpG-19 but the association with plasma ApoE levels was not significant (p > 0.05). Moreover, we found a negative significant association between plasma ApoE levels and CpG-227 (p = 0.008).

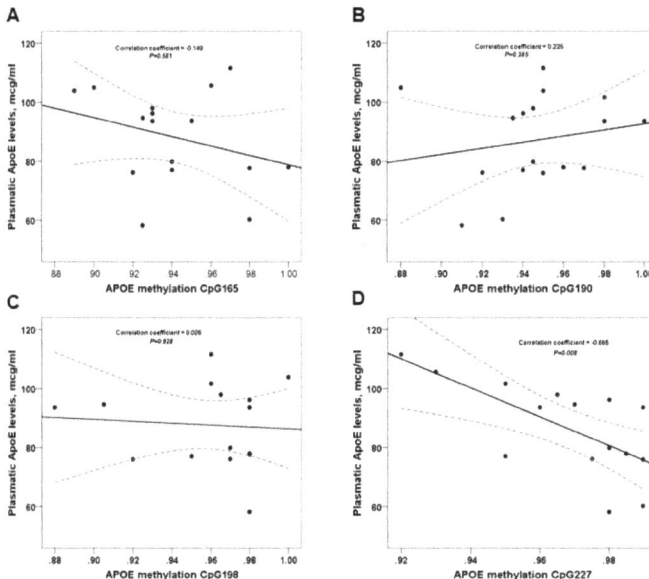

Figure 2. Correlation between Plasmatic ApoE levels and *APOE* Methylation CpGs. The correlation coefficient and p values for (**A–C**) were calculated using Pearson's test; and for (**D**) we used the Spearman's rank correlation.

3. Discussion

In the present study, we examined the association between plasma ApoE levels and *APOE* methylation in 14 CpGs in Chr19, exon IV; from 44,909,188 to 44,909,373 between participants with MCI and control subjects from Bogotá, Colombia, South America. Our key findings were: (i) individuals with MCI had increased plasma ApoE levels in contrast with healthy cognitive controls; (ii) rather than considering global methylation levels, we found that diverse *APOE* CpGs were differentially methylated when comparing participants with MCI and control subjects; (iii) after adjustment by age and sex, increments in ApoE plasma levels and CpG-165, CpG-190 and CpG-198 were found to be associated with increased risk of MCI, whereas lower CpG-227 methylation was related with lower risk; and (iv) only CpG-227 showed a significant correlation with plasma ApoE levels. Although confirmatory studies in larger samples are required, we suggest that assessment of MCI should include plasma ApoE levels and *APOE* methylation levels in order to identify individuals at high risk of developing dementia [28].

Previous studies have shown that decreased serum ApoE levels [28] and hypomethylation in the CpG-252 [26] (cg18799241) are risk factors for the development of dementia. We also found an inverse relationship in which higher plasma levels of ApoE were associated with MCI risk. Our findings might be due to differences in methylation patterns between cell types, with neurons holding higher global levels of DNA methylation [27] and with methylation variations in peripheral blood mononuclear cells related with shortening telomere length [29]. On the other hand, it should be noted that our study examined the possible pathophysiological process involved in a pre-dementia phase and thus our findings may suggest that high concentrations of ApoE would generate a more significant burden of amyloid beta deposition. This is supported by the fact that ApoE protein has a removal effect on amyloid beta [30]. However, this hypothesis cannot be verified with our current research model.

We report that serum ApoE and CpG regions were differentially methylated in MCI patients in contrast with the control group. We found both decreased and increased DNA methylation associated with MCI. Whether the increased DNA methylation of the APOE CpG-165, CpG-190 and CpG-198 is a cause or a consequence of cognitive decline remains to be studied. Foraker et al. [27] suggested an enhancer role of CGI that can be altered by DNA methylation and can modulate gene expression of both *APOE* and *TOMM40* with possible implications in ApoE expression and mitochondrial function. Additionally, these alterations in DNA methylation within genes that are essential for the mitochondrial function could contribute to structural changes in protein and mRNA instability [31]. Our findings support this view, as we found that CpG-227 was correlated with plasma ApoE levels. Despite no statistically significant association, CpG-165 and CpG-190 showed a tendency in relation to ApoE plasma levels.

Liu et al. [32] suggested that hypermethylation levels at multiple CpGs in the *APOE* genomic region are associated with delayed recall during cognitive aging. A previous study of our group [33] found an absence of differences in global *LINE-1* DNA methylation in LOAD subjects; however, this does not imply lack of alterations in DNA methylation for specific loci and their contribution to exonization events and lately in the epigenetic modifications of the landscape [31]. Consequently, global APOE DNA methylation can be useful to complement locus-specific subanalysis. In the same way, the ability to detect DNA methylation in patients with MCI could be enhanced by new approaches focused on specific cell-analysis, such as distinct cerebral cortex layers [27] and correlation with in vivo brain flow biomarkers [34–38].

Yu et al. [39] found that *APOE* CGI exhibits transcriptional enhancer or silencer activity, the mean percentage of methylation *APOE* CGI tends to be directly proportional with *APOE* expression, although it did not reach the cutoff value of statistical significance.

To the best of our knowledge, we found no previous studies analyzing the methylation pattern in this specific locus in subjects with MCI and its correlation with *APOE* transcriptional activity; therefore, the underlying mechanisms of transcriptional regulation of *APOE* and correlation with CpG 227 will need to be studied in a larger sample of patients with higher statistical significance.

Developing countries do not usually have advanced diagnostic methods that can be implemented to identify patients at high risk for MCI. Thus, the study of peripheral blood DNA methylation promises to be a useful pre-clinical biomarker of MCI.

4. Materials and Methods

4.1. Study Design and Population Sample

Participants from a cohort of Colombian patients enrolled at the Memory Clinic of the National University of Colombia agreed to participate in this research. Inclusion criteria were: (i) individuals free of dementia at baseline assessment; and (ii) available data of plasma ApoE or APOE methylation levels. Exclusion criteria were: a history of schizophrenia, manic-depressive disorders, schizoaffective disorder, drug/dependence abuse, severe brain trauma or significant disability or unstable medical conditions (i.e., chronic renal failure, chronic hepatic disease, or severe pulmonary disease) and thyroid disease with no hormonal substitution. From a total of 100 participants, 41 had plasma ApoE levels, and 59 had APOE methylation data available (only 18 participants had both genetic phenotypes). Informed consent was obtained from both the participants and their closest relatives. This study was approved by the ethics committee of the National University of Colombia Act 011-107-15 (01/07/2015). All participants, or their closest relatives, gave written informed consent before participating in this study.

4.2. Medical Evaluation

A clinical neurological assessment was performed and all available data were registered, such as personal clinical history, mental and neurological examination, cognitive screening tests, neuropsychiatric inventory, functional scales and blood tests, e.g., lipid profile, glucose, thyroid function, vitamin B12 and folate levels, hepatic and renal function, serology VDRL, and complete metabolic panel. For those participants with abnormal cognitive tests, brain magnetic resonance image (MRI) was obtained and reviewed in consultation with our multidisciplinary team during follow-up.

4.3. Neuropsychological (NP) Evaluation

We used the Neuronorma Colombia (Neuronorm-Col) diagnostic NP battery for cognitive assessment [40–48]. Neuronorm-Col consists of language tests (Boston Naming Test and Token Test), visuoconstructive skills (Rey–Osterrieth Complex Figure), attention and executive functions (WAIS-III Digit Retention Tests, Corsi Cubes, Trail Making Tests A and B (TMT A and B), Digit–Symbol Test (SDMT), Stroop Color–Word Test, Tower of London test, Win-Dingo Card Sorting Test and Verbal Fluency), and memory (Free and Cued Selective Reminding Test) [40–48].

4.4. Diagnostic Classification of the Participants

4.4.1. Mild Cognitive Impairment

MCI was diagnosed by consensus of a multidisciplinary group that included neurologists, neuropsychologists and neuroscientists, according to the criteria of Petersen et al. [4] modified from the Cognitive assessment test study described by Estrada-Orozco et al. [40]. Differential diagnosis of other related cognitive disorders was based on information from complete neuropsychological testing, medical and social history, activities of daily living, reported cognitive symptoms, and neuroimaging findings. Global cognitive functioning was assessed with the Neuronorm-Col diagnostic neuropsychological battery [40–46] and other functional scales [47,48]. NP criteria for MCI included scores 1.5–2.0 SD below education- and age-corrected values on at least two individual tests within a cognitive domain.

4.4.2. Normal Performance in Healthy Subjects

Criteria for normal performance were: (1) no more than one test score lower than expected within a cognitive domain; and (2) no more than two scores lower than expected across domains, with the threshold corresponding to 1.0 standard deviation (SD) below age-adjusted control means. Moreover, medical and social history, activities of daily living, reported cognitive symptoms, and neuroimaging findings were reviewed to classify the subjects as healthy normal controls.

The control group was composed of cognitively healthy subjects who were selected based on the performances obtained in the screening scales and in the Neuronorma-Colombia battery, a neuropsychological battery normalized to our population in the context of the Spanish Multicenter Normative Studies (NEURONORMA project) [40–46] The cognitive domains evaluated in this battery were: attention (TMT A, TMT B, SDMT, Stroop Test, and Verbal and Direct Visual Span), memory (Grober & Buschke Test, Rey–Osterrieth Complex Figure Evocation), language (Boston Naming Test, Token Test, Verbal Fluency), visual-constructional skills (Rey–Osterrieth Complex Figure Copy), and executive functions (Verbal Phonological Fluency, Tower of London, Interference Stroop Test, and Wisconsin Card Sorting Test).

The following scores were considered normal for the screening tests: Montreal Cognitive Assessment (MOCA) [49], >24; INECO Frontal Screening (IFS) [50], >17.5; Yesavage scale [51], <5; Neuropsychiatric inventory (NPI) [52], <4; and Modified Lawton scale [47], 14–14. In the Neuropsychological tests [40–47,49–54], the cutoff point was one standard deviation (<1 SD) below the mean according to Petersen criteria [4]. Therefore, subjects with scores below the mean or <1 SD in two or more tests evaluating the same cognitive domain were discarded as controls [4].

4.5. DNA Extraction and Bisulfite Treatment

Genomic DNA was isolated from blood from patients and controls using the kit ReliaPrep Blood gDNA Miniprep System™ (A5082-PROMEGA, Fitchburg, WI, USA) following the protocol recommended by the company. DNA was Quantified in a spectrophotometer NanoDrop2000c ThermoScientific and then saved at 4 °C. Subsequently, the isolated DNA was bisulfite-converted using the EZ DN Methylation-Direct Kit (D5021–ZymmoResearch, Irvine, CA, USA). We then evaluated the methylation status of the APOE-CGI (APOE-ExonVI-CpG118 to CpG252) located in Chr19:44,909,188–44,909,373.

4.6. Bisulfite Sequencing PCR (BSP)

The APOE-CGI primers sequences from APOE-F-(5′-TGGAGAAGGTGTAGGTT-3′) and APOE-R-(5′-TTATTAAACTAAAATCCACCCC-3′) were designed following the parameters proposed by Clark et al. [55]. and modified from Tusnady et al. [56]. Each amplification reaction contained 200 ng of DNA, 20 pmol of each primer, 10% dimethyl sulfoxide, two mM dNTP, and 0.125 U of Taq DNA. Both primers were used in a final concentration of 200 nmol/L. Specificity of the assay for converted DNA was verified with the inclusion of unconverted genomic DNA as a control (non-converted DNA Human Methylated & Non-Methylated DNA set D5014 Zymo Research) [57]. Conditions for BSP assay were 95 °C for 5 min, 40 cycles of 95 °C for 30 s, 58.2 °C for 30 s, and 72 °C for 45 s and standardized in a thermocycler C1000 touch (Bio-Rad, Hercules, CA, USA) Then, the products were purified and sequencing in a ABI PRISM 3500 (Applied Biosystem, Foster City, CA, USA). Methylation percentage for each sample was calculated by analysis with ESME (Epigenetic Sequencing Methylation analysis Software, USA) [33,58].

4.7. Apolipoprotein Plasma Levels

Genotyping for APOE ε2, ε/3 and /ε4 allelic variants was determined as previously described, and ApoE plasma levels [26] were measured [28] using ELISA technique (Thermofisher-Invitrogen Human Apo E (AD2) ELISA Kit, CA, USA).

4.8. Statistical Analysis

The continuous variables are presented as mean and standard deviation ($\pm$) while categorical ones are summarized as frequencies and percentages (%). Global methylation level was calculated averaging each of the CpG sites. We compared the ApoE plasma and APOE methylation levels between participants with MCI and control group. Those APOE methylation CpGs following a non-parametric distribution were analyzed using the U-Mann–Whitney test for determining statistically significant differences; for the remaining traits, we used a t-student test. For categorical variables, we used Chi-square test. The plasma ApoE and APOE methylation levels were compared according to the APOE allelic variants. To determine the risk association of plasma ApoE levels and APOE methylation with MCI, we performed multivariate-adjusted models accounting for age and sex. Regression models were performed in those genetic phenotypes with significant average differences. Finally, CpG sites found as risk factors for MCI were correlated with plasma ApoE levels using Pearson's correlation or Spearman's rank tests when appropriated. Data management and statistical analysis were performed using SPSS version 23 (statistical package for social science). Statistical significance was accepted at $p < 0.05$ for two-tailed tests.

5. Limitations

The present study must be interpreted within the context of its potential limitations. First, the population sample presents a risk of selection bias because analyzed individuals attended a specialized care center for patients with memory complaints. However, selection bias was minimized by local radio and television announcements in an effort to recruit healthy subjects. Second, the small sample size limits the generalization of the findings. Third, although unlikely, it is possible that peripheral cellular populations with normal DNA methylation levels could mask the detection of more substantial methylation changes [59].

6. Conclusions

We found that, depending on the CpG region, decreased or increased DNA methylation levels, as well as increased plasma ApoE levels, are potential biomarkers for MCI. These findings might have implications for clinical practice given that these peripheral blood genetic phenotypes could be used for the early diagnosis of MCI. Moreover, if a high-risk profile for vascular cognitive impairment is identified [14], clinical intervention strategies to treat and control modifiable risk factors [16] associated with MCI progression can be intensively implemented to prevent or delay the development of dementia [14,16]. Further studies are needed to confirm these findings and to clarify the risk-association of DNA methylation from different tissues with MCI and neuropsychological profile, and to determine whether the clinical intervention of controlling modifiable risk factors found in dementia can modify the DNA methylation pattern and reduce the risk for MCI progression.

Author Contributions: Conceptualization, G.C.R., O.M.-P., R.P.-T. and H.A.; Methodology, O.M.-P., K.E.-O., J.D.M. and E.G.; Validation, H.A., O.M.-P. and R.P.-T.; Formal Analysis, M.F.M., E.G., J.O.-R., D.S.-T. and J.D.M.; Investigation, O.M.-P., R.P.-T., K.B.-V., K.E.-O., F.C. and H.A.; Resources, H.A.; Data Curation, O.M.-P., K.B.-V., F.C., K.E.-O., E.V., H.A. and N.S.; Writing—Original Draft Preparation, O.M.-P., K.E.-O., J.D.M., D.S.-T., R.P.-T. and H.A.; Writing—Review and Editing, O.M.-P., H.A., K.E.-O., K.B.-V., D.S.-T., G.C.R, and R.P.-T.; Supervision, G.C.R., O.M.-P., R.P.-T. and H.A.; Project Administration, H.A.; and Funding Acquisition, G.C.R., O.M.-P. and H.A.

Funding: This work was supported by Colciencias (Code: 110171149904), grant agreement No. 848-2015 and by the Faculty of Medicine National University of Colombia, grant agreement No. 33350.

Acknowledgments: Mancera-Páez acknowledges the mentorship of Gustavo Román for this work. Mancera-Páez also acknowledges David Cabello for his support of the neurology education program at Houston Methodist Hospital. We thank Juan Camilo Vargas for methodological advice. Jesus Melgarejo acknowledges the support of the International Brain Research Organization (IBRO) for his research fellowship at the National University of Colombia's Institute of Genetics through IBRO's Latin America Regional Committee (LARC) short research training grant between July and December 2018. The authors would like to thank Colciencias for constant support to clinical research in Colombia.

Conflicts of Interest: The authors declare no conflict of interests.

References

1. Gauthier, S.; Reisberg, B.; Zaudig, M.; Petersen, R.C.; Ritchie, K.; Broich, K.; Belleville, S.; Brodaty, H.; Bennett, D.; Chertkow, H. Mild cognitive impairment. *Lancet* **2006**, *367*, 1262–1270. [CrossRef]
2. Henao-Arboleda, E.; Aguirre-Acevedo, D.; Munoz, C.; Pineda, D.; Lopera, F. Prevalence of mild cognitive impairment, amnestic-type, in a Colombian population. *Rev. Neurol.* **2008**, *46*, 709–713. [PubMed]
3. Hänninen, T.; Hallikainen, M.; Tuomainen, S.; Vanhanen, M.; Soininen, H. Prevalence of mild cognitive impairment: A population-based study in elderly subjects. *Acta Neurol. Scand.* **2002**, *106*, 148–154. [CrossRef] [PubMed]
4. Petersen, R.C.; Roberts, R.O.; Knopman, D.S.; Boeve, B.F.; Geda, Y.E.; Ivnik, R.J.; Smith, G.E.; Jack, C.R. Mild cognitive impairment: Ten years later. *Arch. Neurol.* **2009**, *66*, 1447–1455. [CrossRef]
5. Manly, J.J.; Tang, M.X.; Schupf, N.; Stern, Y.; Vonsattel, J.P.G.; Mayeux, R. Frequency and course of mild cognitive impairment in a multiethnic community. *Ann. Neurol.* **2008**, *63*, 494–506. [CrossRef] [PubMed]
6. Kaduszkiewicz, H.; Eisele, M.; Wiese, B.; Prokein, J.; Luppa, M.; Luck, T.; Jessen, F.; Bickel, H.; Mösch, E.; Pentzek, M. Prognosis of mild cognitive impairment in general practice: Results of the German AgeCoDe study. *Ann. Family Med.* **2014**, *12*, 158–165. [CrossRef] [PubMed]
7. Farrer, L.A.; Cupples, L.A.; Haines, J.L.; Hyman, B.; Kukull, W.A.; Mayeux, R.; Myers, R.H.; Pericak-Vance, M.A.; Risch, N.; Van Duijn, C.M. Effects of age, sex, and ethnicity on the association between apolipoprotein E genotype and Alzheimer disease: A meta-analysis. *JAMA* **1997**, *278*, 1349–1356. [CrossRef]
8. Maestre, G.E.; Mena, L.J.; Melgarejo, J.D.; Aguirre-Acevedo, D.C.; Pino-Ramírez, G.; Urribarrí, M.; Chacon, I.J.; Chávez, C.A.; Falque-Madrid, L.; Gaona, C.A. Incidence of dementia in elderly Latin Americans: Results of the Maracaibo Aging Study. *Alzheimer's Dement.* **2018**, *14*, 140–147. [CrossRef]
9. Petersen, R.C.; Roberts, R.O.; Knopman, D.S.; Geda, Y.E.; Cha, R.H.; Pankratz, V.; Boeve, B.; Tangalos, E.; Ivnik, R.; Rocca, W. Prevalence of mild cognitive impairment is higher in men The Mayo Clinic Study of Aging. *Neurology* **2010**, *75*, 889–897. [CrossRef]
10. Plassman, B.L.; Langa, K.M.; Fisher, G.G.; Heeringa, S.G.; Weir, D.R.; Ofstedal, M.B.; Burke, J.R.; Hurd, M.D.; Potter, G.G.; Rodgers, W.L. Prevalence of cognitive impairment without dementia in the United States. *Ann. Intern. Med.* **2008**, *148*, 427–434. [CrossRef]
11. Sosa, A.L.; Albanese, E.; Stephan, B.C.; Dewey, M.; Acosta, D.; Ferri, C.P.; Guerra, M.; Huang, Y.; Jacob, K.; Jimenez-Velazquez, I.Z. Prevalence, distribution, and impact of mild cognitive impairment in Latin America, China, and India: A 10/66 population-based study. *PLoS Med.* **2012**, *9*, e1001170. [CrossRef] [PubMed]
12. Schargrodsky, H.; Hernández-Hernández, R.; Champagne, B.M.; Silva, H.; Vinueza, R.; Ayçaguer, L.C.S.; Touboul, P.-J.; Boissonnet, C.P.; Escobedo, J.; Pellegrini, F. CARMELA: Assessment of cardiovascular risk in seven Latin American cities. *Am. J. Med.* **2008**, *121*, 58–65. [CrossRef]
13. Melgarejo, J.D.; Maestre, G.E.; Thijs, L.; Asayama, K.; Boggia, J.; Casiglia, E.; Hansen, T.W.; Imai, Y.; Jacobs, L.; Jeppesen, J. Prevalence, treatment, and control rates of conventional and ambulatory hypertension across 10 populations in 3 continents. *Hypertension* **2017**, *70*, 50–58. [CrossRef] [PubMed]
14. Knopman, D.; Boland, L.; Mosley, T.; Howard, G.; Liao, D.; Szklo, M.; McGovern, P.; Folsom, A. Atherosclerosis Risk in Communities (ARIC) Study Investigators. Cardiovascular risk factors and cognitive decline in middle-aged adults. *Neurology* **2001**, *56*, 42–48. [CrossRef]
15. Kivipelto, M.; Ngandu, T.; Fratiglioni, L.; Viitanen, M.; Kåreholt, I.; Winblad, B.; Helkala, E.-L.; Tuomilehto, J.; Soininen, H.; Nissinen, A. Obesity and vascular risk factors at midlife and the risk of dementia and Alzheimer disease. *Arch. Neurol.* **2005**, *62*, 1556–1560. [CrossRef] [PubMed]
16. Román, G.C.; Mancera-Páez, O.; Bernal, C. Epigenetic Factors in Late-Onset Alzheimer's disease: *MTHFR* and *CTH* Gene Polymorphisms, Metabolic Trans-sulfuration and Methylation Pathways, and B Vitamins. *Int. J. Mol. Sci.* **2019**, *20*, 319. [CrossRef] [PubMed]
17. Tang, M.-X.; Stern, Y.; Marder, K.; Bell, K.; Gurland, B.; Lantigua, R.; Andrews, H.; Feng, L.; Tycko, B.; Mayeux, R. The APOE-ε4 allele and the risk of Alzheimer disease among African Americans, whites, and Hispanics. *JAMA* **1998**, *279*, 751–755. [CrossRef]
18. Taddei, K.; Clarnette, R.; Gandy, S.E.; Martins, R.N. Increased plasma apolipoprotein E (apoE) levels in Alzheimer's disease. *Neurosci. Lett.* **1997**, *223*, 29–32. [CrossRef]

19. Sullivan, P.; Han, B.; Liu, F.; Mace, B.; Ervin, J.; Wu, S.; Koger, D.; Paul, S.; Bales, K. Reduced levels of human apoE4 protein in an animal model of cognitive impairment. *Neurobiol. Aging* **2011**, *32*, 791–801. [CrossRef]

20. Song, F.; Poljak, A.; Crawford, J.; Kochan, N.A.; Wen, W.; Cameron, B.; Lux, O.; Brodaty, H.; Mather, K.; Smythe, G.A. Plasma apolipoprotein levels are associated with cognitive status and decline in a community cohort of older individuals. *PLoS ONE* **2012**, *7*, e34078. [CrossRef]

21. Corder, E.H.; Saunders, A.M.; Strittmatter, W.J.; Schmechel, D.E.; Gaskell, P.C.; Small, G.; Roses, A.D.; Haines, J.; Pericak-Vance, M.A. Gene dose of apolipoprotein E type 4 allele and the risk of Alzheimer's disease in late onset families. *Science* **1993**, *261*, 921–923. [CrossRef] [PubMed]

22. Liu, N.; Zhang, K.; Zhao, H. Haplotype-association analysis. *Adv. Genet.* **2008**, *60*, 335–405. [PubMed]

23. Chartier-Hariln, M.-C.; Parfitt, M.; Legrain, S.; Pérez-Tur, J.; Brousseau, T.; Evans, A.; Berr, C.; Vldal, O.; Roques, P.; Gourlet, V. Apolipoprotein E, ε4 allele as a major risk factor for sporadic early and late-onset forms of Alzheimer's disease: Analysis of the 19q13. 2 chromosomal region. *Hum. Mol. Genet.* **1994**, *3*, 569–574. [CrossRef]

24. Ciceri, F.; Rotllant, D.; Maes, T. Understanding epigenetic alterations in Alzheimer's and Parkinson's disease: Towards targeted biomarkers and therapies. *Curr. Pharm. Des.* **2017**, *23*, 839–857. [CrossRef] [PubMed]

25. Bae, M.G.; Kim, J.Y.; Choi, J.K. Frequent hypermethylation of orphan CpG islands with enhancer activity in cancer. *BMC Med. Genomics* **2016**, *9*, 38. [CrossRef] [PubMed]

26. Shao, Y.; Shaw, M.; Todd, K.; Khrestian, M.; D'Aleo, G.; Barnard, P.J.; Zahratka, J.; Pillai, J.; Yu, C.-E.; Keene, C.D. DNA methylation of TOMM40-APOE-APOC2 in Alzheimer's disease. *J. Hum. Genet.* **2018**, *63*, 459. [CrossRef] [PubMed]

27. Foraker, J.; Millard, S.P.; Leong, L.; Thomson, Z.; Chen, S.; Keene, C.D.; Bekris, L.M.; Yu, C.-E. The APOE gene is differentially methylated in Alzheimer's disease. *J. Alzheimer's Dis.* **2015**, *48*, 745–755. [CrossRef] [PubMed]

28. Rasmussen, K.L.; Tybjærg-Hansen, A.; Nordestgaard, B.G.; Frikke-Schmidt, R. Plasma levels of apolipoprotein E and risk of dementia in the general population. *Ann. Neurol.* **2015**, *77*, 301–311. [CrossRef] [PubMed]

29. Paul, L.; Cattaneo, M.; D'Angelo, A.; Sampietro, F.; Fermo, I.; Razzari, C.; Fontana, G. Telomere length in peripheral blood mononuclear cells is associated with folate status in men in me. *J. Nutr.* **2009**, *139*, 1273–1278. [CrossRef]

30. Cramer, P.E.; Cirrito, J.R.; Wesson, D.W.; Lee, C.D.; Karlo, J.C.; Zinn, A.E.; Casali, B.T.; Restivo, J.L.; Goebel, W.D.; James, M.J. ApoE-directed therapeutics rapidly clear β-amyloid and reverse deficits in AD mouse models. *Science* **2012**, *335*, 1503–1506. [CrossRef]

31. Larsen, P.A.; Lutz, M.W.; Hunnicutt, K.E.; Mihovilovic, M.; Saunders, A.M.; Yoder, A.D.; Roses, A.D. The Alu neurodegeneration hypothesis: A primate-specific mechanism for neuronal transcription noise, mitochondrial dysfunction, and manifestation of neurodegenerative disease. *Alzheimer's Dement.* **2017**, *13*, 828–838. [CrossRef] [PubMed]

32. Liu, J.; Zhao, W.; Ware, E.B.; Turner, S.T.; Mosley, T.H.; Smith, J.A. DNA methylation in the APOE genomic region is associated with cognitive function in African Americans. *BMC Med. Genomics* **2018**, *11*, 43. [CrossRef] [PubMed]

33. Hernández, H.G.; Mahecha, M.F.; Mejía, A.; Arboleda, H.; Forero, D.A. Global long interspersed nuclear element 1 DNA methylation in a Colombian sample of patients with late-onset Alzheimer's disease. *Am. J. Alzheimer's Dis. Other Dement.* **2014**, *29*, 50–53. [CrossRef] [PubMed]

34. Du, A.; Jahng, G.; Hayasaka, S.; Kramer, J.; Rosen, H.; Gorno-Tempini, M.; Rankin, K.; Miller, B.; Weiner, M.; Schuff, N. Hypoperfusion in frontotemporal dementia and Alzheimer disease by arterial spin labeling MRI. *Neurology* **2006**, *67*, 1215–1220. [CrossRef] [PubMed]

35. Tosun, D.; Schuff, N.; Jagust, W.; Weiner, M.W.; Initiative, A.s.d.N. Discriminative power of arterial spin labeling magnetic resonance imaging and 18F-fluorodeoxyglucose positron emission tomography changes for amyloid-β-positive subjects in the Alzheimer's disease continuum. *Neurodegener. Dis.* **2016**, *16*, 87–94. [CrossRef] [PubMed]

36. Fällmar, D.; Haller, S.; Lilja, J.; Danfors, T.; Kilander, L.; Tolboom, N.; Egger, K.; Kellner, E.; Croon, P.M.; Verfaillie, S.C. Arterial spin labeling-based Z-maps have high specificity and positive predictive value for neurodegenerative dementia compared to FDG-PET. *Eur. Radiol.* **2017**, *27*, 4237–4246. [CrossRef] [PubMed]

37. Musiek, E.S.; Chen, Y.; Korczykowski, M.; Saboury, B.; Martinez, P.M.; Reddin, J.S.; Alavi, A.; Kimberg, D.Y.; Wolk, D.A.; Julin, P. Direct comparison of fluorodeoxyglucose positron emission tomography and arterial spin labeling magnetic resonance imaging in Alzheimer's disease. *Alzheimer's Dement.* **2012**, *8*, 51–59. [CrossRef] [PubMed]

38. Wolk, D.A.; Detre, J.A. Arterial spin labeling MRI: An emerging biomarker for Alzheimer's disease and other neurodegenerative conditions. *Curr. Opin. Neurol.* **2012**, *25*, 421. [CrossRef] [PubMed]

39. Yu, C.-E.; Cudaback, E.; Foraker, J.; Thomson, Z.; Leong, L.; Lutz, F.; Gill, J.A.; Saxton, A.; Kraemer, B.; Navas, P.; et al. Epigenetic signature and enhancer activity of the human APOE gene. *Hum. Mol. Genet.* **2013**, *22*, 5036–5047. [CrossRef]

40. Estrada-Orozco, K.; Bonilla-Vargas, K.; Cruz, F.; Mancera, O.; Ruiz, M.; Alvarez, L.; Pardo, R.; Arboleda, H. Cognitive Assessment Test: Validation of a Short Cognitive Test for the Detection of Mild Cognitive Disorder. *Int. J. Alzheimer's Dis.* **2018**, *2018*, 3280621. [CrossRef]

41. Espitia Mendieta, D. Funciones Ejecutivas en el Envejecimiento Normal: Datos Normativos Con La Batería Neuronorma. Master's Thesis, Universidad Nacional de Colombia, Bogotá, Colombia, 2017.

42. Peña-Casanova, J.; Quiñones-Ubeda, S.; Gramunt-Fombuena, N.; Quintana-Aparicio, M.; Aguilar, M.; Badenes, D.; Cerulla, N.; Molinuevo, J.L.; Ruiz, E.; Robles, A.; et al. Spanish Multicenter Normative Studies (NEURONORMA Project): Norms for verbal fluency tests. *Arch. Clin. Neuropsychol.* **2009**, *24*, 395–411. [CrossRef] [PubMed]

43. Sánchez-Benavides, G.; Peña-Casanova, J.; Casals-Coll, M.; Gramunt, N.; Manero, R.M.; Puig-Pijoan, A.; Aguilar, M.; Robles, A.; Antúnez, C.; Frank-García, A.; et al. One-Year Reference Norms of Cognitive Change in Spanish Old Adults: Data from the NEURONORMA Sample. *Arch. Clin. Neuropsychol.* **2016**, *31*, 378–388. [CrossRef] [PubMed]

44. Aranciva, F.; Casals-Coll, M.; Sánchez-Benavides, G.; Quintana, M.; Manero, R.M.; Rognoni, T.; Calvo, L.; Palomo, R.; Tamayo, F.; Peña-Casanova, J. Spanish normative studies in a young adult population (NEURONORMA young adults Project): Norms for the Boston Naming Test and the Token Test. *Neurología* **2012**, *27*, 394–399. [CrossRef] [PubMed]

45. Peña-Casanova, J.; Quiñones-Ubeda, S.; Gramunt-Fombuena, N.; Quintana, M.; Aguilar, M.; Molinuevo, J.L.; Serradell, M.; Robles, A.; Barquero, M.S.; Payno, M.; et al. Spanish Multicenter Normative Studies (NEURONORMA Project): Norms for the Stroop color-word interference test and the Tower of London-Drexel. *Arch. Clin. Neuropsychol.* **2009**, *24*, 413–429. [CrossRef] [PubMed]

46. Cruz-Sanabria, F.; Bonilla-Vargas, K.; Estrada, K.; Mancera, O.; Vega, E.; Guerrero, E.; Ortega-Rojas, J.; Mahecha María, F.; Romero, A.; Montañés, P.; et al. Análisis de desempeños cognitivos y polimorfismos en SORL, PVRL2, CR1, TOMM40, APOE, PICALM, GWAS_14q, CLU y BIN1 en pacientes con trastorno neurocognitivo leve y en sujetos cognitivamente sanos. *Neurología* **2018**. [CrossRef] [PubMed]

47. Gutiérrez, C.A.C.; Eslava, D.L.M.; Gavilán, P.R.; Ríos, P.M. Cambios en las actividades instrumentales de la vida diaria en la Enfermedad de Alzheimer. *Acta Neurol. Colomb.* **2010**, *26*, 112–121.

48. Jefferson, A.L.; Byerly, L.K.; Vanderhill, S.; Lambe, S.; Wong, S.; Ozonoff, A.; Karlawish, J.H. Characterization of activities of daily living in individuals with mild cognitive impairment. *Am. J. Geriatr. Psychiatry* **2008**, *16*, 375–383. [CrossRef] [PubMed]

49. Gil, L.; Ruiz de Sánchez, C.; Gil, F.; Romero, S.J.; Pretelt Burgos, F. Validation of the Montreal Cognitive Assessment (MoCA) in Spanish as a screening tool for mild cognitive impairment and mild dementia in patients over 65 years old in Bogotá, Colombia. *Int. J. Geriatr. Psychiatry* **2015**, *30*, 655–662. [CrossRef] [PubMed]

50. Romero-Vanegas, S.; Romero-Vanegas, S.; Vargas-Gonzalez, J.C.; Arboleda, H.; Lopera, F.; Pardo, R. Validation of the ineco frontal screening in a colombian population. *Alzheimer's Dement.* **2014**, *10*, P726. [CrossRef]

51. Angulo, C.B.G.; Arias, A.C. Escala de Yesavage para Depresión Geriátrica (GDS-15 y GDS-5) estudio de la consistencia interna y estructura factorial. *Univ. Psychol.* **2011**, *10*, 735–743.

52. Cummings, J.L.; Mega, M.; Gray, K.; Rosenberg-Thompson, S.; Carusi, D.A.; Gornbein, J. The Neuropsychiatric Inventory: Comprehensive assessment of psychopathology in dementia. *Neurology* **1994**, *44*, 2308–2314. [CrossRef]

53. Peña-Casanova, J.; Casals-Coll, M.; Quintana, M.; Sánchez-Benavides, G.; Rognoni, T.; Calvo, L.; Palomo, R.; Aranciva, F.; Tamayo, F.; Manero, R.M. Estudios normativos españoles en población adulta joven (Proyecto NEURONORMA jóvenes): Métodos y características de la muestra. *Neurología* **2012**, *27*, 253–260. [CrossRef]
54. Schinka, J.A.; Loewenstein, D.A.; Raj, A.; Schoenberg, M.R.; Banko, J.L.; Potter, H.; Duara, R. Defining Mild Cognitive Impairment: Impact of Varying Decision Criteria on Neuropsychological Diagnostic Frequencies and Correlates. *Am. J. Geriatr. Psychiatry* **2010**, *18*, 684–691. [CrossRef]
55. Clark, S.J.; Statham, A.; Stirzaker, C.; Molloy, P.L.; Frommer, M. DNA methylation: Bisulphite modification and analysis. *Nat. Protoc.* **2006**, *1*, 2353. [CrossRef]
56. Tusnady, G.E.; Simon, I.; Varadi, A.; Aranyi, T. BiSearch: Primer-design and search tool for PCR on bisulfite-treated genomes. *Nucleic Acids Res.* **2005**, *33*, e9. [CrossRef]
57. Hernández, H.G.; Tse, M.Y.; Pang, S.C.; Arboleda, H.; Forero, D.A. Optimizing methodologies for PCR-based DNA methylation analysis. *Biotechniques* **2013**, *55*, 181–197. [CrossRef]
58. Lewin, J.; Schmitt, A.O.; Adorján, P.; Hildmann, T.; Piepenbrock, C. Quantitative DNA methylation analysis based on four-dye trace data from direct sequencing of PCR amplificates. *Bioinformatics* **2004**, *20*, 3005–3012. [CrossRef]
59. Mastroeni, D.; Grover, A.; Delvaux, E.; Whiteside, C.; Coleman, P.D.; Rogers, J. Epigenetic changes in Alzheimer's disease: Decrements in DNA methylation. *Neurobiol. Aging* **2010**, *31*, 2025–2037. [CrossRef]

International Journal of
Molecular Sciences

MDPI

Review

Pin1 Modulation in Physiological Status and Neurodegeneration. Any Contribution to the Pathogenesis of Type 3 Diabetes?

Marzia Bianchi and Melania Manco *

Research Unit for Multi-factorial Diseases, Obesity and Diabetes, Bambino Gesù Children's Hospital IRCCS (Istituto di Ricovero e Cura a Carattere Scientifico), viale di San Paolo 15, 00146 Rome, Italy; marzia.bianchi@opbg.net
* Correspondence: melania.manco@opbg.net

Received: 27 June 2018; Accepted: 6 August 2018; Published: 8 August 2018

Abstract: Prolyl isomerases (Peptidylprolyl isomerase, PPIases) are enzymes that catalyze the isomerization between the *cis/trans* Pro conformations. Three subclasses belong to the class: FKBP (FK506 binding protein family), Cyclophilin and Parvulin family (Pin1 and Par14). Among Prolyl isomerases, Pin1 presents as distinctive feature, the ability of binding to the motif pSer/pThr-Pro that is phosphorylated by kinases. Modulation of Pin1 is implicated in cellular processes such as mitosis, differentiation and metabolism: The enzyme is dysregulated in many diverse pathological conditions, i.e., cancer progression, neurodegenerative (i.e., Alzheimer's diseases, AD) and metabolic disorders (i.e., type 2 diabetes, T2D). Indeed, Pin1 KO mice develop a complex phenotype of premature aging, cognitive impairment in elderly mice and neuronal degeneration resembling that of the AD in humans. In addition, since the molecule modulates glucose homeostasis in the brain and peripherally, Pin1 KO mice are resistant to diet-induced obesity, insulin resistance, peripheral glucose intolerance and diabetic vascular dysfunction. In this review, we revise first critically the role of Pin1 in neuronal development and differentiation and then focus on the in vivo studies that demonstrate its pivotal role in neurodegenerative processes and glucose homeostasis. We discuss evidence that enables us to speculate about the role of Pin1 as molecular link in the pathogenesis of type 3 diabetes i.e., the clinical association of dementia/AD and T2D.

Keywords: Alzheimer's disease; brain glucose metabolism; neuronal differentiation; neuronal degeneration; Prolyl isomerases; Pin1; type 2 diabetes; type 3 diabetes

---

1. Introduction

Prolyl isomerases (Peptidylprolyl isomerase, PPIases) are a class of enzymes that catalyze the *cis/trans* isomerization of the peptide bond between the preceding amino acid and the proline residue (Pro) [1–3]. The presence of Pro's unusual structure lowers the free energy difference and allows these conformational changes [4]. The catalytic activity of PPIases modulates enzyme activity, protein stability and cellular localization by mediating conformational changes of substrates [3,5–9]. Three distinctive groups belong to the class of PPIases: the FKBP (FK506 binding protein), the Cyclophilin and the Parvulin families (Peptidylprolyl *cis/trans* isomerase NIMA-interacting 1 (Pin1) and Par14) [1,3,7,8].

With respect to other PPIases, the distinctive feature of Pin1 is that the Serine (Ser) or the Threonine (Thr) that precedes the Pro residue is phosphorylated accounting for the correct enzyme-substrate recognition [3,9]. The proline-directed kinases phosphorylate Pin1 substrates and induce Pin1-dependent post-phosphorylation conformational changes that impact key proteins. These changes influence cellular functions such as mitosis, differentiation and metabolism. Deregulation of Pin1 substrates has been implied in the onset of various diseases, i.e., cancer, neurodegenerative disorders and metabolic syndromes including type 2 diabetes (T2D) [2–4,10–12].

Expression of Pin1 increases significantly in cell types that undergo active cell division [4,13], depending on cell type, availability of specific substrates, and biological context. Pin1 seems to play a dual role, promoting cell proliferation or even, at the opposite, cell death depending on the biological context. Different capacities result from the adoption of diverse mechanisms of regulation, i.e., gene transcription and protein phosphorylation.

Pin1 serves as molecular "switch" that modulates enzyme activity or an entire signaling pathway [14]. Deeper understanding of function and regulation of Pin1 is essential to appreciate fine mechanisms of signal transduction and integration that are triggered by the propyl isomerase.

In this review, we will discuss the pivotal role of Pin1 in physiological i.e., the neuronal development as well as in neurodegenerative conditions i.e., the Huntinton's, the Parkinson's, the temporal lobe epilepsy (TLE) and the Alzheimer's (AD) diseases.

Then, with regard to AD, important features of the disease phenotype are reduced brain utilization of glucose and impaired glycolytic enzyme activities owing to defective insulin signaling [15]. Therefore AD is deemed as "type 3 diabetes" (T3D) [16] and regarded a metabolic disease. de la Monte defines AD "as a brain form of diabetes in which insulin resistance and deficiency develop either primarily in the brain, or due to systemic insulin resistance disease with secondary involvement of the brain" [17]. Interestingly, in patients with AD, Pin1 seems involved not only in neuronal degeneration but also in brain impaired glucose metabolism. Therefore, Pin1 may be at the crossroad between AD and T2D-associated dementia, contributing to the clinical relationship of these two conditions in the T3D.

2. Pin1 Characterization: Structure, Regulation and Subcellular Localization

Pin1 is a ubiquitous enzyme, with high similarity across species (*Homo sapiens*, *Xenopus laevis*, and *Danio rerio*) that suggests conserved function in vertebrates [4]. X-ray crystallography and solution nuclear magnetic resonance (NMR) studies in extracts of human brain tissues show that Pin1 is a monomeric enzyme of 163 amino acids. Pin1 consists of two functional domains, the WW- at the N-terminus and the PPIase-domains at the C-terminus, covalently fastened by a flexible linker (amino acids 38–53) of 15-residues [9,18]. The linker allows the domains to rotate independently of each other [19,20]. The PPIase domain catalyzes specifically the *cis*/*trans* isomerization of pSer/pThr-peptidyl-prolyl bonds. The WW domain (type IV) binds, but does not catalyze, similar epitopes [20,21]. Upon Pin1 binding of the specific substrate to the previous phosphorylated S/T-P motifs by proline directed kinases, the two functional domains of Pin1 interact partially with each other in a substrate-dependent manner [20,22,23]. Recognition of the substrate determines a loss of flexibility of residue side-chains in the region between the catalytic loop and the inter-domain surface. The reduced flexibility leads to increased contacts between the two functional domains [23].

Both E2F (E2 factor) [24], a protein involved in the cell cycle regulation, and N1ICD (Notch Intracellular Domain) [25], a protein that regulates Pin1 generating a positive loop in breast cancer cells, are able to activate transcription of Pin1 upon binding. On the contrary, AP4 (Activating Enhancer Binding Protein 4), the brain-selective transcription factor [26], down regulates Pin1 transcription, and the proteasome is responsible of the protein degradation following its ubiquitination [4,27].

Protein kinases regulate the capability of Pin1 to bind substrates. Steps for binding are phosphorylation of S/T-P motifs of substrates by kinases first, and secondly, phosphorylation of specific Ser residues in the Pin1 WW domain. Phosphorylation at the Ser16 residue modulates the ability of Pin1 to bind its substrates in a substrate-dependent manner [3]. In particular, phosphorylation by PKA (Protein kinase A) [28] and Aurora A [29] kinases inhibited the binding of Pin1 to its substrate resulting in mitotic block [28] or progression to G2/M phase [29]. Conversely, both Ser/Thr-kinase RSK2 (ribosomal protein S6 kinase 2) [30] and COT (Cancer Osaka Thyroid) [31] increased the binding of Pin1 to substrates leading to tumor progression.

Other kinases induce post-translational modifications of Pin1 in different Ser residues. The phosphorylation on Ser71 by DAPK1 (Death-associated protein kinase 1), a known tumor suppressor, inhibited Pin1 catalytic activity during the cell cycle progression [32]. MLK3 (Mixed Lineage Kinase 3) is a MAP3K (Mitogen-activated protein kinase kinase kinase) phosphorylated Pin1 on Ser138, thus increasing its activity to drive cell cycle progression [33]. Phosphorylation on Ser65 by PLK1 (Polo-like kinase 1), a regulator of mitotic checkpoints, stabilized Pin1 structure, reducing its ubiquitination and consequently degradation [27].

As to the exact subcellular localization of Pin1, this is not unique. Pin1 subcellular localization seems to vary upon post-translational modification of different Ser residues and different cellular contexts. When the Ser16 residue was phosphorylated and Pin1 modulation was associated with mitotic arrest, Pin1 was found in the nucleus [28,29]. When the same residue was phosphorylated in mouse embryonic fibroblasts and breast cancer cells [30,31], Pin1 was detected out of the nucleus. When Pin1 was phosphorylated on Ser138 enhancing tumorigenesis, Pin1 *trans*-located into the nucleus [33].

Subcellular localization of Pin1 seems to vary in different cell types, tissues and health status. Pin1 localized in the nucleus of cell lines such as SH-SY5Y, but also in the cytoplasm of primary neuron cultures or tissues [4]. Pin1 presented a cytoplasmic localization in axons from cultured DRG (dorsal root ganglia) neurons [34], but it was localized in both compartments at comparable levels in primary cultured mouse cortical neurons [35] and embryonic NPCs (neural stem/progenitor cells) [36]. Ibarra and co-workers found that Pin1 was preferentially localized in the cytoplasm of a number of neuronal cells, but there was nuclear localization in some cases or even nuclear exclusion of Pin1 in some other cases (i.e., in embryos and in adult brain of zebrafish) [4]. In brains from autopsies, PIN1 was found to localize in the nucleus of normal neurons, but both in nuclei and cytoplasm in brains from patients suffering frontotemporal dementias (FTD) [37] or AD [38–40]. In AD patients, Pin1 localized also exclusively in the cytoplasm of neurons of certain brain regions i.e., the hippocampus [38–40].

A sound explanation for the varying cell localization of Pin1 is a dynamic regulation of the enzyme in different cell-types, developmental stages and pathological conditions. Such dynamic regulation might depend on phosphorylation of Pin1 and generation of different isovariants [4]. Indeed, Pin1 showed a number of different phosphorylated isovariants [28–33] that were detectable combining bi-dimensional (2D) denaturant gel electrophoresis followed by western blotting analysis in the zebrafish model at different developmental stages [4].

3. Physiological Role of Pin1 in In Vivo Brain Development

There is a bulk of evidence demonstrating the effects of Pin1 on specific substrates and/or signaling pathways in cell lines. On the contrary, in vivo evidence is limited. We focused on studies performed in in vivo models of mouse and zebrafish to highlight the role of Pin1 in neuronal development.

3.1. Pin1 Expression during Embryogenesis in the Zebrafish

The expression of Pin1 is tightly regulated during embryogenesis. Levels of expression vary during stages of embryogenesis and across different regions of the embryo [4]. Pin1 mRNA was already detected at 1–2 cell-stages, indicating a maternal heredity. Its levels decreased during the next developmental steps. As to regional distribution, analysis of Pin1 mRNA/protein level and distribution provided inconsistent results. In whole mount in situ hybridization (WISH), mRNA levels of Pin1 were higher in head regions (cerebellum, ventricular zone of the diencephalon and thelencephalon) and lower in trunk and tail regions. Levels of Pin1 protein were not different between trunk and head regions [4] in immunofluorescence assays.

3.2. Pin1 Regulates Neuronal Cortical Differentiation: Modulation of the Wnt/β-Catenin Pathway

In 2012, it was demonstrated that Pin1 regulates differentiation of cortical neurons by affecting the Wnt/β-catenin pathway [36]. Pivotal in neurogenesis [36,41], the Wnt/β-catenin pathway works on the two main steps of the neural stem/progenitor cells (NPCs) development: the expansion phase, stimulating proliferation; and the neurogenic phase, regulating the timing and the area of the neuronal differentiation [36,42]. There was no doubt that the β-catenin functionality is regulated by phosphorylation, but the exact mechanism was unclear for long time until β-catenin was identified as specific substrate of Pin1 in NPCs. The researchers used a proteomic approach, glutathione S-transferase (GST)-pull down strategies followed by Electrospray Ionization-Mass Spectrometry (ESI-MS) analysis [36] to demonstrate that Pin1 regulates β-catenin functionality by modulating its conformation after the phosphorylation in the Ser246-Pro motif [36,43]. This occurred in the later phases of neuronal differentiation both in NPCs and mouse brain (Figure 1).

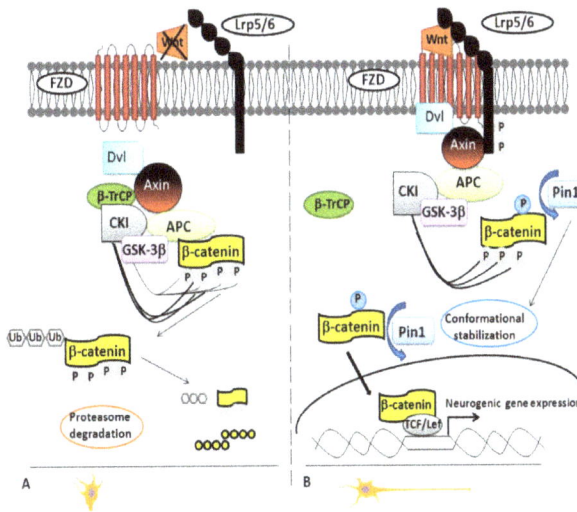

Figure 1. Pin1 modulation of Wnt/β-catenin pathway. **Panel A.** Inactivation of the Wnt/β-catenin pathway. In the absence of Wnt stimulation, levels of β-catenin decrease in the cytoplasm owing to ubiquitination and degradation by the proteasome. Both casein kinase 1α (CK1α) and glycogen synthase kinase-3β (GSK-3β) phosphorylate β-catenin to assemble a complex of proteins (named "destruction complex"). The complex includes scaffold protein axin, adenomatous polyposis coli (APC) and β-transducin repeat-containing E3 ubiquitin protein ligase (β-TrCP). Once phosphorylated, β-catenin is recognized by β-TrCP, ubiquitinated and then degraded. **Panel B.** Activation of the Wnt/β-catenin pathway. A Wnt ligand binds a Frizzled (Fz) receptor and coreceptors LRP5/6 activating the protein Dishevelled (Dvl) mostly by phosphorylation. This modification triggers the recruitment of axin to the phosphorylated tail of LRP and leads to the inhibition of the degradation pathway. Pin1 binds the β-catenin phosphorylated on the Ser246-Pro motif thus downregulating its binding with APC. By doing so, Pin1 catalyzes conformational stabilization of cytoplasmic β-catenin, which gets into the nucleus to regulate the transcription of Wnt target genes.

In NPCs and embryonal mouse cerebral cortex, Pin1 and β-catenin co-localized in both cytosol and nuclei [36].

In Pin1 KO NPCs from embryonal and adult brain, the percentage of differentiated neurons was significantly lower than in WT NPCs [36,44] as evident by using specific markers for neurons and glia cells [36]. Pin1 deficiency did not affect the NPCs proliferation in BrdU assay experiments, but it inhibited specifically cell differentiation into migrating immature neurons at embryonal stage E15.5. These results were consistent with the higher level of Pin1 expression in NPCs at later developmental step. Consistent with this finding, Pin1 KO mice showed an impaired motor activity during the neonatal stage, and this was the result of a specific inhibition of the differentiation of the upper layers neurons in the motor cortex. The authors performed experiments to stimulate/inhibit the signaling in order to confirm the role of Pin1 in the regulation of the Wnt/β-catenin pathway. They observed a rescue of Pin1 KO NPCs phenotype by using constitutively active S33Y β-catenin mutant or NPCs overexpressing Pin1. On the contrary, the truncated TCF4 (DN-TCF4), a dominant-negative mutant of the β-catenin activity, blocked the signaling and resulted in reduced neuronal differentiation [36].

3.3. Pin1 Regulation of Axonal Guidance by Modulation of Microtubule Assembly and Buffering Sema-3A Stimulation

During development of the CNS (central nervous system), axonal growth is tightly regulated by many extracellular mediators that are secreted or bound to cell membranes. The binding of these molecules to their receptors at the active growth cones triggers signaling cascades that modify dynamics of microtubules and result in axonal growth, turn, stop, or retraction [34,45]. Mechanisms of the signaling cascade are not fully understood.

Collapsin response mediator protein 2 (CRMP2), a tubulin heterodimer-binding protein that supports microtubule assembly and axon growth, plays a pivotal role in the cascade [46–48]. The binding affinity of CRMP2 to tubulin regulates the dynamic equilibrium of microtubule assembly-disassembly through phosphorylation by CDK5/GSK-3β (cyclin-dependent kinase 5/glycogen synthase kinase-3β) or Rho kinase [48,49]. In turn, stimulation of growing axons by Sema3A (semaphorin-3A) activates CDK5 [50]. The gene encoding CMRP2 presents two alternative splicing isoforms that differ in their N-terminus: CRMP2B and CRMP2A, a ~100 amino acids longer isoform. The latter was localized in axons rather than dendrites [51,52] and is likely modulated by conformational changes [53]. Balastik and co-authors contributed to the comprehension of this tightly regulated pathway, demonstrating that Pin1 binds and stabilizes CRMP2A. They observed that Pin1 is driven towards the growth cone after stimulation with Sema3A both in vitro and in vivo conditions [34]. Using a proteomic approach, combining GST-pull down followed by SDS-PAGE and tandem mass spectrometry (LC-MS/MS) experiments, they were the first to identify CRMP2A as major target of Pin1 in postnatal neurons. Pin1 stabilized CRMP2A previously phosphorylated by CDK5 selectively in distal axons [34]. Then, they found reduced level of CRMP2A at the growth cone in different experimental models: primary cortical neurons derived from Pin1 KO mice, cell lines knocked down (KD) for Pin1, or after using the specific Pin1 inhibitor (Juglone). In these experiments, Pin1 KO neurons had significantly shorter axons and this phenotype was completely reverted by over expression of CRMP2A. Therefore, they provided robust evidence on the relationship between Pin1 and CRMP2A [34].

Treatment with different concentrations of Sema-3A, and not with lysophosphatidic acid (LPA), a bioactive phospholipid in the Rho-kinase pathway, induced CRMP2 signaling via phosphorylation [54] and collapse of the growth cone both in WT and Pin1 KO primary dorsal root ganglia (DRG) neurons. This observation was associated with a change of Pin1 level and distribution in the growth cone of Pin1 WT. The change was dependent on the stimulation by Sema-3A. These observations confirmed that the catalytic action of Pin1 is specific for the Sema-3A signaling pathway in the vicinity of the growth cone where it co-localizes with CRMP2A [34]. Balastik et al. reported also that Pin1 KO embryos present selective defects of the axon growth that affect several regions of peripheral and CNS, like stunted neurite process and lack of arborization, probably owing to the impaired CRMP2A signaling [55,56]. The authors hypothesized that this uneven neuronal phenotype characterized by axonal defects might be due to compensatory mechanisms put in place for Pin1 deficiency by other members of the CRMP family to rescue reduction of CRMP2A. Indeed, Pin1 expression levels were negatively correlated with susceptibility to neurofibrillary degeneration in different regions of mouse and human brain [34,44].

To confirm in vivo the interplay between Pin1 and Sema-3A signaling, Balastick et al. used a zebrafish model of motor neuron development. Silencing of Sema-3A signaling by using KD of Neuropilin1 (NRP1) induced defects of the motor neuron growth [57,58]. Defects of the motor neuron growth were partially rescued in the simultaneous KD of NRP1 and Pin1. In Pin1 KD silencing of Sema-3A (NRP1 KD) produced a milder phenotype of motor neuron growth defects owing to the reduced stabilization of phosphorylated CRMP2A by Pin1. This would further support the notion that Pin1 regulates the Sema3A-driven axonal guidance. Indeed, Pin1 stabilizes CMRP2A selectively in distal axons and buffers low-level Sema3A stimulation both in vitro and in vivo [34].

4. Role of Pin1 in Neurodegenerative Disorders

The control of mitotic entry and progression is accompanied by the formation of specific phosphoepitopes such as MPM-2 (mitotic protein monoclonal 2), that are formed on Ser or Thr residues next to a Pro residue. Pin1 modulates function and/or dephosphorylation of some of these phosphoproteins, that are mostly recognized by the specific monoclonal antibody MPM-2 [59]. The induction of MPM-2 epitopes is a common feature in a number of neurodegenerative disorders (i.e., AD, FTD with Parkinsonism, progressive supranuclear palsy, corticobasal degeneration, Down's syndrome, and Pick's disease) [44]. Therefore, the presence of MPM-2 epitopes suggests the likely involvement of Pin1's catalytic activity in the pathogenesis of these heterogeneous conditions.

4.1. Huntington's Disease

Pin1 was reported to contribute also to the neurodegeneration seen in a mouse model of Huntington's disease. The expression of mutant Huntingtin (mHtt) determined the phosphorylation of p53 on a Ser46 residue that made it a target site for binding and modulation by Pin1 [60]. The authors hypothesized that this interaction caused the dissociation of p53 from the apoptosis inhibitor iASPP, promoting the p53 activation cascade in striatal neurons. The authors demonstrated that inhibition of Pin1, by using the specific inhibitor PiB, protected in vitro neuronal cells from mHtt-induced apoptosis. Therefore, inhibition of Pin1 might represent a therapeutic target for the treatment of Huntington's disease [60].

4.2. Parkinson's Disease

Lewy bodies (LBs) are aggregates of proteins that represent the histological hallmark of the Parkinson's disease (PD). α-synuclein, a presynaptic neuronal protein of unknown function, is the major constituent of LBs. Post-mortem histochemical analysis of patients' brain revealed the detection of Pin1 in the 50–60% of LBs [61]. Pin1 interacted indirectly and co-localized with α-synuclein in intracellular inclusions. Indeed, Pin1 bound the phosphorylated form of synphilin-1 mediated by casein kinase II (CKII) (player in cell cycle progression) and modulated the interaction between α-synuclein and synphilin-1. In neurons from substantia nigra or locus ceruleus of patients with PD, Pin1 had higher affinity for α-synuclein-synphilin-1 complex than for tau protein (in patients with AD it has the opposite affinity) and did not co-localize with the latter in LBs. Again, we face an example of the diverse modulation of Pin1 depending upon cellular types and biological contexts [61].

4.3. Temporal Lobe Epilepsy

In epileptic mice and patients with TLE, Pin1 was down-regulated as well, suggesting the involvement of the protein also in this disease [62]. Pin1 modulates an important neuronal protein, gephyrin [63], a postsynaptic scaffolding protein that favors the clustering of GABA(A) receptors at inhibitory synapses and that is down expressed in TLE patients [64]. Based on immunofluorescence, Pin1 localized in cytoplasm and cytoplasmic membranes of neurons from hippocampus and neocortex of epileptic patients and pilocarpine-induced epileptic mice [62]. Pin1 expression was down regulated in the hippocampus and cortex of mice with spontaneous recurrent seizures (SRS), compared to controls following the epileptic seizures. The authors performed immunoprecipitation experiments that demonstrated the interaction of Pin1 with NMDAR subunits 2A/2B (NR2A/2B) containing NMDA receptors (NMDARs) and not α-amino-3-hydroxy-5-methyl-4-isoxazole propionic acid receptors (AMPARs). The authors speculated that Pin1 influenced competitively the formation of synapse-associated protein-95/NR2B (PSD95/NR2B) complex thus, negatively affecting NMDAR-mediated synaptic transmission and spine morphology. The reduced expression of Pin1 caused the decreased surface trafficking of NMDARs by promoting NMDARs internalization with the net result of reduced neuronal hyper-excitability. Nevertheless, the fine mechanism by which Pin1 modulates NMDARs internalization remains unclear (Figure 2).

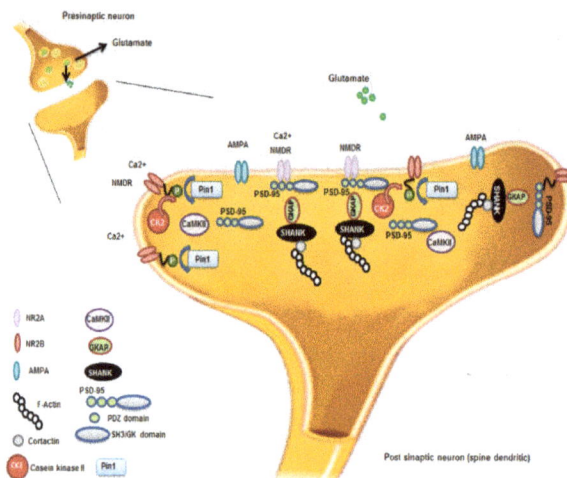

Figure 2. Pin1 regulation of NMDAR complex and synaptic plasticity. Synaptic activation of NMDAR stimulates Ca^{2+}/Calmodulin-dependent protein Kinase II (CaMKII) and casein kinase II (CK2) activity. CK2 phosphorylates the PDZ (postsynaptic density-95(PSD-95)/Discs large/zona occludens-1) ligand of NR2B. Pin1 binds and stabilizes probably the conformational change of NR2B phosphorylated disrupting the interaction between NMDAR on the cell surface and the PDZ domains of PSD-95. This leads to destabilization and internalization of surface NMDAR that influences, in turn, synaptic transmission and spine morphology. NR2A; NR2B; AMPA; F-Actin; Contarctin; CaMKII, Ca^{2+}-Calmodulin dependent protein Kinase II; CK2, casein Kinase 2; PSD-95 presents three PDZ domains, SH3 (Sarc homology 3 domain) and GK (guanylate kinase-like domain) domains; Pin1; GKAP, Guanylate kinase associate protein; Shank, shank protein.

In this disease too, Pin1 with its modulation might represent a target to modulate neuronal hyper-excitability [62].

4.4. Pin1 and Alzheimer's Diseases (AD)

AD is the most frequent form of dementia in the elderly causing progressive cognitive decline and memory loss. AD is characterized by widespread apoptosis of neurons and increased deposition of extracellular plaques and intracellular neurofibrillary tangles (NFTs) within the brain. NFTs are aggregates of microtubules that result from the hyperphosphorylation of tau protein. They are markers of AD: the amount of NFT deposits is correlated with the degree of neurodegeneration [65]. The plaques are primarily comprised of aggregated amyloid-β-peptide (Aβ) derived from the increased processing of the amyloid precursor protein (APP) [66].

Elderly Pin1 KO mice develop a complex phenotype of premature aging characteristics: namely, reduced body size, telomere instability, decreased germ-cell proliferation rate, cognitive impairment and neuronal degeneration that resembles human AD [44,67–70]. Of note, Pin1 KO mice show anomalous behavior and enhanced tau accumulation in the brain. These evidences support the argument that Pin1 dysfunction/deficiency is a pivotal determinant of the AD progression [3].

Level of Pin1 expression increased during neuronal differentiation and remained high during the lifespan [14,44]. In neuronal cells of AD patients, Pin1 was downregulated and the degree of downregulation was inversely correlated with the neuronal loss, since loss the protective ability of the enzyme against degeneration [44]. Pin1 regulated the *cis/trans* conformational changes of tau and APP proteins after their phosphorylation by kinases, such as GSK-3β [71]. Pin1 catalyzed the conformational switch of tau and APP proteins from the dysfunctional *cis*-toward the functional *trans*-structure. As consequence, tau could be dephosphorylated and degraded [39,70,71]. The accumulation of phosphorylated tau in Thr231-Pro motif was an early event in patients affected by mild cognitive impairment (MCI) [71]. In fact, Pin1 co-localized with phosphorylated tau and modulated the assembly of tau and tubulin into microtubules [72]. Pastorino and co-workers demonstrated that Pin1 drove the processing of APP and the formation of nonamyloidogenic βamyloid (Aβ) plaques. Indeed, the over-expression of Pin1 reduced Aβ secretion from cell cultures [70].

Pin1 modulated the APP processing toward the healthy nonamyloidogenic form by catalyzing *cis*-to *trans*-isomerization of the phosphorylated Thr668-Pro motif of APP [70] both directly and inhibiting the phosphorylation of APP in that motif induced by GSK-3β activity [73].

Pin1 inhibited GSK-3β activity trough the binding to a phosphorylated Thr330-Pro motif and catalyzing substrate isomerization [73]. This became evident in H4 neuroglioma cells transfected with WT or mutated construct GSK-3β T330A that did not affect the basal activity of the enzyme. In H4 cells, transfected with the mutated T330A GSK-3β, Pin1 was not able to bind and inhibit GSK-3β activity, thereby increasing the level of toxic amyloidogenic APP processing [73]. It has been speculated that the initial steps of tau phosphorylation have a neuronal protective role against the toxicity exerted by Aβ deposits [74,75], but they become detrimental when the excessive phosphorylation of tau causes the accumulation of NFTs [75]. Thus, the fine regulation of this post-translational modification is central in the pathogenesis of AD.

Alternatively, Pin1 may modulate APP processing to inhibit the phosphorylation of Thr668-Pro motif of APP performed by GSK-3β [73,76] Processing was observed in Pin1 WT or KD cells transfected both with WT or the mutated T330A GSK-3β constructs [73]. When Pin1 KD cells were transfected with both WT and mutated T330A GSK-3β, they showed comparable levels of pThr668, while Pin1 WT cells showed reduced levels of pThr668 only when transfected with WT GSK-3β construct. Therefore, the presence of Pin1 was needed for APP degradation following its binding to the Thr330-Pro motif of GSK-3β, resulting in its inhibition that then reduced phosphorylation of Thr668 [73].

In a molecular model, Pin1 was critical to regulate the active and stable conformation of Akt protein [77], that is an important player in cell survival, growth, migration and proliferation. The Akt signaling cascade is one of the survival pathways activated by neurotrophins through the binding to the tyrosine kinase (Trk) family of receptors. In *post mortem* studies of AD brains, Pin1 and neurotrophins expression levels were reduced in parallel [14]. As Angelucci and Hort hypothesized, another way for Pin1 to determine the fate of neuronal cell survival/death in neurodegenerative disorders might be related to the balance between expression of Pin1 and of neurotrophins (i.e., the brain derived growth factor, BDNF). Pin1 and neurotrophins can modulate the response of p53 [14,78,79], that acts not only as tumor suppressor, but also as a player in neuronal differentiation process [14,80], toward some transcriptional targets. The authors hypothesized that Pin1, modulating Akt and p53 conformation stability, influenced the signal transduction pathways activated by neurotrophins through the Trk receptors binding [14]. In neurodegenerative conditions, the reduced Pin1 expression level determined the reduced activity of neurotrophins-Trk signal transduction pathway causing an increased induction by p53 of neuronal death [14].

Furthermore, Pin1 influenced synaptic plasticity by regulating protein ubiquitination and degradation of postsynaptic density (PSD) proteins. By using a proteomic approach, Xu and collaborators demonstrated the association between Pin1 and Shank proteins at dendritic rafts of neuronal cells and PSD isolated from the synaptosome fractions obtained from the frontal cortical tissues of AD patients and controls [81]. Shank proteins have an "organizing role" in dendritic rafts and PSD. In cultured cortical neurons from Pin1 KO and control mice, Xu and collaborators verified the role of Pin1 in synaptic plasticity, inhibiting both Pin1 (with PiB treatment) and proteasome activity (by using MG132, carbobenzoxy-Leu-Leu-leucinal peptide). Ubiquitinated proteins in the PSD increased after the simultaneous treatment with PiB and MG132. This deregulated pathway determined an enhanced degradation of Shank3 and other PSD proteins, with a consequent alteration in PSD structure from AD brains. Pin1 modulated the NMDA receptor-mediated turnover of Shank proteins. In this regard, Tang and co-workers who studied Pin1 modulation in TLE patients [62], hypothesized that reduced Pin1 activity caused not only misfolded proteins, but also the generation of aberrant synapses contributing to the progression of pre-clinical AD [81].

Despite the role of Pin1 in the pathogenesis of AD, none of the PIN1 polymorphisms have been conclusively associated with delayed AD onset [26,65].

5. Pin1 Links Brain Impaired Glucose Metabolism and Neuronal Degeneration in AD

We speculate that post-translational modulation of Pin1 links AD among neurodegenerative disorders with peripheral and brain impaired glucose metabolism. Pin1 KO mice developed neuronal characteristics of premature aging similar to those observed in human AD i.e., age-related cognitive decline [44,68–70], but also peripheral (in both liver and muscle) and brain impairment of glucose metabolism and altered insulin signaling that resulted in overt glucose intolerance [3].

Autopsy studies in AD brains and intracerebral injected mice with streptozocin (STZ) demonstrated that brain insulin resistance and impaired insulin signaling occur early in the natural history of the disease, even before main clinical and histological characteristics develop. Main features in AD brains and STZ treated mice were cognitive impairment, structural deficits of neuronal cytoskeleton, loss of synaptic connections and increased neuronal apoptosis. Therefore, AD emerged as metabolic brain disease characterized by neuroinflammation and impaired energetic metabolism that lead to neuronal damage [16].

5.1. Type 3 Diabetes

The presence of defective insulin signaling and reduced glucose intake in the brain might be therefore early features that precede the diagnosis of overt cognitive deficits [16].

In *post mortem* brains (i.e., cerebral cortex, hippocampus and hypothalamus), the expression levels of insulin, insulin-like growth factor (IGF) 1, and IGF1 receptors were in AD patients significantly lower than in healthy age-matched individuals [82]. mRNA expression of IGF-2 tended to be different as well. Expression levels were correlated with the degree of cognitive impairment (Braak stages) in AD patients. Differential regional distribution of insulin, IGFs and their receptors was confirmed in primary neuronal cultures from rat fetal brains. Impairment of brain insulin signaling was induced in fetal rat by gestational exposure to ethanol. In this model, there was reduced production of insulin and IGFs as well as reduced expression of their receptors in different areas of the CNS. Reduced synthesis of growth factors in the brain could account for the increased neuronal death. Steen and de la Monte hypothesized that brain insulin resistance manifested as the inability to compensate for reduced secretion of insulin and IGFs with overexpression of their receptors [82]. They speculated a new type of diabetes that they named "type 3 diabetes" [16,17]. T3D would be characteristic of AD patients, affecting glucose metabolism exclusively in the brain or more broadly [75]. The reduced expression of insulin and IGF-1 receptors might explain the insulin resistance registered in the AD brain, thereby affecting negatively the insulin transduction pathway. Insulin/IGF-1 signaling upon binding to receptors activates the auto phosphorylation of receptors at Tyr residues and the activation of docking proteins, named insulin receptor substrates (IRS-1-4) and responsible for signal transduction. The intracellular signal determines the activation of mitogenic functions through MAPK/extracellular signal-regulated kinase

(MAPK/ERK) and metabolic functions through the phosphatidyl-inositol-3-kinase/Akt (PI3K/Akt) pathway that phosphorylates and inactivates GSK-3β. Given the impaired insulin/IGF-1 signal transduction, the expression level of IRS-1 was reduced in tissues from AD brains with respect to controls [82]. As consequence, the decrease of phospho-Akt and the increase of active form of GSK-3β, responsible for tau hyperphosphorylation, increased the apoptotic stimulus and caused mitochondrial dysfunction, thereby increasing mitochondrial-mediated apoptosis and oxidative stress [82].

Insulin and IGFs levels were reduced in brain tissues of AD patients and this was associated with the reduced synthesis of tau [82,83]. Growth factors stimulated tau protein expression in neuroblastoma cells [83]. The condition of brain insulin resistance was characterized by reduced levels of insulin and IGFs, no compensatory hyper-expression of their receptors and therefore a reduced transduction pathway [16,17,82]. de la Monte and collaborators speculated that the brain insulin resistance accounts for most of the molecular, histological and biochemical damages found in AD patients and develops before the onset of the clinical AD phenotype [17].

Therefore, insulin resistance that manifests in patients with T2D as peripheral hyperinsulinemia and increased release of IGF1, defective binding to receptors, impaired insulin signaling in muscle, adipose tissue and liver, could also be characterized in the brain by reduced release of insulin and IGFs and results as well in impaired insulin signaling in the brain. In both cases, impaired signaling causes reduced glucose utilization as determined by clamp and positron emission tomography studies. To further support commonalities in the pathogenesis of T2D and T3D, there is evidence from clinical trials using anti-diabetic dugs (namely analogs of glucagon like peptide 1, GLP1) in patients with T2D and/or AD that demonstrate the amelioration of the cognitive performance in parallel with the improvement of the glucose metabolism [17].

5.2. Pin1 and Insulin Pathways

Pin1 modulated peripheral glucose metabolism influencing independently both insulin secretion and sensitivity. β-cell-specific Pin1 KO (βPin1 KO) mice [84] developed glucose intolerance owing to reduced insulin secretion but preserved peripheral insulin sensitivity. βPin1 KO mice presented reduced β-cell mass as compared to controls suggesting that Pin1 affects β-cells proliferation. Cultured Pin1 KO β-cells had reduced intracellular $Ca2^+$ response to glucose- and KCl, despite preserved cellular ATP and insulin content. The mechanism by which Pin 1 influenced $Ca2^+$ response implied salt inducible kinase 2 (SIK2) and p35 protein. Pin1 interacted with SIK2, a protein belonging to the AMP-activated protein kinase (AMPK) family that is a key player for insulin secretion. The binding of Pin1 with SIK2 enhanced SIK2 kinase activity that, in turn, promoted p35 protein degradation and down-regulation of the p35-CDK5 (Cyclin-dependent kinase-5) complex activity, a negative regulator of $Ca2^+$ influx [84].

Experiments in KO mice, primary human endothelial cells and peripheral blood monocytes (PBMCs) of T2D patients demonstrated that hyperglycemia caused Pin1 upregulation and the latter, in turn, mediated vascular-damage occurring in diabetic patients [85]. Hyperglycemia was associated with reduced methylation of the *Pin1* gene promoter [85], and that is another finding commonly observed also in patients with AD. Indeed, Arosio and collaborators found upregulated *Pin1* gene expression in PBMCs of late onset AD patients and significantly reduced percentage methylation of *Pin1* gene promoter. In PBMCs of these patients, Pin1 quantity and activity tended to be reduced as well [86]. To explain this apparent divergent behavior, Wang [87] suggested that Pin1 reduction contributes first to the accumulation of hyperphosphorylated tau in AD patients, but successively Pin1 is over expressed to compensate for the increased formation of Aβ plaques.

AMPK is the major sensor for cellular energy status. Phosphorylation/dephosphorylation cycles of the α subunit control the activation state of AMPK. In presence of nutrients deficiency, the AMP/ADP molecules bind CBS domain of the γ subunit determining a conformational change of AMPK that protects the α subunit from dephosphorylation by protein-phosphatase 2C (PP2C). Pin1 regulated negatively AMPK, binding its CBS domain in the γ subunit. By doing so, it inhibited the binding of AMP/ADP to the γ subunit of AMPK. Therefore, Pin1 KO mice were resistant to high-fat-diet (HFD) induced obesity [88,89], with the activity level of AMPK in muscle significantly higher than in WT mice [88]. In that interaction with AMPK, Pin1 may represent a therapeutic target also to treat obesity and diabetes.

Indirect evidence in diabetic db/db treated by GLP1-analog demonstrated that PIN1 modulates insulin signaling also at the central level [72,90], being both an insulin receptor substrate-1 (IRS-1) and Akt binding partner [77,89]. About the role of Pin1 on IRS-1 signaling cascades, two hypotheses were formulated. Pin1 isomerase-activity may favor IRS-1 phosphorylation [91], or alternatively the isomerization of IRS-1 may prevent its dephosphorylation by the protein Tyr phosphatase [3]. Although Pin1 interacted with IRS-2, it did not influence its phosphorylation, likely because of a less efficient binding [89,92]. Pin1 isomerization activity induced a reduced response to insulin (insulin resistance), as demonstrated by its association with stress-induced c-Jun N-terminal kinase (JNK) and/or ribosomal protein S6 kinase (S6K), through the modulation of Ser-phosphorylation of IRSs [3,69,93]. In particular, Pin1 increased JNK kinase activity and the phosphorylation level of S6K. In HFD conditions, the Pin1 expression level increased determining the hyper-activation of JNK and S6K. Thus, in this metabolic condition the positive effect of Pin1 on IRS-1 and insulin cascade was abolished causing, instead insulin resistance.

Together with upstream kinases and protein phosphatases, Pin1 influenced the equilibrium between active and stable Akt versus the inactive and unstable form [77]. This modulation represents another way for Pin1 to influence the insulin transduction signaling. In fact, insulin promoted neuronal cytoskeletal dynamics via Akt phosphorylation [94]. Activation of insulin signaling leads usually to the activation of PI3K/Akt pathway. Akt, after being activated through its phosphorylation mainly at Thr308 residue, phosphorylates GSK-3β at Ser9 and inactivates GSK-3β kinase activity [95]. GSK-3β is one of the first enzymes involved in the regulation of glycogen synthase but is also the major protein kinase regulating tau phosphorylation in the brain [96,97] (Figure 3). GSK-3β is inactivated if phosphorylated in Ser9 residue by PI3K/Akt pathway. Its activity was found to be increased in insulin resistance conditions, suggesting that this kinase is pivotal in the regulation of peripheral and brain glucose utilization [75].

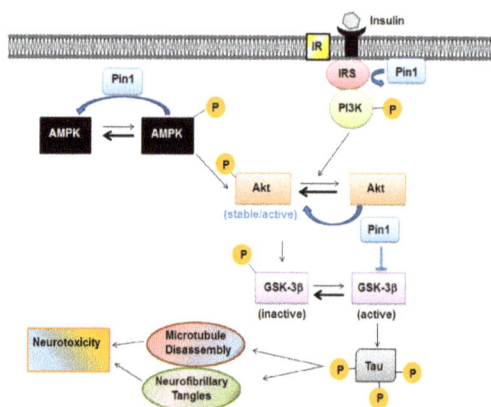

Figure 3. Pin1 in the metabolic pathway acts as a negative regulator of AMPK and modulates the Akt active form. Insulin promotes neuronal cytoskeletal dynamics via Akt phosphorylation. Activation of insulin signaling leads usually to activation of phosphatidyl-inositol-3-kinase/Akt (PI3K/Akt) pathway. After being activated through its phosphorylation mainly at Thr308 residue, Akt phosphorylates glycogen synthase kinase-3β (GSK-3β) at Ser9 and inactivates GSK-3β kinase activity. GSK-3β is also the major protein kinase regulating tau phosphorylation in the brain. Pin1 modulates all these players involved in the signaling cascade.

5.3. Pin1 and GSK-3β(Glycogen Synthase Kinase-3β) Modulation

Hence, Pin1 modulates insulin signaling by acting on Akt [77] and GSK-3β [73] activities. In brain extracts of elderly diabetic db/db mice, Akt activity was down-regulated and GSK-3β activity increased leading to the enhanced phosphorylation of tau protein and formation of NFTs [90]. It may be speculated that, more generally, the oxidative damage occurring in the diabetic brain results in impaired phosphorylation of tau protein and Pin1 modulation is implied. Oxidative modification of Pin1 at Cys113 residue was associated with reduced catalytic activity and expression in hippocampus of patients with MCI and AD too [72,98–100].

The mutual interaction between GSK-3β and Pin1 become evident also in experimental models of hereditary hemochromatosis, whereas homozygous patients carrying the HFE hemocromatosis mutation have enhanced risk of diabetes ("bronze diabetes") if untreated. In SH-SY5Y cells carrying H63D mutation of the HFE gene and in H67D transgenic mice, phosphorylation of Pin1 on Ser16 was increased and resulted in decreased enzyme activity. The treatment with Trolox, an antioxidant, rescued Pin1 activity in both models [101]. In the in vitro model, tau phosphorylation was increased paralleling GSK-3β activity [99], while Pin1 activity was consequently reduced [101]. Conversely, the inhibition of GSK-3β activity by lithium was associated with the increase of Pin1 activity, reinforcing the notion that GSK-3β and Pin1 interact with each other' to influence the ability of Pin1 to modulate its substrates such as tau protein [101,102].

6. Conclusions

The prolyl isomerase Pin1 plays a central role in the switching of proline-directed phosphorylation signaling: it induces the isomerization of the *cis/trans* configuration of protein substrates that are phosphorylated. As molecular "switch" that turns on or off enzyme activities or entire signaling pathways, Pin1 modulates many cellular functions both in physiological processes and pathological conditions (i.e., cancer, neurodegenerative and metabolic diseases). Pin1 exerts different and even opposite effects in vivo that are cell specific, depend upon specific protein substrates and are associated with phosphorylation of distinctive Pin1 binding sites.

Among the various regulatory functions, Pin1 controls neuronal differentiation during brain development by stabilizing β-catenin and synaptic plasticity/degradation of postsynaptic density proteins. Pin1 protects against oxidative damage of neurons, thereby protecting them from neurodegeneration. Indeed, impaired Pin1 function is associated with neurodegeneration and cognitive dysfunction in patients with AD. But Pin1 is also a master regulator of neuronal energy metabolism and interferes particularly with the cell insulin signaling. Because of these dual major roles on neuroprotection and metabolism, we and others speculate that impaired Pin1 function is one of the pivotal molecular links between neurodegeneration and impaired glucose metabolism, both manifesting in patients with dementia. Therefore, we report evidence supporting this hypothesis. There is large overlap between Pin1 and insulin signaling pathways within the brain, with Pin1 influencing insulin signaling.

Therefore, Pin1 may also represent a treatment target to prevent the onset of brain impaired glucose metabolism that is deemed by some researchers as one of the first hints in the pathogenesis of the AD. It may be effective to prevent the onset of or delay the neurodegenerative process, since acting as isomerase, Pin1 catalyzes the conformational switch of tau and APP proteins from the dysfunctional *cis*-toward the functional *trans*-structure and enhances tau dephosphorylation and degradation.

Author Contributions: M.B. and M.M. conceived the review; M.B. write the draft and M.M. revised for intellectual content. Both authors actively contributed in generating figures.

Funding: M.B. is supported by a grant from the Italian Ministry of Health (RC201802P004271).

Conflicts of Interest: The authors declare no conflicts of interest.

References

1. Gothel, S.F.; Marahiel, M.A. Peptidyl-prolyl *cis-trans* isomerases, a superfamily of ubiquitous folding catalysts. *Cell. Mol. Life Sci.* **1999**, *55*, 423–436. [CrossRef] [PubMed]
2. Reimer, U.; Scherer, G.; Drewello, M.; Kruber, S.; Schutkowski, M.; Fischer, G. Side-chain effects on peptidyl-prolyl *cis/trans* isomerisation. *J. Mol. Biol.* **1998**, *279*, 449–460. [CrossRef] [PubMed]
3. Nakatsu, Y.; Matsunaga, Y.; Yamamotoya, T.; Ueda, K.; Inoue, Y.; Mori, K.; Sakoda, H.; Fujishiro, M.; Ono, H.; Kushiyama, A.; et al. Physiological and Pathogenic Roles of Prolyl Isomerase Pin1 in Metabolic Regulations via Multiple Signal Transduction Pathway Modulations. *Int. J. Mol. Sci.* **2016**, *17*, 1495. [CrossRef] [PubMed]
4. Ibarra, M.S.; Borini Etichetti, C.; di Benedetto, C.; Rosano, G.L.; Margarit, E.; del Sal, G.; Mione, M.; Girardini, J. Dynamic regulation of Pin1 expression and function during zebrafish development. *PLoS ONE* **2017**, *12*, e0175939. [CrossRef] [PubMed]
5. Davies, T.H.; Ning, Y.M.; Sanchez, E.R. A new first step in activation of steroid receptors: Hormone-induced switching of FKBP51 and FKBP52 immunophilins. *J. Biol. Chem.* **2002**, *277*, 4597–4600. [CrossRef] [PubMed]
6. Liu, J.; Farmer, J.D., Jr.; Lane, W.S.; Friedman, J.; Weissman, I.; Schreiber, S.L. Calcineurin is a common target of cyclophilin-cyclosporin A and FKBP-FK506 complexes. *Cell* **1991**, *66*, 807–815. [CrossRef]
7. Lu, K.P.; Finn, G.; Lee, T.H.; Nicholson, L.K. Prolyl *cis-trans* isomerization as a molecular timer. *Nat. Chem. Biol.* **2007**, *3*, 619–629. [CrossRef] [PubMed]
8. Lu, K.P.; Zhou, X.Z. The prolyl isomerase PIN1: A pivotal new twist in phosphorylation signalling and disease. *Nat. Rev. Mol. Cell Biol.* **2007**, *8*, 904–916. [CrossRef] [PubMed]
9. Ranganathan, R.; Lu, K.P.; Hunter, T.; Noel, J.P. Structural and functional analysis of the mitotic rotamase Pin1 suggests substrate recognition is phosphorylation dependent. *Cell* **1997**, *89*, 875–886. [CrossRef]
10. Kim, E.K.; Choi, E.J. Pathological roles of MAPK signaling pathways in human diseases. *Biochim. Biophys. Acta* **2010**, *1802*, 396–405. [CrossRef] [PubMed]
11. MacAulay, K.; Doble, B.W.; Patel, S.; Hansotia, T.; Sinclair, E.M.; Drucker, D.J.; Nagy, A.; Woodgett, J.R. Glycogen synthase kinase 3α-specific regulation of murine hepatic glycogen metabolism. *Cell Metab.* **2007**, *6*, 329–337. [CrossRef] [PubMed]
12. Patrick, G.N.; Zukerberg, L.; Nikolic, M.; de la Monte, S.; Dikkes, P.; Tsai, L.H. Conversion of p35 to p25 deregulates Cdk5 activity and promotes neurodegeneration. *Nature* **1999**, *402*, 615–622. [CrossRef] [PubMed]
13. Bao, L.; Kimzey, A.; Sauter, G.; Sowadski, J.M.; Lu, K.P.; Wang, D.G. Prevalent overexpression of prolyl isomerase Pin1 in human cancers. *Am. J. Pathol.* **2004**, *164*, 1727–1737. [CrossRef]
14. Angelucci, F.; Hort, J. Prolyl isomerase Pin1 and neurotrophins: A loop that may determine the fate of cells in cancer and neurodegeneration. *Ther. Adv. Med. Oncol.* **2017**, *9*, 59–62. [CrossRef] [PubMed]
15. Talbot, K.; Wang, H.Y.; Kazi, H.; Han, L.Y.; Bakshi, K.P.; Stucky, A.; Fuino, R.L.; Kawaguchi, K.R.; Samoyedny, A.J.; Wilson, R.S.; et al. Demonstrated brain insulin resistance in Alzheimer's disease patients is associated with IGF-1 resistance, IRS-1 dysregulation, and cognitive decline. *J. Clin. Investig.* **2012**, *122*, 1316–1338. [CrossRef] [PubMed]
16. De la Monte, S.M.; Wands, J.R. Alzheimer's disease is type 3 diabetes-evidence reviewed. *J. Diabetes Sci. Technol.* **2008**, *2*, 1101–1113. [CrossRef] [PubMed]
17. De la Monte, S.M. Insulin Resistance and Neurodegeneration: Progress Towards the Development of New Therapeutics for Alzheimer's Disease. *Drugs* **2017**, *77*, 47–65. [CrossRef] [PubMed]
18. Verdecia, M.A.; Bowman, M.E.; Lu, K.P.; Hunter, T.; Noel, J.P. Structural basis for phosphoserine-proline recognition by group IV WW domains. *Nat. Struct. Biol.* **2000**, *7*, 639–643. [CrossRef] [PubMed]
19. Bayer, E.; Goettsch, S.; Mueller, J.W.; Griewel, B.; Guiberman, E.; Mayr, L.M.; Bayer, P. Structural analysis of the mitotic regulator hPin1 in solution: Insights into domain architecture and substrate binding. *J. Biol. Chem.* **2003**, *278*, 26183–26193. [CrossRef] [PubMed]
20. Eichner, T.; Kutter, S.; Labeikovsky, W.; Buosi, V.; Kern, D. Molecular Mechanism of Pin1-Tau Recognition and Catalysis. *J. Mol. Biol.* **2016**, *428 Pt A*, 1760–1775. [CrossRef]
21. Labeikovsky, W.; Eisenmesser, E.Z.; Bosco, D.A.; Kern, D. Structure and dynamics of pin1 during catalysis by NMR. *J. Mol. Biol.* **2007**, *367*, 1370–1381. [CrossRef] [PubMed]
22. Jacobs, D.M.; Saxena, K.; Vogtherr, M.; Bernado, P.; Pons, M.; Fiebig, K.M. Peptide binding induces large scale changes in inter-domain mobility in human Pin1. *J. Biol. Chem.* **2003**, *278*, 26174–26182. [CrossRef] [PubMed]

23. Namanja, A.T.; Peng, T.; Zintsmaster, J.S.; Elson, A.C.; Shakour, M.G.; Peng, J.W. Substrate recognition reduces side-chain flexibility for conserved hydrophobic residues in human Pin1. *Structure* **2007**, *15*, 313–327. [CrossRef] [PubMed]

24. Ryo, A.; Liou, Y.C.; Wulf, G.; Nakamura, M.; Lee, S.W.; Lu, K.P. *PIN1* is an E2F target gene essential for Neu/Ras-induced transformation of mammary epithelial cells. *Mol. Cell. Biol.* **2002**, *22*, 5281–5295. [CrossRef] [PubMed]

25. Rustighi, A.; Tiberi, L.; Soldano, A.; Napoli, M.; Nuciforo, P.; Rosato, A.; Kaplan, F.; Capobianco, A.; Pece, S.; Di Fiore, P.P.; et al. The prolyl-isomerase Pin1 is a Notch1 target that enhances Notch1 activation in cancer. *Nat. Cell Biol.* **2009**, *11*, 133–142. [CrossRef] [PubMed]

26. Ma, S.L.; Tang, N.L.; Tam, C.W.; Lui, V.W.; Lam, L.C.; Chiu, H.F.; Driver, J.A.; Pastorino, L.; Lu, K.P. A PIN1 polymorphism that prevents its suppression by AP4 associates with delayed onset of Alzheimer's disease. *Neurobiol. Aging* **2012**, *33*, 804–813. [CrossRef] [PubMed]

27. Eckerdt, F.; Yuan, J.; Saxena, K.; Martin, B.; Kappel, S.; Lindenau, C.; Kramer, A.; Naumann, S.; Daum, S.; Fischer, G.; et al. Polo-like kinase 1-mediated phosphorylation stabilizes Pin1 by inhibiting its ubiquitination in human cells. *J. Biol. Chem.* **2005**, *280*, 36575–36583. [CrossRef] [PubMed]

28. Lu, P.J.; Zhou, X.Z.; Liou, Y.C.; Noel, J.P.; Lu, K.P. Critical role of WW domain phosphorylation in regulating phosphoserine binding activity and Pin1 function. *J. Biol. Chem.* **2002**, *277*, 2381–2384. [CrossRef] [PubMed]

29. Lee, Y.C.; Que, J.; Chen, Y.C.; Lin, J.T.; Liou, Y.C.; Liao, P.C.; Liu, Y.P.; Lee, K.H.; Lin, L.C.; Hsiao, M.; et al. Pin1 acts as a negative regulator of G2/M transition by interacting with the Aurora-A-Bora complex. *J. Cell Sci.* **2013**, *126 Pt 21*, 4862–4872. [CrossRef]

30. Cho, Y.S.; Park, S.Y.; Kim, D.J.; Lee, S.H.; Woo, K.M.; Lee, K.A.; Lee, Y.J.; Cho, Y.Y.; Shim, J.H. TPA-induced cell transformation provokes a complex formation between Pin1 and 90 kDa ribosomal protein S6 kinase 2. *Mol. Cell. Biochem.* **2012**, *367*, 85–92. [CrossRef] [PubMed]

31. Kim, G.; Khanal, P.; Kim, J.Y.; Yun, H.J.; Lim, S.C.; Shim, J.H.; Choi, H.S. COT phosphorylates prolyl-isomerase Pin1 to promote tumorigenesis in breast cancer. *Mol. Carcinogen.* **2015**, *54*, 440–448. [CrossRef] [PubMed]

32. Lee, T.H.; Chen, C.H.; Suizu, F.; Huang, P.; Schiene-Fischer, C.; Daum, S.; Zhang, Y.J.; Goate, A.; Chen, R.H.; Zhou, X.Z.; et al. Death-associated protein kinase 1 phosphorylates Pin1 and inhibits its prolyl isomerase activity and cellular function. *Mol. Cell* **2011**, *42*, 147–159. [CrossRef] [PubMed]

33. Rangasamy, V.; Mishra, R.; Sondarva, G.; Das, S.; Lee, T.H.; Bakowska, J.C.; Tzivion, G.; Malter, J.S.; Rana, B.; Lu, K.P.; et al. Mixed-lineage kinase 3 phosphorylates prolyl-isomerase Pin1 to regulate its nuclear translocation and cellular function. *Proc. Natl. Acad. Sci. USA* **2012**, *109*, 8149–8154. [CrossRef] [PubMed]

34. Balastik, M.; Zhou, X.Z.; Alberich-Jorda, M.; Weissova, R.; Ziak, J.; Pazyra-Murphy, M.F.; Cosker, K.E.; Machonova, O.; Kozmikova, I.; Chen, C.H.; et al. Prolyl Isomerase Pin1 Regulates Axon Guidance by Stabilizing CRMP2A Selectively in Distal Axons. *Cell Rep.* **2015**, *13*, 812–828. [CrossRef] [PubMed]

35. Baik, S.H.; Fane, M.; Park, J.H.; Cheng, Y.L.; Yang-Wei Fann, D.; Yun, U.J.; Choi, Y.; Park, J.S.; Chai, B.H.; Park, J.S.; et al. Pin1 promotes neuronal death in stroke by stabilizing Notch intracellular domain. *Ann. Neurol.* **2015**, *77*, 504–516. [CrossRef] [PubMed]

36. Nakamura, K.; Kosugi, I.; Lee, D.Y.; Hafner, A.; Sinclair, D.A.; Ryo, A.; Lu, K.P. Prolyl isomerase Pin1 regulates neuronal differentiation via beta-catenin. *Mol. Cell. Biol.* **2012**, *32*, 2966–2978. [CrossRef] [PubMed]

37. Thorpe, J.R.; Mosaheb, S.; Hashemzadeh-Bonehi, L.; Cairns, N.J.; Kay, J.E.; Morley, S.J.; Rulten, S.L. Shortfalls in the peptidyl-prolyl *cis-trans* isomerase protein Pin1 in neurons are associated with frontotemporal dementias. *Neurobiol. Dis.* **2004**, *17*, 237–249. [CrossRef] [PubMed]

38. Holzer, M.; Gartner, U.; Stobe, A.; Hartig, W.; Gruschka, H.; Bruckner, M.K.; Arendt, T. Inverse association of Pin1 and tau accumulation in Alzheimer's disease hippocampus. *Acta Neuropathol.* **2002**, *104*, 471–481. [CrossRef] [PubMed]

39. Lu, P.J.; Wulf, G.; Zhou, X.Z.; Davies, P.; Lu, K.P. The prolyl isomerase Pin1 restores the function of Alzheimer-associated phosphorylated tau protein. *Nature* **1999**, *399*, 784–788. [CrossRef] [PubMed]

40. Ramakrishnan, P.; Dickson, D.W.; Davies, P. Pin1 colocalization with phosphorylated tau in Alzheimer's disease and other tauopathies. *Neurobiol. Dis.* **2003**, *14*, 251–264. [CrossRef]

41. Clevers, H. Wnt/β-catenin signaling in development and disease. *Cell* **2006**, *127*, 469–480. [CrossRef] [PubMed]

42. Hirabayashi, Y.; Itoh, Y.; Tabata, H.; Nakajima, K.; Akiyama, T.; Masuyama, N.; Gotoh, Y. The Wnt/β-catenin pathway directs neuronal differentiation of cortical neural precursor cells. *Development* **2004**, *131*, 2791–2801. [CrossRef] [PubMed]
43. Ryo, A.; Nakamura, M.; Wulf, G.; Liou, Y.C.; Lu, K.P. Pin1 regulates turnover and subcellular localization of β-catenin by inhibiting its interaction with APC. *Nat. Cell Biol.* **2001**, *3*, 793–801. [CrossRef] [PubMed]
44. Liou, Y.C.; Sun, A.; Ryo, A.; Zhou, X.Z.; Yu, Z.X.; Huang, H.K.; Uchida, T.; Bronson, R.; Bing, G.; Li, X.; et al. Role of the prolyl isomerase Pin1 in protecting against age-dependent neurodegeneration. *Nature* **2003**, *424*, 556–561. [CrossRef] [PubMed]
45. Culotti, J.G.; Kolodkin, A.L. Functions of netrins and semaphorins in axon guidance. *Curr. Opin. Neurobiol.* **1996**, *6*, 81–88. [CrossRef]
46. Fukata, Y.; Itoh, T.J.; Kimura, T.; Menager, C.; Nishimura, T.; Shiromizu, T.; Watanabe, H.; Inagaki, N.; Iwamatsu, A.; Hotani, H.; et al. CRMP-2 binds to tubulin heterodimers to promote microtubule assembly. *Nat. Cell Biol.* **2002**, *4*, 583–591. [CrossRef] [PubMed]
47. Inagaki, N.; Chihara, K.; Arimura, N.; Menager, C.; Kawano, Y.; Matsuo, N.; Nishimura, T.; Amano, M.; Kaibuchi, K. CRMP-2 induces axons in cultured hippocampal neurons. *Nat. Neurosci.* **2001**, *4*, 781–782. [CrossRef] [PubMed]
48. Yoshimura, T.; Kawano, Y.; Arimura, N.; Kawabata, S.; Kikuchi, A.; Kaibuchi, K. GSK-3beta regulates phosphorylation of CRMP-2 and neuronal polarity. *Cell* **2005**, *120*, 137–149. [CrossRef] [PubMed]
49. Arimura, N.; Menager, C.; Kawano, Y.; Yoshimura, T.; Kawabata, S.; Hattori, A.; Fukata, Y.; Amano, M.; Goshima, Y.; Inagaki, M.; et al. Phosphorylation by Rho kinase regulates CRMP-2 activity in growth cones. *Mol. Cell. Biol.* **2005**, *25*, 9973–9984. [CrossRef] [PubMed]
50. Sasaki, Y.; Cheng, C.; Uchida, Y.; Nakajima, O.; Ohshima, T.; Yagi, T.; Taniguchi, M.; Nakayama, T.; Kishida, R.; Kudo, Y.; et al. Fyn and Cdk5 mediate semaphorin-3A signaling, which is involved in regulation of dendrite orientation in cerebral cortex. *Neuron* **2002**, *35*, 907–920. [CrossRef]
51. Quinn, C.C.; Chen, E.; Kinjo, T.G.; Kelly, G.; Bell, A.W.; Elliott, R.C.; McPherson, P.S.; Hockfield, S. TUC-4b, a novel TUC family variant, regulates neurite outgrowth and associates with vesicles in the growth cone. *J. Neurosci.* **2003**, *23*, 2815–2823. [CrossRef] [PubMed]
52. Yuasa-Kawada, J.; Suzuki, R.; Kano, F.; Ohkawara, T.; Murata, M.; Noda, M. Axonal morphogenesis controlled by antagonistic roles of two CRMP subtypes in microtubule organization. *Eur. J. Neurosci.* **2003**, *17*, 2329–2343. [CrossRef] [PubMed]
53. Schmidt, E.F.; Strittmatter, S.M. The CRMP family of proteins and their role in Sema3A signaling. *Adv. Exp. Med. Biol.* **2007**, *600*, 1–11. [CrossRef] [PubMed]
54. Goshima, Y.; Nakamura, F.; Strittmatter, P.; Strittmatter, S.M. Collapsin-induced growth cone collapse mediated by an intracellular protein related to UNC-33. *Nature* **1995**, *376*, 509–514. [CrossRef] [PubMed]
55. Arimura, N.; Inagaki, N.; Chihara, K.; Menager, C.; Nakamura, N.; Amano, M.; Iwamatsu, A.; Goshima, Y.; Kaibuchi, K. Phosphorylation of collapsin response mediator protein-2 by Rho-kinase. Evidence for two separate signaling pathways for growth cone collapse. *J. Biol. Chem.* **2000**, *275*, 23973–23980. [CrossRef] [PubMed]
56. Tan, F.; Thiele, C.J.; Li, Z. Collapsin response mediator proteins: Potential diagnostic and prognostic biomarkers in cancers (Review). *Oncol. Lett.* **2014**, *7*, 1333–1340. [CrossRef] [PubMed]
57. Behar, O.; Golden, J.A.; Mashimo, H.; Schoen, F.J.; Fishman, M.C. Semaphorin III is needed for normal patterning and growth of nerves, bones and heart. *Nature* **1996**, *383*, 525–528. [CrossRef] [PubMed]
58. Gu, C.; Rodriguez, E.R.; Reimert, D.V.; Shu, T.; Fritzsch, B.; Richards, L.J.; Kolodkin, A.L.; Ginty, D.D. Neuropilin-1 conveys semaphorin and VEGF signaling during neural and cardiovascular development. *Dev. Cell* **2003**, *5*, 45–57. [CrossRef]
59. Escargueil, A.E.; Larsen, A.K. Mitosis-specific MPM-2 phosphorylation of DNA topoisomerase IIα is regulated directly by protein phosphatase 2A. *Biochem. J.* **2007**, *403*, 235–242. [CrossRef] [PubMed]
60. Grison, A.; Mantovani, F.; Comel, A.; Agostoni, E.; Gustincich, S.; Persichetti, F.; Del Sal, G. Ser46 phosphorylation and prolyl-isomerase Pin1-mediated isomerization of p53 are key events in p53-dependent apoptosis induced by mutant huntingtin. *Proc. Natl. Acad. Sci. USA* **2011**, *108*, 17979–17984. [CrossRef] [PubMed]

61. Ryo, A.; Togo, T.; Nakai, T.; Hirai, A.; Nishi, M.; Yamaguchi, A.; Suzuki, K.; Hirayasu, Y.; Kobayashi, H.; Perrem, K.; et al. Prolyl-isomerase Pin1 accumulates in lewy bodies of parkinson disease and facilitates formation of α-synuclein inclusions. *J. Biol. Chem.* **2006**, *281*, 4117–4125. [CrossRef] [PubMed]

62. Tang, L.; Zhang, Y.; Chen, G.; Xiong, Y.; Wang, X.; Zhu, B. Down-regulation of Pin1 in Temporal Lobe Epilepsy Patients and Mouse Model. *Neurochem. Res.* **2017**, *42*, 1211–1218. [CrossRef] [PubMed]

63. Zita, M.M.; Marchionni, I.; Bottos, E.; Righi, M.; del Sal, G.; Cherubini, E.; Zacchi, P. Post-phosphorylation prolyl isomerisation of gephyrin represents a mechanism to modulate glycine receptors function. *EMBO J.* **2007**, *26*, 1761–1771. [CrossRef] [PubMed]

64. Fang, M.; Shen, L.; Yin, H.; Pan, Y.M.; Wang, L.; Chen, D.; Xi, Z.Q.; Xiao, Z.; Wang, X.F.; Zhou, S.N. Downregulation of gephyrin in temporal lobe epilepsy neurons in humans and a rat model. *Synapse (N. Y.)* **2011**, *65*, 1006–1014. [CrossRef] [PubMed]

65. Harris, R.A.; Tindale, L.; Cumming, R.C. Age-dependent metabolic dysregulation in cancer and Alzheimer's disease. *Biogerontology* **2014**, *15*, 559–577. [CrossRef] [PubMed]

66. Zhang, H.; Ma, Q.; Zhang, Y.W.; Xu, H. Proteolytic processing of Alzheimer's beta-amyloid precursor protein. *J. Neurochem.* **2012**, *120*, 9–21. [CrossRef] [PubMed]

67. Atchison, F.W.; Capel, B.; Means, A.R. Pin1 regulates the timing of mammalian primordial germ cell proliferation. *Development* **2003**, *130*, 3579–3586. [CrossRef] [PubMed]

68. Atchison, F.W.; Means, A.R. Spermatogonial depletion in adult Pin1-deficient mice. *Biol. Reprod.* **2003**, *69*, 1989–1997. [CrossRef] [PubMed]

69. Lee, N.Y.; Choi, H.K.; Shim, J.H.; Kang, K.W.; Dong, Z.; Choi, H.S. The prolyl isomerase Pin1 interacts with a ribosomal protein S6 kinase to enhance insulin-induced AP-1 activity and cellular transformation. *Carcinogenesis* **2009**, *30*, 671–681. [CrossRef] [PubMed]

70. Pastorino, L.; Sun, A.; Lu, P.J.; Zhou, X.Z.; Balastik, M.; Finn, G.; Wulf, G.; Lim, J.; Li, S.H.; Li, X.; et al. The prolyl isomerase Pin1 regulates amyloid precursor protein processing and amyloid-beta production. *Nature* **2006**, *440*, 528–534. [CrossRef] [PubMed]

71. Driver, J.A.; Zhou, X.Z.; Lu, K.P. Regulation of protein conformation by Pin1 offers novel disease mechanisms and therapeutic approaches in Alzheimer's disease. *Discov. Med.* **2014**, *17*, 93–99. [PubMed]

72. Butterfield, D.A.; Abdul, H.M.; Opii, W.; Newman, S.F.; Joshi, G.; Ansari, M.A.; Sultana, R. Pin1 in Alzheimer's disease. *J. Neurochem.* **2006**, *98*, 1697–1706. [CrossRef] [PubMed]

73. Ma, S.L.; Pastorino, L.; Zhou, X.Z.; Lu, K.P. Prolyl isomerase Pin1 promotes amyloid precursor protein (APP) turnover by inhibiting glycogen synthase kinase-3β (GSK3β) activity: Novel mechanism for Pin1 to protect against Alzheimer disease. *J. Biol. Chem.* **2012**, *287*, 6969–6973. [CrossRef] [PubMed]

74. Ittner, A.; Chua, S.W.; Bertz, J.; Volkerling, A.; van der Hoven, J.; Gladbach, A.; Przybyla, M.; Bi, M.; van Hummel, A.; Stevens, C.H.; et al. Site-specific phosphorylation of tau inhibits amyloid-β toxicity in Alzheimer's mice. *Science (N. Y.)* **2016**, *354*, 904–908. [CrossRef] [PubMed]

75. Zhang, Y.; Huang, N.Q.; Yan, F.; Jin, H.; Zhou, S.Y.; Shi, J.S.; Jin, F. Diabetes mellitus and Alzheimer's disease: GSK-3β as a potential link. *Behav. Brain Res.* **2018**, *339*, 57–65. [CrossRef] [PubMed]

76. Aplin, A.E.; Gibb, G.M.; Jacobsen, J.S.; Gallo, J.M.; Anderton, B.H. In vitro phosphorylation of the cytoplasmic domain of the amyloid precursor protein by glycogen synthase kinase-3β. *J. Neurochem.* **1996**, *67*, 699–707. [CrossRef] [PubMed]

77. Liao, Y.; Hung, M.C. Physiological regulation of Akt activity and stability. *Am. J. Transl. Res.* **2010**, *2*, 19–42. [PubMed]

78. Brynczka, C.; Labhart, P.; Merrick, B.A. NGF-mediated transcriptional targets of p53 in PC12 neuronal differentiation. *BMC Genom.* **2007**, *8*, 139. [CrossRef] [PubMed]

79. Hu, H.; Wulf, G.M. The amplifier effect: How Pin1 empowers mutant p53. *Breast Cancer Res. BCR* **2011**, *13*, 315. [CrossRef] [PubMed]

80. Tedeschi, A.; Di Giovanni, S. The non-apoptotic role of p53 in neuronal biology: Enlightening the dark side of the moon. *EMBO Rep.* **2009**, *10*, 576–583. [CrossRef] [PubMed]

81. Xu, L.; Ren, Z.; Chow, F.E.; Tsai, R.; Liu, T.; Rizzolio, F.; Boffo, S.; Xu, Y.; Huang, S.; Lippa, C.F.; et al. Pathological Role of Peptidyl-Prolyl Isomerase Pin1 in the Disruption of Synaptic Plasticity in Alzheimer's Disease. *Neural Plast.* **2017**, *2017*, 3270725. [CrossRef] [PubMed]

82. Steen, E.; Terry, B.M.; Rivera, E.J.; Cannon, J.L.; Neely, T.R.; Tavares, R.; Xu, X.J.; Wands, J.R.; de la Monte, S.M. Impaired insulin and insulin-like growth factor expression and signaling mechanisms in Alzheimer's disease—Is this type 3 diabetes? *J. Alzheimer's Dis.* **2005**, *7*, 63–80. [CrossRef]

83. De la Monte, S.M.; Chen, G.J.; Rivera, E.; Wands, J.R. Neuronal thread protein regulation and interaction with microtubule-associated proteins in SH-Sy5y neuronal cells. *Cell. Mol. Life Sci.* **2003**, *60*, 2679–2691. [CrossRef] [PubMed]

84. Nakatsu, Y.; Mori, K.; Matsunaga, Y.; Yamamotoya, T.; Ueda, K.; Inoue, Y.; Mitsuzaki-Miyoshi, K.; Sakoda, H.; Fujishiro, M.; Yamaguchi, S.; et al. The prolyl isomerase Pin1 increases beta-cell proliferation and enhances insulin secretion. *J. Biol. Chem.* **2017**, *292*, 11886–11895. [CrossRef] [PubMed]

85. Paneni, F.; Costantino, S.; Castello, L.; Battista, R.; Capretti, G.; Chiandotto, S.; D'Amario, D.; Scavone, G.; Villano, A.; Rustighi, A.; et al. Targeting prolyl-isomerase Pin1 prevents mitochondrial oxidative stress and vascular dysfunction: Insights in patients with diabetes. *Eur. Heart J.* **2015**, *36*, 817–828. [CrossRef] [PubMed]

86. Arosio, B.; Bulbarelli, A.; Bastias Candia, S.; Lonati, E.; Mastronardi, L.; Romualdi, P.; Candeletti, S.; Gussago, C.; Galimberti, D.; Scarpini, E.; et al. Pin1 contribution to Alzheimer's disease: Transcriptional and epigenetic mechanisms in patients with late-onset Alzheimer's disease. *Neuro-Degener. Dis.* **2012**, *10*, 207–211. [CrossRef] [PubMed]

87. Wang, S.; Simon, B.P.; Bennett, D.A.; Schneider, J.A.; Malter, J.S.; Wang, D.S. The significance of Pin1 in the development of Alzheimer's disease. *J. Alzheimer's Dis.* **2007**, *11*, 13–23. [CrossRef]

88. Nakatsu, Y.; Iwashita, M.; Sakoda, H.; Ono, H.; Nagata, K.; Matsunaga, Y.; Fukushima, T.; Fujishiro, M.; Kushiyama, A.; Kamata, H.; et al. Prolyl isomerase Pin1 negatively regulates AMP-activated protein kinase (AMPK) by associating with the CBS domain in the gamma subunit. *J. Biol. Chem.* **2015**, *290*, 24255–24266. [CrossRef] [PubMed]

89. Nakatsu, Y.; Sakoda, H.; Kushiyama, A.; Zhang, J.; Ono, H.; Fujishiro, M.; Kikuchi, T.; Fukushima, T.; Yoneda, M.; Ohno, H.; et al. Peptidyl-prolyl cis/trans isomerase NIMA-interacting 1 associates with insulin receptor substrate-1 and enhances insulin actions and adipogenesis. *J. Biol. Chem.* **2011**, *286*, 20812–20822. [CrossRef] [PubMed]

90. Ma, D.L.; Chen, F.Q.; Xu, W.J.; Yue, W.Z.; Yuan, G.; Yang, Y. Early intervention with glucagon-like peptide 1 analog liraglutide prevents tau hyperphosphorylation in diabetic db/db mice. *J. Neurochem.* **2015**, *135*, 301–308. [CrossRef] [PubMed]

91. Zhang, J.; Nakatsu, Y.; Shinjo, T.; Guo, Y.; Sakoda, H.; Yamamotoya, T.; Otani, Y.; Okubo, H.; Kushiyama, A.; Fujishiro, M.; et al. Par14 protein associates with insulin receptor substrate 1 (IRS-1), thereby enhancing insulin-induced IRS-1 phosphorylation and metabolic actions. *J. Biol. Chem.* **2013**, *288*, 20692–20701. [CrossRef] [PubMed]

92. Rabiee, A.; Kruger, M.; Ardenkjaer-Larsen, J.; Kahn, C.R.; Emanuelli, B. Distinct signalling properties of insulin receptor substrate (IRS)-1 and IRS-2 in mediating insulin/IGF-1 action. *Cell. Signal.* **2018**, *47*, 1–15. [CrossRef] [PubMed]

93. Park, J.E.; Lee, J.A.; Park, S.G.; Lee, D.H.; Kim, S.J.; Kim, H.J.; Uchida, C.; Uchida, T.; Park, B.C.; Cho, S. A critical step for JNK activation: Isomerization by the prolyl isomerase Pin1. *Cell Death Differ.* **2012**, *19*, 153–161. [CrossRef] [PubMed]

94. Simpson, J.E.; Ince, P.G.; Shaw, P.J.; Heath, P.R.; Raman, R.; Garwood, C.J.; Gelsthorpe, C.; Baxter, L.; Forster, G.; Matthews, F.E.; et al. Microarray analysis of the astrocyte transcriptome in the aging brain: Relationship to Alzheimer's pathology and APOE genotype. *Neurobiol. Aging* **2011**, *32*, 1795–1807. [CrossRef] [PubMed]

95. Beale, E.G. Insulin signaling and insulin resistance. *J. Investig. Med.* **2013**, *61*, 11–14. [CrossRef] [PubMed]

96. Hernandez, F.; Lucas, J.J.; Avila, J. GSK3 and tau: Two convergence points in Alzheimer's disease. *J. Alzheimer's Dis.* **2013**, *33* (Suppl. 1), S141–S144. [CrossRef] [PubMed]

97. Takashima, A. GSK-3 is essential in the pathogenesis of Alzheimer's disease. *J. Alzheimer's Dis.* **2006**, *9*, 309–317. [CrossRef]

98. Butterfield, D.A.; Poon, H.F.; St Clair, D.; Keller, J.N.; Pierce, W.M.; Klein, J.B.; Markesbery, W.R. Redox proteomics identification of oxidatively modified hippocampal proteins in mild cognitive impairment: Insights into the development of Alzheimer's disease. *Neurobiol. Dis.* **2006**, *22*, 223–232. [CrossRef] [PubMed]

99. Hall, E.C., 2nd; Lee, S.Y.; Mairuae, N.; Simmons, Z.; Connor, J.R. Expression of the HFE allelic variant H63D in SH-SY5Y cells affects tau phosphorylation at serine residues. *Neurobiol. Aging* **2011**, *32*, 1409–1419. [CrossRef] [PubMed]
100. Chen, C.H.; Li, W.; Sultana, R.; You, M.H.; Kondo, A.; Shahpasand, K.; Kim, B.M.; Luo, M.L.; Nechama, M.; Lin, Y.M.; et al. Pin1 cysteine-113 oxidation inhibits its catalytic activity and cellular function in Alzheimer's disease. *Neurobiol. Dis.* **2015**, *76*, 13–23. [CrossRef] [PubMed]
101. Hall, E.C., 2nd; Lee, S.Y.; Simmons, Z.; Neely, E.B.; Nandar, W.; Connor, J.R. Prolyl-peptidyl isomerase, Pin1, phosphorylation is compromised in association with the expression of the HFE polymorphic allele, H63D. *Biochim. Biophys. Acta* **2010**, *1802*, 389–395. [CrossRef] [PubMed]
102. Min, S.H.; Cho, J.S.; Oh, J.H.; Shim, S.B.; Hwang, D.Y.; Lee, S.H.; Jee, S.W.; Lim, H.J.; Kim, M.Y.; Sheen, Y.Y.; et al. Tau and GSK3β dephosphorylations are required for regulating Pin1 phosphorylation. *Neurochem. Res.* **2005**, *30*, 955–961. [CrossRef] [PubMed]

International Journal of
Molecular Sciences

MDPI

Article

miRNA-34c Overexpression Causes Dendritic Loss and Memory Decline

Yu-Chia Kao [1,2], I-Fang Wang [1,3] and Kuen-Jer Tsai [1,4,*]

[1] Institute of Clinical Medicine, College of Medicine, National Cheng Kung University, Tainan 704, Taiwan; yukanomail2006@yahoo.com.tw (Y.-C.K.); e_fung0207@hotmail.com (I.-F.W.)
[2] Department of Pediatrics, E-Da Hospital, Kaohsiung 824, Taiwan
[3] Institute of Molecular Biology, Academia Sinica, Taipei 115, Taiwan
[4] Research Center of Clinical Medicine, National Cheng Kung University Hospital, College of Medicine, National Cheng Kung University, Tainan 704, Taiwan
* Correspondence: kjtsai@mail.ncku.edu.tw; Tel.: +886-6-235-3535-4254; Fax: +886-6-275-8781

Received: 30 June 2018; Accepted: 3 August 2018; Published: 8 August 2018

Abstract: Microribonucleic acids (miRNAs) play a pivotal role in numerous aspects of the nervous system and are increasingly recognized as key regulators in neurodegenerative diseases. This study hypothesized that miR-34c, a miRNA expressed in mammalian hippocampi whose expression level can alter the hippocampal dendritic spine density, could induce memory impairment akin to that of patients with Alzheimer's disease (AD) in mice. In this study, we showed that miR-34c overexpression in hippocampal neurons negatively regulated dendritic length and spine density. Hippocampal neurons transfected with miR-34c had shorter dendrites on average and fewer filopodia and spines than those not transfected with miR-34c (control mice). Because dendrites and synapses are key sites for signal transduction and fundamental structures for memory formation and storage, disrupted dendrites can contribute to AD. Therefore, we supposed that miR-34c, through its effects on dendritic spine density, influences synaptic plasticity and plays a key role in AD pathogenesis.

Keywords: miR-34c; dendritic spine; Alzheimer's disease

---

1. Introduction

Alzheimer's disease (AD) is a neurodegenerative disorder characterized by memory loss and cognitive decline due to extracellular accumulation of beta-amyloid peptide and intracellular accumulation of tau; it is also a consequence of dysfunction and loss of synapses [1]. From a pathological perspective, although neurofibrillary tangles and extracellular amyloid accumulations—defined as neuritic plaques—are the main hallmarks of AD, synaptic loss is the best predictor of clinical symptoms of AD [1].

Microribonucleic acids (miRNAs) are small noncoding RNAs composed of 19–25 nucleotides, which mediate the post-transcriptional regulation of gene expression. These non-coding RNAs function by binding to the 3' untranslated regions of target messenger RNAs (mRNAs) (within the RNA-induced silencing complex), silencing their target mRNAs, and downregulating protein translation [2]. Each miRNA contains seed sequences crucial for recognizing and binding to target mRNAs. Some miRNAs participate in key neuronal functions, such as cell signaling, transcription, and axonal guidance [3]. More than three miRNAs, including miR-124, miR-34, and miR-132, are key to hippocampal function [4,5]. miR-124 expression in the medial prefrontal cortex could partially rescue the behavioral deficits associated with frontotemporal dementia [6]. Zovoilis et al. found that miR-34c was upregulated in AD patients [7] Bhatnagar et al. showed that there was a strong correlation between the expression level of miR-34c and scores of mini-mental state examination [8]. Other studies have observed that miR-132 is downregulated in patients with mild cognitive impairment [9], and the miR-132 level is reduced in AD-affected brains [10]. Therefore, miR-124, miR-34, and miR-132 are regarded as memory miRNAs, which play a key role in neurodegenerative diseases such as AD, Parkinson's disease, and Huntington's disease [5].

miR-34 is a markedly conserved miRNA of the let-7 family, with an identical seed sequence of orthologues in flies, *Caenorhabditis elegans*, mice, and humans [11]. The miR-34 family consists of three members: miR-34a, miR-34b, and miR-34c. Of these, miR-34a and miR-34c have identical seed sequences, whereas the seed sequence of miR-34b is similar, but not identical. This suggests that miR-34a and miR-34c share similar mRNA targets, whereas those of miR-34b might be slightly different [12]. Although miR-34b and miR-34c are co-transcribed from the same chromosome, they differ in amount and may regulate different targets in a particular brain region [13]. In mice, miR-34a is ubiquitously expressed, with its highest levels appearing in the brain, whereas miR-34b and miR-34c are expressed mainly in the brain, lungs, testes, and ovaries [14].

The miR-34 family is essential for normal brain development [14]; miR-34 has been linked to neurogenesis, spine morphology, neurite outgrowth, neurodegeneration, and hippocampal memory formation [15]. In zebrafish, the repression of miR-34 led to developmental defects in the neuronal system with an enlargement of the hindbrain during early embryonic development [16]. Moreover, miR-34/449 controls mitotic spindle orientation during mammalian cortex development [17]. Furthermore, miR-34 has been implicated in brain disorders and aging; in *Drosophila*, loss of miR-34 decreased survival and accelerated brain aging and degeneration, whereas miR-34 upregulation extended survival and diminished neurodegeneration due to human pathogenic polyglutamine disease protein [11]. In addition to being present in neurons, miR-34 is expressed in glial cells [4].

Zovoilis et al. showed that miR-34c was elevated in the hippocampi of AD patients, aging mice, and APPPS1-21, a model of amyloid pathology linked to AD. The researchers demonstrated that miR-34c constrained memory consolidation, and miR-34c-mediated memory impairment was regulated, at least in part, by decreased hippocampal sirtuin-1 (SIRT1) levels [7]. Similarly, the expression of another miR-34 family member—miR-34a—was elevated in APPswe/PSΔE9 (AD mouse model) mice compared with control mice [18]. However, a contradictory correlation between miR-34 and cognitive function was also observed. A study that used Adeno-associated virus-delivered sponges in mice revealed that miR-34 inhibition impaired reference memory during the Morris water maze [4]. In *Drosophila*, loss of miR-34 accelerated brain aging and late-onset brain degeneration. miR-34 mutant flies were born with typical brains but showed dramatic vacuolization with age, whereas miR-34 upregulation extended their median lifespan and mitigated neurodegeneration [11]. Because AD is a neurodegenerative disorder associated with aging, we were unsure why loss of miR-34 accelerated brain aging in *Drosophila* while its overexpression caused memory impairment in mammals. Dickson et al. observed that miR-34a inhibited expression of endogenous tau, a crucial intraneuronal aggregate of AD [19]. In addition, Wu et al. found that overexpression of miR-34c downregulated tau expression in gastric cancer cells [20], although the researchers did not perform experiments to test cognitive function or analyze tau proteins in neural cells, the results were confusing in that the decreased tau expression associated with miR-34a and miR-34c overexpression contradicted AD pathogenesis. Therefore, miR-34 seems to play a mutual role in neurodegeneration; it can be protective or causative. For this reason, we wanted to study the role of miR-34c in memory function in mammals.

Dendritic spines are tiny protrusions along dendrites that constitute major postsynaptic sites for synaptic transmission [21]. These spines are highly dynamic structures that can undergo remodeling even in adults. Spine remodeling and new synapse formation, termed "plasticity", underlie the basis for learning and memory [22]. Loss or alteration of these structures was described in patients with neurodegenerative disorders such as AD [23], and synaptic reduction is the feature most closely related with decline in memory and cognition [24]. In one study, the density of neocortical synapses revealed highly powerful correlations with cognitive alterations in AD—the lower the mental status scores, the greater was the loss of synapses [25]. In another study, only weak correlations were observed between psychometric indices and plaques and tangles [26]. Thus, AD is primarily considered to be a synaptopathy [27].

Based on previous reports regarding the importance of particular hippocampal miRNAs in memory function, we studied the influence of miR-34c on memory function and compared dendritic spines between hippocampal neurons overexpressing miR-34c and controls, given that dendritic spines are the fundamental elements of synapses that form the basis of learning and memory.

2. Results

2.1. Overexpression of miR-34c Caused Memory Impairment

Using the Morris water maze task, we found that mice overexpressing miR-34c exhibited markedly increased latency compared with vector-transfected (control) mice (Figure 1A), indicating that the mice overexpressing miR-34c exhibited memory impairment. We used quantitative polymerase chain reaction (PCR) to measure the expression level of miR-34c in the hippocampi of miR-34c and control mice. The miR-34c expression level of the miR-34c-transfected mice was five times higher than that of the control mice (Figure 1B).

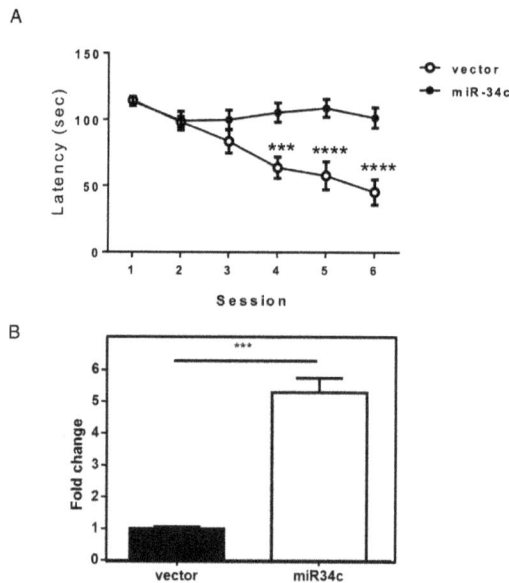

Figure 1. Overexpression of miR-34c caused memory impairment. (**A**) Escape latencies of mice overexpressing miR-34c and control vector (vec) mice ($n = 16$ in each group) in the Morris water maze. The asterisks represent the level of statistical significance calculated by a two-way analysis of variance (*** $p < 0.001$, **** $p < 0.0001$); (**B**) Hippocampal miR-34c expression levels were determined by quantitative PCR in mice transfected with miR-34c construct or mice transfected with the control vector and calculated using a two-tailed Student's *t*-test (*** $p < 0.001$).

2.2. Expression Level of miR-34c in the Hippocampus

The miR-34c construct was transfected into primary hippocampal cells to investigate the role of miR-34c in neuronal morphology. The transfection efficiency of miR-34c was identified on the seventh day of culture (DIV7) through immunofluorescence staining. Cells that co-expressed green fluorescence (from green fluorescent protein GFP$^{+/-}$ mice) and red fluorescence (from miR-34c-transfected cells) were selected, and neural dendrites were confirmed using the microtubule-associated protein 2 (Map2) antibody, which stained the soma and dendrites, but not the axons (Figure 2A). The expression levels of miR-34c in the transfected and control cells were detected using reverse transcription PCR (RT-PCR).

Of the hippocampal primary cells, the expression of miR-34c was nearly four times higher in the miR-34c-transfected cells than in the control cells (Figure 2B).

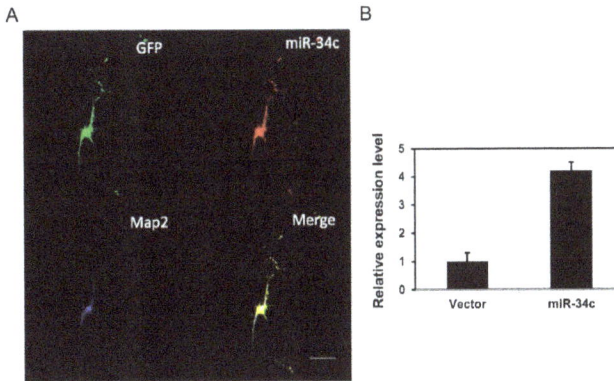

Figure 2. We observed that miR-34c was expressed in cultured hippocampal neurons. (**A**) Transfection of miR-34c was identified on DIV7 through immunofluorescence staining. Cells co-expressing green fluorescence (from GFP$^{+/-}$ mice) and red fluorescence (from miR-34c-transfected cells) were selected, and neural dendrites were confirmed using the Map2 antibody. Scale bar = 50 μm; (**B**) The expression level of miR-34c in the cultured hippocampal cells transfected with the miR-34c construct or control vector were detected using RT-PCR and calculated using a two-tailed Student's *t*-test.

2.3. Overexpression of miR-34c Reduced Dendritic Length and Branch Numbers

To investigate the effect of miR-34c on neuronal morphology, we measured the total dendritic length, dendritic main shaft, and dendritic branch numbers of the hippocampi two days after transfection. In the earlier stage—DIV 7—it appeared that miR-34c suppressed the outgrowth and branching of dendrites; however, this result was nonsignificant (Figure 3A). In the later stage—at DIV 14—the total dendritic length of neurons with miR-34c overexpression had a shorter dendritic length than did the neurons transfected with the control vector (Figure 3B). The numbers of dendritic shafts and branches were slightly decreased in the miR-34c-overexpression group; however, this result was nonsignificant. Therefore, miR-34c partially inhibited dendritic outgrowth but did not affect dendritic branching.

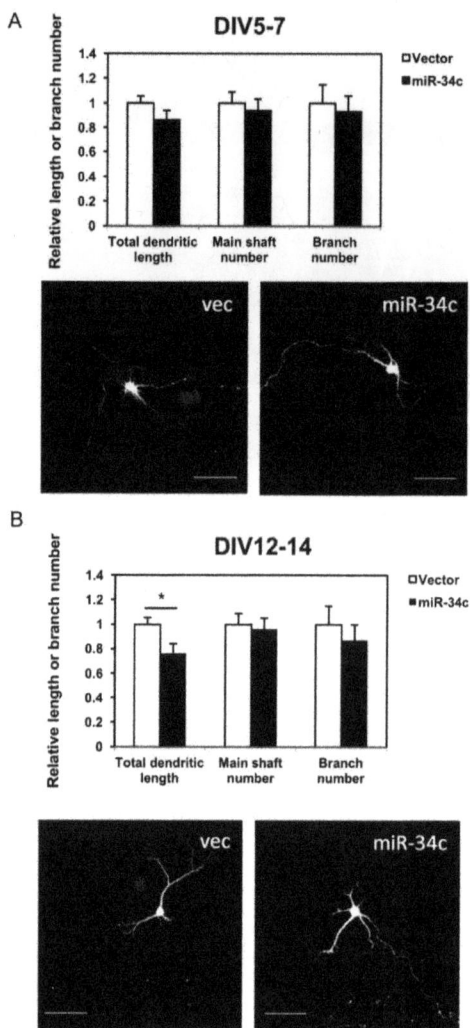

Figure 3. We observed that miR-34c affected the dendritic length of hippocampal neurons. Immunostaining with GFP, red fluorescent protein, and the Map2 antibody was performed on (**A**) DIV 7 (Scale bar = 100 μm) and (**B**) DIV 14. The total dendritic length, shaft number, and branch number of miR-34c hippocampal neurons and vector-transfected (vec) hippocampal neurons (*n* = 18 in each group) were counted under a fluorescence and confocal microscope (20× magnification) and calculated using a two-tailed Student's *t*-test (* *p* < 0.05). Scale bar = 100 μm.

2.4. Effects of miR-34c Overexpression on Neurite Protrusion Density

Protrusions of neurites develop at an early stage of neuronal development and later differentiate into mature spines. This process requires electrical signals from other neurons. The density of protrusions reflects not only the morphological state of the neurons, but also the physiological function. Therefore, we investigated the status of these protrusions by categorizing them into spines and filopodia, and measured the total protrusions, defined as the summation of spines and filopodia. Spines were defined as short extensions from the membranes of dendrite, with mushroom-like heads or extensions with an obscure neck. Filopodia were slender protrusions mostly twice longer as spines, and were identified as protrusions without an apparent shallow head; they were considered the immature form of protrusions. On DIV7, when the protrusions had not developed completely, neurons transfected with miR-34c had a significantly lower spine density and total number of protrusions (Figure 4A). On DIV14, the neurons transfected with miR-34c exhibited declines in spines and filopodia, as well as in total protrusions compared with the controls (Figure 4B).

2.5. Effects of miR-34c Overexpression on Protrusion Density in Various Dendritic Areas

To gain a better understanding of the effects of miR-34c on dendritic protrusion, each dendrite was divided into three equal segments of 40 μm for further analysis, starting from the soma as the proximal segment and followed by the middle and distal segments. The results revealed various density patterns in the spines and filopodia (Figure 5A). Spine density and total protrusions in all three segments of the miR-34c-transfected cells remained lower than those in the controls (Figure 5B,D). In addition, the reductions in spine density and total protrusions were proportional to their distance from the soma, meaning that in the miR-34c-overexpressing neurons, the farther the dendrites were from the soma, the lower was the number of spines. By contrast, no such change was observed in the filopodia along the dendrites (Figure 5C). Based on these findings, we assumed that miR-34c participated in the formation (turnover) of dendritic protrusions and may have acted as a negative regulator in the maturation of dendritic protrusions from filopodia into spines. While the spine density increased with the distance of dendrites from the soma in the control group, the miR-34c-transfected cells exhibited reduced spine density away from the soma.

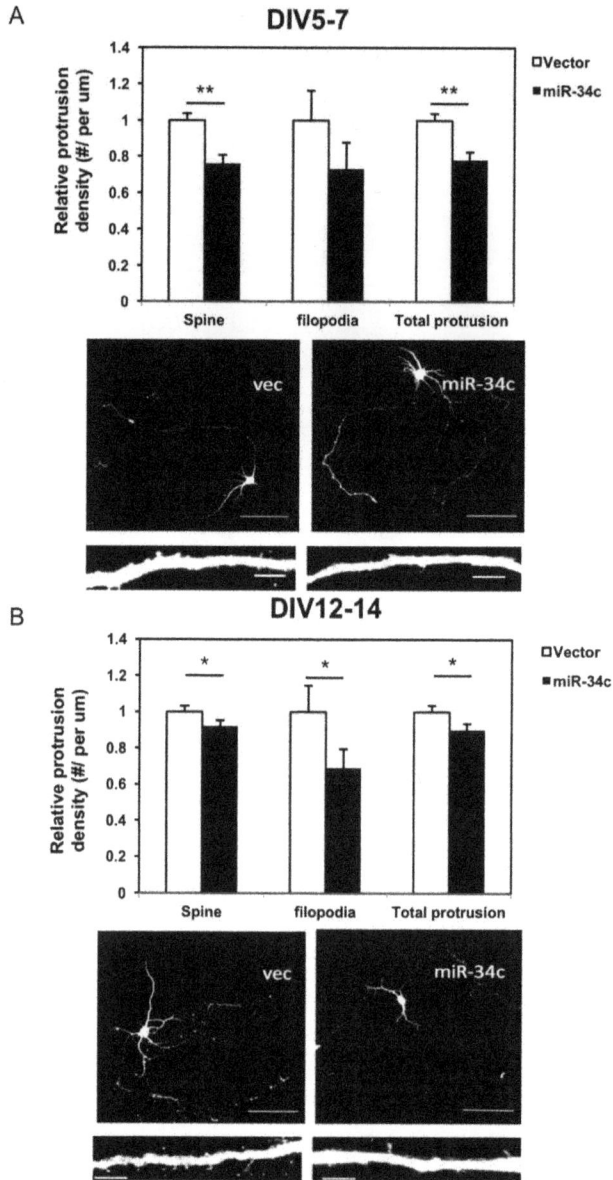

Figure 4. We observed that miR-34c lowered dendritic spine density and total protrusion density. Quantification of the protrusion density of neurites including mature spines, immature filopodia, and the summation of both using immunostaining on the (**A**) DIV 7 (Scale bar of upper panel = 100 μm; lower panel = 5 μm) and (**B**) DIV 14 in miR-34c and vector-transfected (vec) cells (*n* = 28 in each group) was conducted under 100× magnification and calculated using a two-tailed Student's *t*-test (* *p* < 0.05, ** *p* < 0.01). Scale bar of upper panel = 100 μm; lower panel = 5 μm.

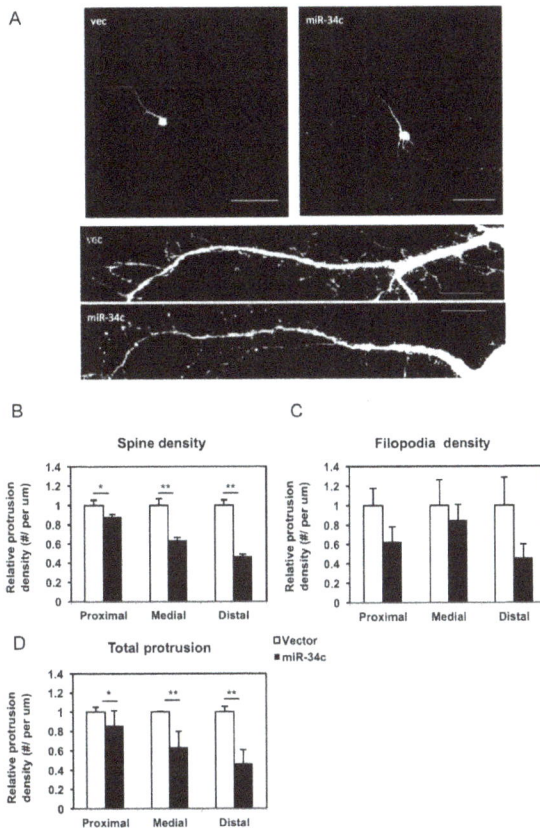

Figure 5. We observed that miR-34c exerted a negative effect on spine density along the dendrites. (**A**) Low- and high-power fields (20× in upper panels and 100× in lower panels) were used for counting. Scale bar of upper panel = 100 μm, lower panel = 20 μm. Quantification of the densities of (**B**) spines, (**C**) filopodia, and (**D**) total protrusions was conducted by dividing each dendrite into three equal segments of 40-μm in length, starting from the soma as the proximal segment (0–40 μm) and followed by the middle (40–80 μm) and distal (80–120 μm) segments, and calculated using a two-tailed Student's *t*-test (* $p < 0.05$, ** $p < 0.01$).

3. Discussion

The present study showed that mice overexpressing miR-34c exhibited significant memory impairment and hippocampal dendritic spine loss. In our study, miR-34c partially inhibited dendritic outgrowth as the total dendritic length was shorter. In addition, miR-34c resulted in reduced numbers of dendritic spines and total protrusions at an early stage (DIV 7). Subsequently, the miR-34c-transfected cells exhibited loss of filopodia and spines; in these cells, the farther the dendrites were from the soma, the lower was the number of spines; however, no such change was observed among filopodia. A study on the acoustic and visual cortices of autopsied AD patients revealed a marked decrease in the number of dendritic spines and loss of distal spines. However, the branches of the apical part of the dendritic tree and basal arborizations appeared to be equally affected by spine depletion [28]. The hippocampus contains two types of dendrites on pyramidal cells: apical and basal dendrites. Apical dendrites can be further divided into distal and proximal parts. Pyramidal cells segregate their inputs by using proximal apical dendrites, which project radially to local pyramidal cells and interneurons, while distal apical dendrites form non-local synapses by receiving inputs from distant cortical and thalamic locations [29]. CA1 neurons receive inputs at the distal tuft from the entorhinal cortex through the perforant path and from the thalamic nucleus reuniens, whereas the basal and apical dendrites receive inputs from the CA3 through the Schaffer collaterals. Based on the finding of more prominent distal dendritic spinal loss in the miR-34c cells in our study, we assumed that miR-34c might disrupt the perforant path from the entorhinal cortex of the memory circuit in mammalian brains.

Dendritic abnormalities in AD are widespread and occur in early stages of the disease; such abnormalities can be divided into (1) dystrophic neurites, (2) dendritic complexity reduction, and (3) dendritic spine loss. Marked dendritic spine loss is the final major dendritic abnormality found in AD patients, and the cortex and hippocampus are the two areas most affected by spine loss [30,31]. In APPxPS1-KI (an AD mouse model), an alteration of spine morphology occurred before reduction of synapses and neuronal density. The researchers observed a reduction in spine length and enlargement its neck, giving the spines a more "stubby" appearance [32].

Filopodia are the smallest structures protruding from dendrites [33]. Filopodia-like protrusions are highly unstable and can form and be removed within hours, whereas larger mushroom-type spines are more likely to remain stable for months to years. In both cell culture and in vivo, filopodia can form, stabilize, and grow into larger spines; this suggests that they represent an early stage of spine formation [34]. Transitions in spine size and stability—from immature and unstable filopodia to mature and stable spines—accompany the maturation and strengthening of synaptic contacts on spines [33]. An in vivo imaging study on the rodent neocortex demonstrated that small spines were NMDAR-dominated and highly motile, whereas large spines were AMPAR-dominated and highly stable. These findings led to the hypothesis that small thin spines are learning spines, whereas larger mushroom spines are memory spines [35]. Changes in the shapes of spine heads (morphogenesis) were actin-dependent and likely regulated by synaptic stimulation in response to a variety of stimuli [36]. In the present study, in the miR-34c-transfected mice, the number of spines were decreased compared with the control mice at an early stage and the filopodia remained unaffected, whereas immature filopodia and mature spines were lost at a later stage; thus, we speculated that miR-34c might disrupt the maturation of spines at an earlier stage and eventually destroy all dendritic spines, thereby affecting learning and memory consolidation.

Studies by Zovoilis and Bhatnagar et al. showed that miR-34c was increased in AD patients' brains and they pointed out one specific target—SIRT1—that contributed to memory decline [7,8]. Codocedo et al. later proved that SIRT1 overexpression was sufficient to change dendritic morphogenesis in the hippocampi and offer certain resistance against the cytotoxic damage induced by Aβ and avoid neuritic dystrophy [37]. Hence, the causal relationship between miR-34c and changes in dendritic spines in our study can be verified. Future research is required to investigate the pathogenic mechanism underlying the association between miR-34c and dendritic spine loss and to search for other possible miR-34c targets.

Structural changes in spines were driven by remodeling of the actin cytoskeleton [38]. Small Rho GTPases were deemed as central regulators of cell cytoarchitecture and played key roles in modulating cell migration, neurite outgrowth, survival, and synapse formation in neurons [39]. Cell division cycle 42 (Cdc42), ras-related C3 botulinum toxin substrate 1 (Rac1), and ras homolog family member A (RhoA) are the most frequently studied members of the small Rho GTPase family [40]. Rac and Rho are crucial to the maintenance of dendritic spines and branches in hippocampal pyramidal neurons [41] and appear to be critical in the formation and maintenance of memory [42]. Activation of Rac1 facilitated the formation of dendritic spines and increased spine head volume. By contrast, RhoA activation prevented spine formation and induced spine shortening [43]. As mentioned, miR-124, miR-34, and miR-132 are regarded as memory miRNAs. Both miR-124 and miR-132 have been reported to be associated with changes in dendritic spines through Rho GTPases, which underlie the stabilization of memory. In one study, miR-124 reduced the expression of Rho GTPs, thereby inhibiting axonal and dendritic branching via Rac1 and Cdc42 signaling, respectively [44]. Moreover, miR-132 activated Rac1 and promoted spine formation [45]. Thus far, no relationship between miR-34 and Rho GTPases in neurons has been reported. Since SIRT1 activators (RES) could modulates the dendritic arborization through the inhibition of Rho-associated protein kinase (ROCK) [37]—downstream effector of Rho GTPases—this served as indirect evidence for the association between miR-34c and Rho GTPases. Besides, one study investigated the roles of miR-34 and Rho GTPases in cancer cells; Huang et al. demonstrated that two key mechanisms involved in cancer metastasis—epithelial-to-mesenchymal transition and mesenchymal-to-amoeboid transition—were coupled through two miRNAs, namely miR-200 and miR-34, both of which inhibited RhoA and Rac1 [46]. This implies that Rho GTPase-dependent cytoskeletal changes might occur in the dendritic spines of miR-34-expressing hippocampal neurons, and thus underlie memory impairment.

Because postsynaptic density protein 95 (PSD-95) is the hallmark of a mature and stable glutamatergic synapse associated with memory consolidation [33], miR-34c might be associated with PSD-95. Bustos et al. designed an epigenetic editing strategy by using a zinc finger construct to control Dlg4/PSD-95 expression in the hippocampus and validate PSD-95 as a key player in plasticity and memory. PSD-95-6ZF-VP64 transduction could increase PSD-95 levels and recover learning and memory deficits in aged and AbPPswe/PS-1 (AD model) mice [47]. Whether synaptic changes in memory miRNAs are also related to PSD-95 requires further investigation.

4. Materials and Methods

4.1. Animals

C57BL/6JNarl (B6) mice were kept in a ventilated room under controlled conditions with a 12-h/12-h light/dark cycle and a maintained temperature of 22 ± 2 °C. The animals were given access to food and water ad libitum. This study was approved by the University Institutional Animal Care and Use Committee of National Cheng Kung University (NCKU) (Project code-106073, date of approval-23/12/2016). The experimental procedures for handling the mice were in accordance with the guidelines of the Institutional Animal Care and Use Committee of NCKU. Transgenic mice with $GFP^{+/-}$ expression and B6 mice were provided by the national laboratory animal center.

4.2. Deoxyribonucleic Acid Constructions

To produce the miRNA constructs, the genomic deoxyribonucleic acid (DNA) from a B6-wild type mouse was used as the complementary template to amplify the precursor of miR-34c by PCR, with the forward primer: 5'-GCA GTG TAA TTA GCT GAT TGT AGT G-3', and reverse primer: 5'-ATA TTA GGA AAC CAG CTG GT TTT AA-3'. After PCR amplification, the precursor miR-34c product was separated by electrophoresis and eluted from the agarose gel, ligated into a lentiviral construct-pFUGW, driven by a ubiquitin promoter, with a red fluorescent protein gene (dsRed) for co-expression.

4.3. Lentivirus Production

The lentivirus was produced by the RNAi core laboratory of Academia Sinica, Taipei, Taiwan with a 100× concentration.

4.4. Stereotaxic Hippocampal Injections

Male B6 mice were placed in the induction chambers. The oxygen flowmeter was adjusted to approximately 0.1–0.3 L/min. Isoflurane vaporizer was adjusted to 5% for anesthesia induction and 1.5–3.5% for maintenance. The anesthetized mice were then mounted in a stereotactic apparatus for intra-hippocampal injections. Concentrated lentivirus (1×10^7 in 2 μL) was injected into the dentate gyrus of bilateral hippocampus (−2.0 mm anterior–posterior, 1.8 mm medial–lateral, and −2.3 mm dorsal–ventral relative to the bregma) using a Hamilton micro-syringe (Hamilton Company, Reno, Nevada). After each injection, the needle was left in situ for 8 min to prevent regurgitation of the virus during removal. Control groups were injected with the same volume of control lentivirus. A Morris Water Maze test was performed 30 days after the injection.

4.5. Morris Water Maze Test

miR-34c-expressing and control vector-infected male B6 mice, $n = 16$ in each group, were tested 30 days after hippocampal injections of viral particles. The swimming pool is divided into four quadrants and contains an escape platform placed beneath the water. The mice were trained with four consecutive trials per day for six days. When released, the mice swam around the pool in search of an exit and on subsequent trials the mice were able to locate the platform more rapidly. Each trial lasted until the mice found the platform or for a maximum of 2 min. The time to locate the platform was recorded and the average latency calculated from the values of four trials each day. The swimming speed and probe test were recorded on video for further analysis.

4.6. Primary Hippocampal Culture

The embryos from B6 wild-type pregnant female mice were used for hippocampal primary culture. Pregnant B6 wild-type mice were dissected at embryonic day 16.5, and the hippocampi were separated from embryonic brains. The hippocampal tissues were washed with 1× Hank's balanced salt solution (HBSS) twice and digested with papain for about 10 min at 37 °C. Then, the aggregated tissues were washed with 1× HBSS twice, followed by resuspension in 1 ml Neurobasal medium with 1× B27 (Invitrogen, Thermo Fisher Scientific, Waltham, MA, USA), 1× Penicillin Streptomycin (P/S) antibiotics (Invitrogen), and 1× GlutaMax (Invitrogen). Cell viability was counted by an Invitrogen countess automatic cell counter. Cells were plated at 2×10^5 cells/well on poly-D-lysine-coated coverslips in six-well plates and incubated at 37 °C/5% CO_2. After 2 h, the medium was changed to a new Neurobasal medium to prevent contamination. Cells were cultured until DIV5 and DIV12 to perform miRNA transfection.

4.7. Transfection

Cells plated in six-well plates were transfected at DIV5 or DIV12. Thirty minutes before transfection, the medium was replaced with P/S antibiotics-free Neurobasal medium. The ubiquitin promoter-driven miR-34c plasmid, or vector only, were transfected into hippocampal primary culture with Lipofectamine 2000 (Invitrogen) for 2 h. The transfected cells were cultured for two days and analyzed by confocal microscopy for neuronal morphology or collected to extract RNA samples.

4.8. Immunofluorescence Staining

The transfected cells were washed twice with phosphate-buffered saline (PBS) and fixed with 4% paraformaldehyde (Electron Microscopy Sciences, Hatfield, UK). The cells were blocked for 1 h (blocking solution: 10% goat serum/PBS) at room temperature. The primary antibody was Map2 (Invitrogen) with 1% goat serum. All images were visualized using a Zeiss laser confocal microscope (LSM510, Zeiss, Oberkochen, Germany) and a fluorescence microscope (DeltaVision, GE Healthcare Life Sciences, London, UK), with a single optical section.

4.9. Image Analysis

Neural cells emitted green fluorescence because of the inherited GFP protein. All images were at the same resolution of 1024 pixels × 1024 pixels. Confocal z sectioning was used, and images were overlapped to have full signals. Each neuron was captured under the low-power field (20× objective, NA 1.4, respectively) for both soma and neurites. These low power images were used to measure the neuritic length and branch numbers, owing to the whole cell coverage with this larger visual field. Dendritic spines were focused on under high magnification (40× and 100× objective) in order to count spine numbers and observe spine morphology, specifically for measuring (1) total dendritic length, defined as the summation of all dendritic length in a neuron; (2) main dendritic shaft number, defined as the numbers of dendritic outgrowth from one soma; (3) dendritic branch number, defined as the total branch numbers from the primary dendrites; and (4) dendritic spine density, expressed as the average number of spines per μm of dendritic length. The total dendritic length and branch numbers were measured using the software NeuronJ (provided by Dr. Erik Meijering, Biomedical Imaging Group Rotterdam of Erasmus University Medical Center, Rotterdam, The Netherlands), whereas the numbers of different protrusions (spines, filopodia, and total protrusions) were counted using ImageJ (provided by Dr. Erik Meijering, Biomedical Imaging Group Rotterdam of Erasmus University Medical Center, Rotterdam, The Netherlands) in various areas divided by their distance from the soma (proximal: 0–40 μm, middle: 40–80 μm, distal: 80–120 μm).

4.10. Reverse Transcription PCR (RT-PCR)

The total RNA products extracted from the hippocampal primary culture or hippocampi in the brain were then reverse-transcribed to complementary DNA (cDNA) by SuperScript II RT (Invitrogen) following the Universal ProbeLibrary (UPL; Roche, Basel, Switzerland) protocol, and the expression levels of miR-34c were then detected by UPL probe #21, according to the manufacturer's protocol. Primer sequences used were as followed, miR-34c UPL RT primer: GTT GGC TCT GGT GCA GGG TCC GAG GTA TTC GC ACCAGAGCCAACGCAATC, miR-34c forward primer: 5'-CGG CGA GGC AGT GTA GTT AGC T-3', universal reverse primer: 5'-GTG CAG GGT CCG AGG T-3'.

4.11. Statistics

Data were analyzed using SPSS and the Morris water maze was analyzed using a two-way analysis of variance. Mean values were analyzed using a two-tailed Student's *t*- test. All results are represented as mean ± standard error of the mean and were first examined using an *f*-test to identify the homogeneity of variance. A probability level of <0.05 was accepted as statistically significant and degrees of significance are presented as * $p < 0.05$, ** $p < 0.01$, *** $p < 0.001$, **** $p < 0.0001$.

5. Conclusions

We established miR-34c expression constructs and transfected miR-34c into primary hippocampal cells. The present study revealed that miR-34c overexpression resulted in memory decline, accompanied by a decrease in dendritic spine density. Our data suggested that miR-34c plays a pivotal role in cognitive decline and therapies targeting miR-34c may restore synaptic defects and shed further light on AD.

Author Contributions: Conceptualization, K.-J.T.; Data curation, Y.-C.K. and I.-F.W.; Funding acquisition, K.-J.T.; Investigation, K.-J.T.; Methodology, Y.-C.K. and I.-F.W.; Project administration, K.-J.T.; Resources, K.-J.T.; Supervision, K.-J.T.; Validation, Y.-C.K. and I.-F.W.; Writing—original draft, Y.-C.K.; Writing—review & editing, K.-J.T.

Funding: This research was funded by the Ministry of Science and Technology, Taiwan (grant numbers MOST-105-2628-B-006-016-MY3 and MOST-106-2628-B-006-001-MY4).

Conflicts of Interest: The authors declare no conflict of interest.

References

1. Forner, S.; Baglietto-Vargas, D.; Martini, A.C.; Trujillo-Estrada, L.; LaFerla, F.M. Synaptic impairment in Alzheimer's disease: A dysregulated symphony. *Trends Neurosci.* **2017**, *40*, 347–357. [CrossRef] [PubMed]
2. Pillai, R.S.; Bhattacharyya, S.N.; Filipowicz, W. Repression of protein synthesis by mirnas: How many mechanisms? *Trends Cell Biol.* **2007**, *17*, 118–126. [CrossRef] [PubMed]
3. Malmevik, J.; Petri, R.; Klussendorf, T.; Knauff, P.; Akerblom, M.; Johansson, J.; Soneji, S.; Jakobsson, J. Identification of the mirna targetome in hippocampal neurons using rip-seq. *Sci. Rep.* **2015**, *5*, 12609. [CrossRef] [PubMed]
4. Malmevik, J.; Petri, R.; Knauff, P.; Brattas, P.L.; Akerblom, M.; Jakobsson, J. Distinct cognitive effects and underlying transcriptome changes upon inhibition of individual mirnas in hippocampal neurons. *Sci. Rep.* **2016**, *6*, 19879. [CrossRef] [PubMed]
5. Hernandez-Rapp, J.; Rainone, S.; Hebert, S.S. MicroRNAs underlying memory deficits in neurodegenerative disorders. *Prog. Neuro-Psychopharmacol. Biol. Psychiatry* **2017**, *73*, 79–86. [CrossRef] [PubMed]

6. Gascon, E.; Lynch, K.; Ruan, H.; Almeida, S.; Verheyden, J.M.; Seeley, W.W.; Dickson, D.W.; Petrucelli, L.; Sun, D.; Jiao, J.; et al. Alterations in microRNA-124 and ampa receptors contribute to social behavioral deficits in frontotemporal dementia. *Nat. Med.* **2014**, *20*, 1444–1451. [CrossRef] [PubMed]

7. Zovoilis, A.; Agbemenyah, H.Y.; Agis-Balboa, R.C.; Stilling, R.M.; Edbauer, D.; Rao, P.; Farinelli, L.; Delalle, I.; Schmitt, A.; Falkai, P.; et al. MicroRNA-34c is a novel target to treat dementias. *EMBO J.* **2011**, *30*, 4299–4308. [CrossRef] [PubMed]

8. Bhatnagar, S.; Chertkow, H.; Schipper, H.M.; Yuan, Z.; Shetty, V.; Jenkins, S.; Jones, T.; Wang, E. Increased microRNA-34c abundance in Alzheimer's disease circulating blood plasma. *Front. Mol. Neurosci.* **2014**, *7*, 2. [CrossRef] [PubMed]

9. Xie, B.; Zhou, H.; Zhang, R.; Song, M.; Yu, L.; Wang, L.; Liu, Z.; Zhang, Q.; Cui, D.; Wang, X.; et al. Serum mir-206 and mir-132 as potential circulating biomarkers for mild cognitive impairment. *J. Alzheimers Dis.* **2015**, *45*, 721–731. [CrossRef] [PubMed]

10. Lau, P.; Frigerio, C.S.; de Strooper, B. Variance in the identification of microRNAs deregulated in Alzheimer's disease and possible role of lincrnas in the pathology: The need of larger datasets. *Ageing Res. Rev.* **2014**, *17*, 43–53. [CrossRef] [PubMed]

11. Liu, N.; Landreh, M.; Cao, K.; Abe, M.; Hendriks, G.J.; Kennerdell, J.R.; Zhu, Y.; Wang, L.S.; Bonini, N.M. The microRNA mir-34 modulates ageing and neurodegeneration in drosophila. *Nature* **2012**, *482*, 519–523. [CrossRef] [PubMed]

12. Rokavec, M.; Li, H.; Jiang, L.; Hermeking, H. The p53/mir-34 axis in development and disease. *J. Mol. Cell Biol.* **2014**, *6*, 214–230. [CrossRef] [PubMed]

13. Jauhari, A.; Singh, T.; Singh, P.; Parmar, D.; Yadav, S. Regulation of mir-34 family in neuronal development. *Mol. Neurobiol.* **2018**, *55*, 936–945. [CrossRef] [PubMed]

14. Wu, J.; Bao, J.; Kim, M.; Yuan, S.; Tang, C.; Zheng, H.; Mastick, G.S.; Xu, C.; Yan, W. Two mirna clusters, mir-34b/c and mir-449, are essential for normal brain development, motile ciliogenesis, and spermatogenesis. *Proc. Nat.Acad. Sci. USA* **2014**, *111*, E2851–E2857. [CrossRef] [PubMed]

15. Agostini, M.; Tucci, P.; Steinert, J.R.; Shalom-Feuerstein, R.; Rouleau, M.; Aberdam, D.; Forsythe, I.D.; Young, K.W.; Ventura, A.; Concepcion, C.P.; et al. MicroRNA-34a regulates neurite outgrowth, spinal morphology, and function. *Proc. Nat.Acad. Sci. USA* **2011**, *108*, 21099–21104. [CrossRef] [PubMed]

16. Soni, K.; Choudhary, A.; Patowary, A.; Singh, A.R.; Bhatia, S.; Sivasubbu, S.; Chandrasekaran, S.; Pillai, B. Mir-34 is maternally inherited in drosophila melanogaster and danio rerio. *Nucleic Acids Res.* **2013**, *41*, 4470–4480. [CrossRef] [PubMed]

17. Fededa, J.P.; Esk, C.; Mierzwa, B.; Stanyte, R.; Yuan, S.; Zheng, H.; Ebnet, K.; Yan, W.; Knoblich, J.A.; Gerlich, D.W. MicroRNA-34/449 controls mitotic spindle orientation during mammalian cortex development. *EMBO J.* **2016**, *35*, 2386–2398. [CrossRef] [PubMed]

18. Wang, X.; Liu, P.; Zhu, H.; Xu, Y.; Ma, C.; Dai, X.; Huang, L.; Liu, Y.; Zhang, L.; Qin, C. miR-34a, a microRNA up-regulated in a double transgenic mouse model of Alzheimer's disease, inhibits bcl2 translation. *Brain Res. Bull.* **2009**, *80*, 268–273. [CrossRef] [PubMed]

19. Dickson, J.R.; Kruse, C.; Montagna, D.R.; Finsen, B.; Wolfe, M.S. Alternative polyadenylation and mir-34 family members regulate tau expression. *J. Neurochem.* **2013**, *127*, 739–749. [CrossRef] [PubMed]

20. Wu, H.; Huang, M.; Lu, M.; Zhu, W.; Shu, Y.; Cao, P.; Liu, P. Regulation of microtubule-associated protein tau (mapt) by mir-34c-5p determines the chemosensitivity of gastric cancer to paclitaxel. *Cancer Chemother. Pharmacol.* **2013**, *71*, 1159–1171. [CrossRef] [PubMed]

21. Uchizono, K. Characteristics of excitatory and inhibitory synapses in the central nervous system of the cat. *Nature* **1965**, *207*, 642–643. [CrossRef] [PubMed]

22. Attardo, A.; Fitzgerald, J.E.; Schnitzer, M.J. Impermanence of dendritic spines in live adult ca1 hippocampus. *Nature* **2015**, *523*, 592–596. [CrossRef] [PubMed]

23. Takashima, S.; Ieshima, A.; Nakamura, H.; Becker, L.E. Dendrites, dementia and the down syndrome. *Brain Dev.* **1989**, *11*, 131–133. [CrossRef]

24. Spires-Jones, T.; Knafo, S. Spines, plasticity, and cognition in Alzheimer's model mice. *Neural Plast.* **2012**, *2012*, 319836. [CrossRef] [PubMed]

25. DeKosky, S.T.; Scheff, S.W. Synapse loss in frontal cortex biopsies in Alzheimer's disease: Correlation with cognitive severity. *Ann. Neurol.* **1990**, *27*, 457–464. [CrossRef] [PubMed]

26. Terry, R.D.; Masliah, E.; Salmon, D.P.; Butters, N.; DeTeresa, R.; Hill, R.; Hansen, L.A.; Katzman, R. Physical basis of cognitive alterations in Alzheimer's disease: Synapse loss is the major correlate of cognitive impairment. *Ann. Neurol.* **1991**, *30*, 572–580. [CrossRef] [PubMed]

27. Selkoe, D.J. Alzheimer's disease is a synaptic failure. *Science* **2002**, *298*, 789–791. [CrossRef] [PubMed]

28. Baloyannis, S.J. Dendritic pathology in Alzheimer's disease. *J. Neurol. Sci.* **2009**, *283*, 153–157. [CrossRef] [PubMed]

29. Spruston, N. Pyramidal neurons: Dendritic structure and synaptic integration. *Nat. Rev. Neurosci.* **2008**, *9*, 206–221. [CrossRef] [PubMed]

30. Catala, I.; Ferrer, I.; Galofre, E.; Fabregues, I. Decreased numbers of dendritic spines on cortical pyramidal neurons in dementia. A quantitative golgi study on biopsy samples. *Hum. Neurobiol.* **1988**, *6*, 255–259. [PubMed]

31. Einstein, G.; Buranosky, R.; Crain, B.J. Dendritic pathology of granule cells in Alzheimer's disease is unrelated to neuritic plaques. *J. Neurosci.* **1994**, *14*, 5077–5088. [CrossRef] [PubMed]

32. Androuin, A.; Potier, B.; Nagerl, U.V.; Cattaert, D.; Danglot, L.; Thierry, M.; Youssef, I.; Triller, A.; Duyckaerts, C.; El Hachimi, K.H.; et al. Evidence for altered dendritic spine compartmentalization in Alzheimer's disease and functional effects in a mouse model. *Acta Neuropathol.* **2018**, *135*, 839–854. [CrossRef] [PubMed]

33. Berry, K.P.; Nedivi, E. Spine dynamics: Are they all the same? *Neuron* **2017**, *96*, 43–55. [CrossRef] [PubMed]

34. Zuo, Y.; Lin, A.; Chang, P.; Gan, W.B. Development of long-term dendritic spine stability in diverse regions of cerebral cortex. *Neuron* **2005**, *46*, 181–189. [CrossRef] [PubMed]

35. Kasai, H.; Matsuzaki, M.; Noguchi, J.; Yasumatsu, N.; Nakahara, H. Structure-stability-function relationships of dendritic spines. *Trends Neurosci.* **2003**, *26*, 360–368. [CrossRef]

36. Tada, T.; Sheng, M. Molecular mechanisms of dendritic spine morphogenesis. *Curr. Opin. Neurobiol.* **2006**, *16*, 95–101. [CrossRef] [PubMed]

37. Codocedo, J.F.; Allard, C.; Godoy, J.A.; Varela-Nallar, L.; Inestrosa, N.C. Sirt1 regulates dendritic development in hippocampal neurons. *PLoS ONE* **2012**, *7*, e47073. [CrossRef] [PubMed]

38. Lamprecht, R.; LeDoux, J. Structural plasticity and memory. *Nat. Rev. Neurosci.* **2004**, *5*, 45–54. [CrossRef] [PubMed]

39. Haditsch, U.; Leone, D.P.; Farinelli, M.; Chrostek-Grashoff, A.; Brakebusch, C.; Mansuy, I.M.; McConnell, S.K.; Palmer, T.D. A central role for the small gtpase rac1 in hippocampal plasticity and spatial learning and memory. *Mol. Cell. Neurosci.* **2009**, *41*, 409–419. [CrossRef] [PubMed]

40. Vadodaria, K.C.; Jessberger, S. Maturation and integration of adult born hippocampal neurons: Signal convergence onto small rho gtpases. *Front. Synaptic Neurosci.* **2013**, *5*, 4. [CrossRef] [PubMed]

41. Nakayama, A.Y.; Harms, M.B.; Luo, L. Small gtpases rac and rho in the maintenance of dendritic spines and branches in hippocampal pyramidal neurons. *J. Neurosci.* **2000**, *20*, 5329–5338. [CrossRef] [PubMed]

42. Hedrick, N.G.; Yasuda, R. Regulation of rho gtpase proteins during spine structural plasticity for the control of local dendritic plasticity. *Curr. Opin. Neurobiol.* **2017**, *45*, 193–201. [CrossRef] [PubMed]

43. Nakayama, A.Y.; Luo, L. Intracellular signaling pathways that regulate dendritic spine morphogenesis. *Hippocampus* **2000**, *10*, 582–586. [CrossRef]

44. Schumacher, S.; Franke, K. miR-124-regulated RhoG: A conductor of neuronal process complexity. *Small GTPases* **2013**, *4*, 42–46. [CrossRef] [PubMed]

45. Castaneda, P.; Munoz, M.; Garcia-Rojo, G.; Ulloa, J.L.; Bravo, J.A.; Marquez, R.; Garcia-Perez, M.A.; Arancibia, D.; Araneda, K.; Rojas, P.S.; et al. Association of n-cadherin levels and downstream effectors of rho gtpases with dendritic spine loss induced by chronic stress in rat hippocampal neurons. *J. Neurosci. Res.* **2015**, *93*, 1476–1491. [CrossRef] [PubMed]

46. Huang, B.; Jolly, M.K.; Lu, M.; Tsarfaty, I.; Ben-Jacob, E.; Onuchic, J.N. Modeling the transitions between collective and solitary migration phenotypes in cancer metastasis. *Sci. Rep.* **2015**, *5*, 17379. [CrossRef] [PubMed]

47. Bustos, F.J.; Ampuero, E.; Jury, N.; Aguilar, R.; Falahi, F.; Toledo, J.; Ahumada, J.; Lata, J.; Cubillos, P.; Henriquez, B.; et al. Epigenetic editing of the *Dlg4*/PSD95 gene improves cognition in aged and Alzheimer's disease mice. *Brain* **2017**, *140*, 3252–3268. [CrossRef] [PubMed]

International Journal of
Molecular Sciences

MDPI

Article

Inherited and Acquired Decrease in Complement Receptor 1 (CR1) Density on Red Blood Cells Associated with High Levels of Soluble CR1 in Alzheimer's Disease

Rachid Mahmoudi [1,2,*], Sarah Feldman [3,4], Aymric Kisserli [5,6], Valérie Duret [5,6], Thierry Tabary [5,6], Laurie-Anne Bertholon [1], Sarah Badr [1], Vignon Nonnonhou [1], Aude Cesar [7], Antoine Neuraz [3,4], Jean Luc Novella [1,2] and Jacques Henri Max Cohen [5,6]

[1] Department of Internal Medicine and Geriatrics, Reims University Hospitals, Maison Blanche Hospital, 45 rue cognac Jay, 51092 Reims, France; labertholon@chu-reims.fr (L.-A.B.); sbadr@chu-reims.fr (S.B.); vnonnonhou@chu-reims.fr (V.N.); jlnovella@chu-reims.fr (J.L.N.)

[2] Faculty of Medicine, University of Reims Champagne-Ardenne, EA 3797, 51092 Reims, France

[3] Institut National de la Santé et de la Recherche Médicale (INSERM), Centre de Recherche des Cordeliers, UMR 1138 Equipe 22, Paris Descartes, Sorbonne Paris Cité University, 75006 Paris, France; sarah.feldman@aphp.fr (S.F.); antoine.neuraz@aphp.fr (A.N.)

[4] Department of Medical Informatics, Necker-Enfants Malades Hospital, Assistance Publique des Hôpitaux de Paris (AP-HP), 75015 Paris, France

[5] Department of Immunology, Reims University Hospitals, Robert Debré Hospital, 51092 Reims, France; akisserli@chu-reims.fr (A.K.); vduret@chu-reims.fr (V.D.); ttabary@chu-reims.fr (T.T.); jhmcohen@gmail.com (J.H.M.C.)

[6] Faculty of Medicine, University of Reims Champagne-Ardenne, LRN EA 4682, 51092 Reims, France

[7] Department of Research and Innovation, Reims University Hospitals, Robert Debré Hospital, 51092 Reims, France; acesar@chu-reims.fr

* Correspondence: rmahmoudi@chu-reims.fr; Tel.: +33-326-832-341

Received: 30 June 2018; Accepted: 23 July 2018; Published: 25 July 2018

Abstract: The complement receptor 1 (*CR1*) gene was shown to be involved in Alzheimer's disease (AD). We previously showed that AD is associated with low density of the long CR1 isoform, CR1*2 (S). Here, we correlated phenotype data (CR1 density per erythrocyte (CR1/E), blood soluble CR1 (sCR1)) with genetic data (density/length polymorphisms) in AD patients and healthy controls. CR1/E was enumerated using flow cytometry, while sCR1 was quantified by ELISA. *CR1* polymorphisms were assessed using restriction fragment length polymorphism (RFLP), pyrosequencing, and high-resolution melting PCR. In AD patients carrying the H allele (*Hind*III polymorphism) or the Q allele (Q981H polymorphism), CR1/E was significantly lower when compared with controls carrying the same alleles ($p < 0.01$), contrary to sCR1, which was significantly higher ($p < 0.001$). Using multivariate analysis, a reduction of 6.68 units in density was associated with an increase of 1% in methylation of *CR1* (estimate -6.68; 95% confidence intervals (CIs) -12.37, -0.99; $p = 0.02$). Our data show that, in addition to inherited genetic factors, low density of CR1/E is also acquired. The involvement of CR1 in the pathogenesis of AD might be linked to insufficient clearance of amyloid deposits. These findings may open perspectives for new therapeutic strategies in AD.

Keywords: Alzheimer's disease; complement receptor 1; *CR1* length polymorphism; CR1 density; complement C3b/C4b receptor; complement; dementia; molecular biology; neurosciences; genetic risk

1. Introduction

Alzheimer's disease (AD) is a neurodegenerative disease that depends on both genetic and environmental factors. Genetic studies showed that the determinants of AD are manifold. In fact, while certain early-onset forms of AD are directly linked to mutations in genes that follow traditional Mendelian transmission, it was also established that other genetic risk factors play a role in sporadic forms of the disease. Genome-wide association studies (GWASs) identified variations in over 20 loci that contribute to disease risk, including the complement component (C3b/C4b) receptor 1 (*CR1*) gene [1–7].

The *CR1* gene encodes the complement receptor 1 (*CR1*), which is one of the regulators of complement activity. CR1 is a membrane-bound glycoprotein that binds to the complement proteins C3b, C4b, C3bi, and C1q. In humans, 90% of the total circulating CR1 is found on erythrocytes [8]. On the surface of erythrocytes, CR1 binds to C3b- or C4b-opsonized microorganisms or immune complexes, thus facilitating their clearance from circulation [9–11]. By limiting the deposition of C3b and C4b, CR1 might prevent excessive complement activation. In this way, the presence of CR1 on erythrocytes is viewed as a critical component in protecting tissues against immune-complex deposition and subsequent disease, such as AD [12,13].

CR1 is a glycoprotein of approximately 200 kDa. The extracellular domain of the most common form of CR1 is composed of a series of 30 repeating units named short consensus repeats (SCRs). The SCRs are arranged in tandem groups of seven, known as long homologous regions (LHRs). CR1 is arranged in four LHRs designated as LHR-A, -B, -C, and -D, arising from duplication of a seven-SCR unit [14,15].

CR1 presents three types of polymorphisms: density polymorphism, structural polymorphism (length), and Knops blood group polymorphism [14,16].

The density polymorphism is a stable phenotype that accounts for the constitutive expression level of CR1 on erythrocytes, although acquired deficiency may also occur in some diseases, such as systemic lupus erythematosus (SLE) and acquired immune deficiency syndrome (AIDS) [17]. In Caucasians, erythrocytes from different healthy subjects show up to a 10-fold variation in the number of CR1 molecules per erythrocyte (range: 150–1200 molecules/erythrocyte) [18]. Moreover, previously published data showed that CR1 density was not correlated with age [19,20]. CR1 density on erythrocytes is genetically associated with an autosomal co-dominant bi-allelic system on the *CR1* gene, which is correlated with a *Hind*III restriction fragment length polymorphism (RFLP) [21]. A single-point mutation in intron 27 of the *CR1* gene, which is located between the exons that encode the second SCR in LHR-D, results in the generation of a polymorphic *Hind*III site within this region [22]. Genomic *Hind*III fragments of 7.4 and 6.9 kDa identify alleles associated with high (H allele) or low (L allele) CR1 density, respectively, on erythrocytes [14,16,21]. CR1 density on erythrocytes is also associated with the presence of a nucleotide mutation (G3093T) in exon 19 encoding the polymorphism Q981H in SCR 16 (LHR-C) of *CR1*. CR1 density is higher in individuals who are homozygous for the QQ genotype, and lower in individuals who are homozygous for the HH genotype [15,23,24].

The second CR1 polymorphism is the structural polymorphism (length), related to a variation in LHR number. The most common isoform of *CR1* (CR1*1, also termed F), found in about 87% of Caucasians, is composed of four LHRs. The second most common isoform (CR1*2, also termed S), which is found in about 11% of Caucasians, is composed of five LHRs; thus, this isoform contains additional C3b/C4b sites. The two other, rarer, *CR1* isoforms, CR1*3 (F') and CR1*4 (D), exhibit a deletion of one LHR or the presence of two additional LHRs, respectively [14].

The third CR1 polymorphism is the Knops (KN) polymorphism, whose role in AD remains to be determined [25].

Finally, CR1 is also present in circulation in a soluble form (sCR1) [26], resulting from either the proteolysis of the membrane-bound form of CR1 [27] or exocytosis from erythrocytes (E) [28]. It is hypothesized that sCR1 is a locally active molecule that seems to have highly efficacious complement regulatory and anti-inflammatory activities [16,29]. In fact, sCR1 is a potent local inhibitor that functions in the complement pathways [30]. In addition, increased plasma levels of sCR1 were reported in some autoimmune diseases, such as SLE and glomerulonephritis [31]. In AD, a slight increase in sCR1 was reported in subjects with risk of AD single-nucleotide polymorphisms (SNPs) [32].

The complement's role in AD pathogenesis was highlighted in different studies [33,34], suggesting that AD is associated with increased complement activation [35,36]. Previous studies showed that the AD risk associated with *CR1* can be explained by low density [20] of the long *CR1* isoform, CR1*2 (S) [35–37]. However, the mechanisms underlying the decrease in CR1 density in AD remain to be elucidated. In the current study, we aimed to correlate genetic data (density/length polymorphisms) with phenotypic data (CR1 density per erythrocyte (CR1/E) and soluble CR1 (sCR1)) in patients with AD and control subjects.

2. Results

A total of 187 Caucasian subjects (100 AD patients and 87 controls) were investigated. Their main socio-demographic and clinical characteristics are shown in Table 1.

Table 1. Demographic and clinical characteristics of the study sample.

Variable	AD Patients (*n* = 100)	Controls (*n* = 87)	*p*
Age (years)	81.5 ± 7.2	74.3 ± 6.3	$<10^{-4}$
Female sex	66 (66.0%)	50 (57.4%)	0.23
APOE-ε4+ (*n* = 73)	48 (48.0%)	25 (28.74%)	0.0071
Living at home	88 (88.0%)	83 (95.4%)	0.73
Comorbidities (Charlson)	1.31 ± 1.26	1.12 ± 1.02	0.27
Level of dependence			
IADL	4.81 ± 2.57	7.83 ± 0.86	$<10^{-4}$
ADL	5.38 ± 1.0	5.95 ± 0.25	$<10^{-4}$
Cognitive status			
MMSE	19.2 ± 5.3	28.8 ± 1.3	$<10^{-4}$
AD stage			
Mild (MMSE ≥ 21)	42 (42.00%)	–	–
Moderate (MMSE 10–20)	55 (55.00%)	–	–
Severe (MMSE < 10)	3 (3.00%)	–	–

Notes: AD = Alzheimer's disease; *n* = number of subjects; *APOE*-ε4+ = subject with at least one *APOE*-ε4 allele; IADL = instrumental activities of daily living, values range from 0 (completely dependent) to 8 (completely independent); ADL = activities of daily living, values range from 0 (completely dependent) to 6 (completely independent); MMSE = Mini-Mental State Examination, scores range from 0 to 30, whereby higher scores correspond to better cognitive status.

The average age was 81.5 ± 7.2 years for AD patients, and 74.3 ± 6.3 years for controls. Indeed, there was no correlation between age and *CR1* density, using the Pearson correlation coefficient (r = −0.1, p= 0.17 in the overall population; r = 0.03, p = 0.75 in AD patients; and r = −0.07, p = 0.53 in controls).

Moreover, as expected, the *APOE*-ε4 allele, a high level of dependence, and cognitive disorders were associated with AD in our population (p = 0.0071, $p < 10^{-4}$, and $p < 10^{-4}$, respectively). However, no significant differences in sex (p = 0.23), place of residence (p = 0.73), or comorbidities (p = 0.27) were observed between AD patients and controls, which confirmed that the rationale of this study was valid.

CR1/E density in the overall study population was, on average, 677 ± 288. The average CR1 density among AD patients was significantly lower compared to controls (626 ± 272 vs. 737 ± 297; $p = 0.009$). After adjustment for age, this difference remained statistically significant ($\beta = -106.6 \pm 47.4$; $p = 0.03$).

2.1. Association between the Genetic CR1 Density Polymorphism and the CR1 Density Phenotypic Polymorphism

2.1.1. Association between the Genetic *CR1* Density Polymorphism, *Hind*III, and the CR1 Density Phenotypic Polymorphism

Among the 187 subjects investigated, 114 exhibited the HH genotype, 65 exhibited the HL genotype, and eight exhibited the LL genotype (Table 2).

Table 2. Distribution of *CR1* density and length polymorphisms among AD patients and controls.

CR1 Polymorphisms			Subjects		
			All (*n* = 187), %	AD Patients (*n* = 100), %	Controls (*n* = 87), %
Density polymorphisms	*Hind*III	HH	114 (61.0)	59 (59.0)	55 (63.2)
		HL	65 (34.8)	37 (37.0)	28 (32.2)
		LL	8 (4.3)	4 (4.0)	4 (4.6)
	Q981H	QQ	118 (63.1)	62 (62.0)	56 (64.4)
		QH	60 (32.1)	35 (35.0)	25 (28.7)
		HH	9 (4.8)	3 (3.0%)	6 (6.9)
Length polymorphisms		CR1*1 CR1*1	126 (67.4)	63 (63.0)	63 (72.4)
		CR1*1 CR1*1 + CR1*1 CR1*3	128 (68.5)	65 (65.0)	63 (72.4)
		CR1*1 CR1*2	48 (25.7)	28 (28.0)	20 (23.0)
		CR1*2 CR1*2	10 (5.4)	6 (6.0	4 (4.6)
		CR1*1 CR1*3	2 (1.1)	2 (2.0)	0 (0)
		CR1*1 CR1*2 + CR1*2 CR1*2	58 (31.0)	34 (34.0)	24 (27.6)
		CR1*2 CR1*4	1 (0.5)	1 (1.0)	0 (0)
		CR1*1 CR1*2 + CR1*2 CR1*2 + CR1*2 CR1*4	59 (31.6)	35 (35.0)	24 (27.6)
		CR1*2 CR1*2 + CR1*2 CR1*4	11 (5.9)	7 (7.0)	4 (4.6)

Notes: AD = Alzheimer's disease; *CR1* = complement receptor 1, where numbers following asterisk denote an isoform; HH = individuals homozygous for the H allele (*Hind*III polymorphism); HL = individuals heterozygous for the *Hind*III polymorphism; LL = individuals homozygous for the L allele (*Hind*III polymorphism); QQ = individuals homozygous for the Q allele (Q981H polymorphism); QH = individuals heterozygous for the Q981H polymorphism; HH = individuals homozygous for the H allele (Q981H polymorphism).

Among the AD patients, subjects with the HH genotype had a higher CR1 density (742 ± 262) than subjects carrying the L allele (HL genotype, 486 ± 175, $p < 0.0001$; LL genotype, 210 ± 142, $p < 0.01$), including both the HL and LL genotypes (460 ± 190, $p < 0.0001$), as shown in Figure 1a.

Similar findings were observed in the control subjects (Figure 1a), showing that the genetic criteria were in concordance with the phenotypic criteria (the H allele was associated with a higher density compared to the L allele).

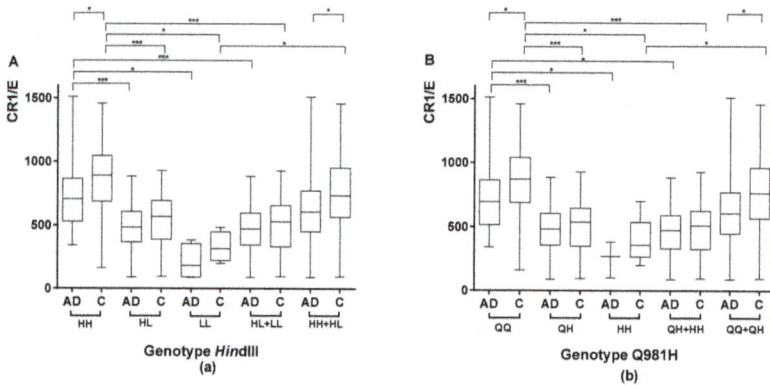

Figure 1. Comparison of the mean number of complement receptor 1 per erythrocyte (CR1/E) according to *CR1* density polymorphisms in Alzheimer's disease (AD) patients and/or controls. Box plots of CR1/E are shown. The upper and lower limits of the boxes, and the middle line across the boxes indicate the 75th and 25th percentiles, and the median, respectively. The upper and lower horizontal bars indicate the maximum and minimum values, respectively. Wilcoxon's rank test was used to compare CR1 density among AD patients or among controls according to genotype for non-normally distributed variables. The Mann-Whitney U test was used to compare AD patients with controls according to genotype for non-normally distributed variables, and the Student's *t*-test was used for normally distributed variables; * $p < 0.01$, ** $p < 0.001$, and *** $p < 0.0001$. (**a**) Comparison of the mean number of CR1/E according to *Hind*III polymorphisms in AD patients and/or controls. (**b**) Comparison of the mean number of CR1/E according to Q981H polymorphisms in AD patients and/or controls.

2.1.2. Association between the *CR1* Density Genetic Polymorphism Encoding Q981H and the CR1 Density Phenotypic Polymorphism

Among the 187 subjects investigated, 118 exhibited the QQ genotype, 60 exhibited the QH genotype, and nine exhibited the HH genotype (Table 2).

Among AD patients, subjects with the QQ genotype had a significantly higher CR1 density (730 ± 262) than subjects carrying the H allele (QH genotype, 474 ± 189, $p < 0.0001$; HH genotype, 251 ± 143, $p < 0.01$), including both the QH and HH genotypes (456 ± 194, $p < 0.01$), as shown in Figure 1b.

Similar findings were observed in the control subjects, again showing that the genetic criteria were in agreement with the phenotypic criteria (the Q allele being associated with a higher density compared to the H allele).

2.1.3. Study of the Agreement between the *Hind*III Genotype and Q981H

Analysis of the agreement between the *Hind*III genotype and Q981H showed excellent results for AD patients (weighted Kappa coefficient: 0.93; 95% confidence intervals (CIs): 0.86, 1.0), control subjects (weighted Kappa coefficient: 0.90; 95% CIs: 0.80, 1.0), and the overall population (AD patients + controls; weighted Kappa coefficient: 0.91; 95% CIs: 0.85, 0.97).

2.1.4. Comparison of CR1 Density Using *Hind*III and Q981H Genotype in AD Patients vs. Controls

The comparison of CR1 density between patients and controls according to the *Hind*III genotype showed a lower density in AD patients homozygous for the H allele compared with controls homozygous for the H allele (742 ± 262 vs. 864 ± 268, respectively; $p < 0.01$). Furthermore, when combining HH and HL subjects, the average CR1 density was significantly lower in AD patients vs. controls (643 ± 263 and 756 ± 289, respectively; $p < 0.01$), as shown in Figure 1a.

A comparison of CR1 density between AD patients and control subjects according to the Q981H density polymorphism showed that density was significantly lower in patients homozygous for the Q allele compared with controls homozygous for the Q allele (730 ± 262 vs. 866 ± 261, respectively; $p < 0.01$). When combining QQ and QH subjects, the average CR1 density was also significantly lower in AD patients vs. controls (638 ± 267 and 762 ± 289, respectively; $p < 0.001$), as shown in Figure 1b.

These findings suggest that, in addition to genetic factors, the low-density phenotype is acquired in AD.

2.2. Evaluation of the CR1 Length Polymorphisms

Table 2 presents the distribution of subjects according to the *CR1* length polymorphism.

2.3. Evaluation of the Serum Levels of sCR1

The average level of serum sCR1 in the overall population enrolled in this study was 27.17 ± 21.55 ng/mL. In AD patients, sCR1 levels were significantly higher than in controls (31.60 ± 22.86 vs. 21.96 ± 18.71 ng/mL, respectively; $p = 0.002$). The difference remained significant after adjustment for age ($\beta = 9.3 \pm 3.53$; $p = 0.009$).

2.3.1. Serum sCR1 Levels According to *CR1* Length Polymorphisms

In subjects with the genotype *CR1*1 CR1*1*, the level of sCR1 was significantly higher in AD patients than in controls (31.69 ± 25.24 vs. 23.38 ± 21.06 ng/mL, respectively; $p = 0.048$). However, when subjects with the *CR1* short alleles (*CR1*1 CR1*1* and *CR1*1 CR1*3*) were pooled, the difference between patients and controls was no longer significant. In subjects with the long *CR1* allele (*CR1*1 CR1*2* and *CR1*2 CR1*2*), the level of sCR1 was significantly higher in AD patients compared to controls (31.08 ± 18.60 vs. 18.35 ± 10.14 ng/mL, respectively; $p = 0.006$). We observed similar findings in heterozygous subjects (*CR1*2 CR1*2* subjects), with a significantly higher level of sCR1 in AD patients vs. controls (30.43 ± 18.37 vs. 19.01 ± 10.74 ng/mL, respectively; $p = 0.027$), as shown in Figure 2.

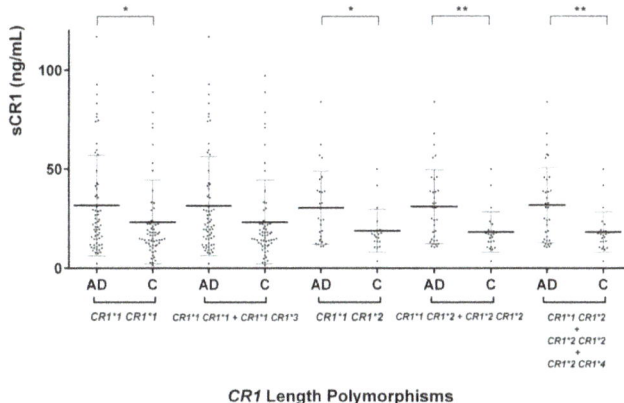

Figure 2. Comparison of soluble complement receptor 1 (sCR1) levels according to *CR1* length polymorphisms in Alzheimer's disease (AD) patients and controls. The middle line indicates the mean value, and the upper and lower horizontal bars indicate the standard deviation values. A Student's *t*-test was used to compare AD patients with controls according to *CR1* length polymorphisms for normally distributed variables; * $p < 0.05$, ** $p < 0.01$.

2.3.2. Evaluation of Serum sCR1 Levels According to the *CR1* Density Polymorphisms *Hind*III and Q981H

There was no significant difference in sCR1 levels according to genotype, either in the overall population or in the AD and control groups separately. The comparison of the levels of sCR1 between AD subjects and controls according to the *Hind*III genotype showed that there was a significantly higher level of sCR1 in AD patients homozygous for the H allele compared to controls (31.67 ± 22.23 vs. 20.65 ± 16.99, respectively; $p = 0.004$). sCR1 levels were also significantly higher in AD patients compared to controls when we pooled subjects who were homozygous for the H allele and subjects who were heterozygous for the allele (HH and HL subjects, respectively; 30.77 ± 21.46 and 21.74 ± 18.57, respectively; $p = 0.003$), as shown in Figure 3a.

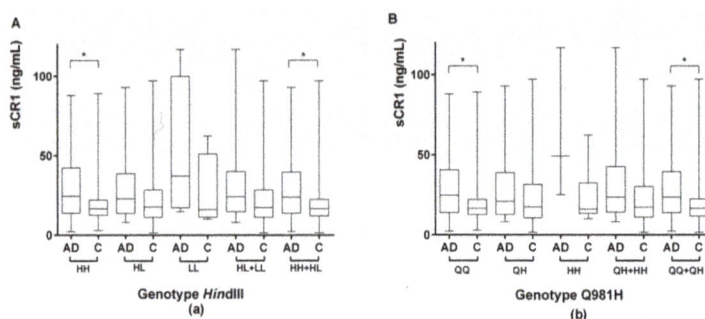

Figure 3. Comparison of sCR1 levels according to *CR1* density polymorphisms, *Hind*III and Q981H (genotype), among AD patients and controls. Box plots of sCR1 are shown. The upper and lower limits of the boxes, and the middle line across the boxes indicate the 75th and 25th percentiles, and the median, respectively. The upper and lower horizontal bars indicate the maximum and minimum values, respectively. A Student's *t*-test was used to compare AD patients with controls according to *CR1* density polymorphisms for normally distributed variables; * $p < 0.005$; sCR1 = soluble CR1. (**a**) Comparison of sCR1 levels according to *Hind*III polymorphisms in AD patients and controls. HH = individuals homozygous for the H allele (*Hind*III polymorphism); HL = individuals heterozygous for the *Hind*III polymorphism; LL = individuals homozygous for the L allele (*Hind*III polymorphism). (**b**) Comparison of sCR1 levels according to Q981H polymorphisms in AD patients and controls. QQ = individuals homozygous for the Q allele (Q981H polymorphism); QH = individuals heterozygous for the Q981H polymorphism; HH = individuals homozygous for the H allele (Q981H polymorphism).

The comparison of sCR1 levels between AD patients and controls according to the Q981H genotypes showed significantly higher levels in patients homozygous for the Q allele vs. controls (31.44 ± 21.71 vs. 20.65 ± 16.83 ng/mL, respectively; $p = 0.0038$). sCR1 levels were also significantly higher in patients compared to controls when we grouped subjects who were homozygous for the Q allele and subjects who were heterozygous for the allele (QQ and QH subjects, respectively; 30.61 ± 21.41 and 21.82 ± 18.79 ng/mL, respectively; $p = 0.0048$), as shown in Figure 3b.

2.4. Association between CR1/E and sCR1 and the Stage of AD

The comparison of CR1/E between AD subjects according to the stage of AD, as assessed by Mini-Mental State Examination (MMSE) scores, showed that CR1/E was significantly lower in patients with moderate or severe AD as compared with mild AD (583.67 ± 238.58 vs. 698.75 ± 295.32, respectively; $p = 0.034$), as shown in Figure S1a. In contrast, there was no significant difference in sCR1 in AD subjects according to the stage of AD (Figure S1b).

In addition, the association between CR1/E and the severity of AD, measured by the MMSE score and tested using univariate linear regression (with MMSE scores alternatively used as a quantitative variable and a categorical variable in two classes) showed that CR1/E density was significantly lower for moderate and severe AD patients than for mild AD patients (estimate: −115.08; 95% CIs: −221.78, −8.39; p = 0.035), as shown in Table S1. In contrast, no association was found between sCR1 and AD severity (estimate: −1.12; 95% CIs: −10.45, −8.21; p = 0.812), as shown in Table S2.

2.5. CR1 Methylation

The second methylation site was associated with a reduction of 6.68 units in density, for an increase of 1% in methylation (estimate: −6.68; 95% CIs: −12.37, −0.99; p = 0.02), independently of AD, age, and density polymorphism. Density decreased with age (estimate: −6.51; 95% CIs: −11.79, −1.23; p = 0.016), as shown in Table 3.

Table 3. Multivariate analysis of factors associated with density.

Variable	Unit	Estimate	95% CIs	p
Alzheimer's disease	–	−63.14	−144.89, 18.6	0.129
Age	1	−6.51	−11.79, −1.23	0.016
Density polymorphism *Hind*III (reference: HH)				<0.001
HL	–	−311.67	−389.16, −234.17	
LL	–	−569.51	−740.02, −398.99	
2nd methylation site	1%	−6.68	−12.37, −0.99	0.022

Notes: 95% CIs = 95% confidence intervals.

2.6. Assessment of Factors Associated with AD

A multivariate analysis identified six factors that were independently related to AD: age, female sex, *APOE*-ε4 carrier, number of CR1 antigenic sites per erythrocyte (density), the level of sCR1, and the density polymorphism Q981H (Table 4).

Table 4. Multivariate analysis of factors independently associated with Alzheimer's disease at different levels of variation in the explanatory variables.

Variable	Unit	OR	95% CIs	p
Age (years)	1	1.182	1.118, 1.260	<0.0001
Sex (female)	–	2.605	1.172, 6.050	0.0215
APOE-ε4+	–	4.745	2.152, 11.199	0.0002
Density (number of CR1 antigenic sites per erythrocyte)	30	0.936	0.894, 0.975	0.0025
	100	0.801	0.689, 0.920	0.0025
	200	0.641	0.475, 0.847	0.0025
	400	0.411	0.225, 0.718	0.0025
	500	0.329	0.155, 0.661	0.0025
Density polymorphism Q981H (Q vs. HH)	–	12.416	1.603, 112.155	0.0193
Serum level of soluble CR1 (ng/mL)	1	1.032	1.013, 1.054	0.0015
	10	1.369	1.139, 1.685	0.0015
	20	1.874	1.298, 2.840	0.0015
	30	2.565	1.479, 4.787	0.0015
	40	3.512	1.684, 8.068	0.0015
	50	4.807	1.919, 13.597	0.0015

Notes: CR1 = complement receptor 1; OR = odds ratio; 95% CIs = 95% confidence intervals.

We failed to show a significant interaction between *CR1* length polymorphisms and *APOE* (p = 0.106), or between CR1 density and *APOE* (p = 0.3795).

To investigate the risk associated with different quantitative variations in the explanatory variables, the results of the multivariate analysis presented in Table 4 modeled the adjusted risk for a range of different variations in the two quantitative variables (density and level of sCR1). This enabled us to explore variations that could be clinically relevant, given the absence of data in the literature.

Regarding the variable "density", since the threshold of biological detection is 30 sites and the average density in the general population is 500 (150–1500), according to our multivariate model, a variation of 30 sites was associated with a 6.4% reduction in the risk of developing AD. However, a variation of 30 sites is neither clinically nor biologically relevant. Conversely, a variation of 200 antigenic sites has greater biological discrimination and, according to our model, it was associated with a 35.9% reduction in the risk of developing AD, which is also more clinically relevant. Furthermore, our multivariate model showed that an increase of 20 ng/mL in serum sCR1 was associated with a 1.8-fold increase in the risk of AD (95% CIs: 1.29, 2.84; Table 4).

The multivariate model fitted the data well (*p*-value for the Hosmer and Lemeshow test = 0.23).

3. Discussion

The originality of our study resides in the combination of phenotypic and genotypic data relating to *CR1* polymorphisms in a well-characterized cohort of AD patients and control subjects. Furthermore, we confirmed findings from previous works showing that AD is associated with the long *CR1* isoform [35,36], and that low CR1 density could explain the association between *CR1* and AD [20], as identified by GWAS [3].

The present study established that abnormally low CR1 density on erythrocytes was associated with AD, independently of genetic factors. The agreement between the two genotypes associated with the *CR1* density polymorphism (*Hind*III and Q981H) was excellent, both in patients and controls. Univariate analysis showed that the presence of the H allele (*Hind*III) or the Q allele (Q981H) was associated with significantly higher CR1 density as compared with the L allele (*Hind*III) or the H allele (Q981H), both in patients and in controls. However, CR1 density was lower in AD patients compared with controls in the presence of both the H allele (*Hind*III) and the Q allele (Q981H). This observation of lower density in carriers of the high-density allele (H for *Hind*III or Q for Q981H) was more pronounced in patients who were homozygous for the allele coding for high density. Altogether, our findings suggest that genetic factors determining CR1 density are indeed present; however, non-genetic factors, such as acquired factors, can be involved in AD, resulting in a CR1 low-density phenotype acquired during the course of AD. This might also contribute to AD development.

With regards to the *CR1* length polymorphism, our results generally followed the same trends as those observed in our previous work [20]. No association between length polymorphism-associated genotype and density polymorphism-associated genotype was observed. Conversely, we previously reported an association between the length polymorphism and the density polymorphism at the protein level. Again, these results support our hypothesis that, in addition to constitutive genetic factors such as the long *CR1* allele, which appears to be linked to regulatory factors probably in the promoter region, other non-genetic factors also have an effect on CR1 expression. This may lead to a lower-density phenotype than that expected from the genotype of a given patient, acquired during the course of the disease. Reciprocally, this could result in a decline in the clearance of the amyloid beta 1-42 (Aβ_{1-42}) peptide, as well as in a lower control of in situ inflammation, in turn leading to a higher risk of developing AD [13,35,36].

In this study, analysis of five methylation sites revealed an increase in methylation at the second site located in the additional segment of the long *CR1* allele (CR1*2 (S)) suggesting that the increase in methylation of the long *CR1* allele (CR1*2 (S)) might be the direct mechanism of the lower expression of that isoform at the protein level.

In the present study, serum levels of sCR1 were assessed. Firstly, we showed that sCR1 levels were higher in AD patients compared with controls. This could be explained by increased proteolysis of CR1, as shown in patients with diseases related to protease production [17,38], and/or its vesiculation (exocytosis) demonstrated in erythrocytes [28]. As described in SLE, this suggests that, at the peripheral level and during the binding or capture of amyloid peptides, C3b molecules are deposited at the cell surface, and removed together with CR1 as the surrounding molecule via vesiculation of the membrane [39]. Accordingly, the CR1-enriched vesicles are taken into account in sCR1 dosages [40]. Secondly, the univariate analysis performed in our study showed that the long *CR1* alleles (CR1*2 and CR1*4) were associated with higher levels of sCR1 in AD patients compared with controls. In this regard, our results may help explain the pathological observations of Hazrati and colleagues [36], who found that the distribution of CR1 in the brain was different between *CR1* length polymorphism-associated genotypes, hypothesizing that the CR1*1 isoform is transported between protein-sorting compartments, whereas the longer CR1*2 isoform accumulates in the membrane of cytoplasmic vesicles [36]. Taken together, our results suggest that the long *CR1* alleles are linked, during the course of AD, to lower CR1 density, probably due to the effect of other genetic or acquired factors, which might partially explain the increase in sCR1 through proteolysis and/or exocytosis.

Lastly, using multivariate analysis, we identified six factors that were independently associated with AD, namely age, female sex, *APOE-ε4* carrier, CR1 density, serum sCR1 level, and the density polymorphism Q981H (Q allele). In fact, age, female sex, and the *APOE-ε4* allele were already described in the literature as risk factors for AD [41]; however, this is the first study showing that CR1 density, serum sCR1 levels, and the density polymorphism Q981H (Q allele) are independent factors related to AD. According to our multivariate model, an increase of 200 CR1 antigenic sites was associated with a 35.9% reduction in the risk of developing AD. This does not imply that subjects with a low density (e.g., 100 CR1/E sites) are at higher risk of developing AD, since the risk of developing AD was not higher in a subject who expressed 200 CR1/E sites than in a subject who expressed 900 CR1/E sites, for example. This can likely be explained by the existence of other factors belonging to other systems. However, when an individual is genetically programmed to display 900 CR1/E sites, yet, under the influence of external factors, expresses a reduced CR1 density of 200 CR1/E, then the risk of developing AD would be increased. Our findings showed a high frequency of the QQ or HH genotype, both in AD patients and in control subjects, which was in line with previous reports [42]. Taken together, our phenotype and genotype findings suggest that the biological pathway of CR1 expression is ruled by both genetic and acquired factors that are intrinsically linked. Thus, an acquired reduction in CR1 density, as opposed to a low-density genotype, seems to be associated with an increased risk of AD. Another finding of our multivariate analysis was that the serum sCR1 level was associated with AD, independently of age, CR1 density, and density polymorphism (Q allele for Q981H).

Our findings suggest that, in addition to genetic factors, a low density of CR1 is also acquired during the course of AD, and that the involvement of CR1 in the pathogenesis of AD might be linked to both insufficient in situ inhibition of complement and/or inflammation, or impaired amyloid protein clearance in the peripheral blood. In fact, an improved understanding of the pathophysiological mechanisms of AD may pave the way toward new therapeutic avenues for this disease. In light of our results, and in view of the physiological role and potential implication of CR1 in the pathogenesis of AD, two avenues deserve to be further explored: the increase in CR1 expression (which requires a better understanding of regulatory factors), and the use of recombinant forms of sCR1 to restore improved control of complement-induced inflammation.

4. Materials and Methods

4.1. Study Population

The study was approved by the regional ethics committee, and written informed consent was obtained from each participant. All AD patients met the diagnostic criteria for probable AD according to the Diagnostic and Statistical Manual of Mental Disorders, fourth edition (DSM IV) [43], and the criteria of the National Institute of Neurological and Communicative Disorders and Stroke and the Alzheimer's Disease and Related Disorders Association (NINCDS-ADRDA) [44]. The controls were subjects without any organic brain or cognitive disorders. Subjects with diseases that were likely to affect the physiology of CR1, such as hemolytic anemia, terminal renal or liver failure, or SLE, were excluded from the study. We also excluded subjects receiving treatments likely to modify sCR1 rates, such as non-steroidal or steroidal anti-inflammatory drugs, as well as those receiving treatments likely to modify CR1/E density, such as blood transfusion.

Blood samples were drawn into Vacutainer tubes containing 0.12 mL of 0.15% ethylenediaminetetraacetic acid (EDTA) and 5-mL Vacutainer dry tubes.

4.2. Quantification of CR1 Density Using Flow Cytometry

The mean CR1 density on erythrocytes was determined using flow cytometry and a J3D3 monoclonal antibody (moAb) [45], as previously described [46]. Moreover, the anti-CR1 moAb J3B11, and the TO5 and E11 moAbs were also used in flow cytometry or in control experiments [20]. A standard curve was obtained from donors of known CR1 antigenic sites, with a density ranging from 180 to 1000 sites per erythrocyte. Flow cytometry was performed on stained cells using a flow cytometer (FACScan; Becton Dickinson, Mountain View, CA, USA). At least 10,000 events were collected for each sample. The mean fluorescence intensity channel was used to quantify the staining of each sample. The detection threshold was 30 CR1 antigenic sites per erythrocyte.

4.3. DNA Extraction

DNA from 2-mL whole blood samples was isolated using the QuickGene-610L (Fujifilm, Asnières, France). The manufacturer's instructions were followed according to the recommended protocol. Briefly, tubes containing 2 mL of EDTA whole blood, 2.5 mL of lysis buffer (containing guanidine hydrochloride), and 300 μL of proteinase K were mixed and incubated for 5 min at 56 °C. After this incubation, 2.5 mL of ethanol (>99%) was added and mixed, and the samples were applied to the Quickgene Cartridge. Cartridges were placed on the instrument and QuickGene-610L's pre-programmed protocol automated the rest of the process, before DNA was finally eluted in 500 μL of elution buffer.

4.3.1. *APOE* Genotyping using Amplification and High-Resolution Melting Analyses

PCR was performed in 10-μL volumes in a LightCycler 480 (Roche) using 96-well plates. The Type-it HRM PCR Kit (Quiagen, Courtaboeuf, France) was used following the LightCycler 480 manufacturer's instructions and specific primer sets (Spot-to-Lab. Montpellier France) for each SNP genotyping rs429358 and rs7412 (primer sequences are available on request from the manufacturer). The temperature-cycling protocol included an initial denaturation step at 95 °C for 10 min, followed by 50 cycles of denaturation at 95 °C for 15 s, annealing for 15 s, and extension at 72 °C with a transition from annealing to extension of 2.2 °C/s. The touchdown program of the annealing step was used with a starting temperature of 65 °C and a progressive decrease of 0.5 °C per cycle to reach the final temperature of 55 °C. The reactions were monitored during PCR at the end of each extension phase. Following the amplification phase, the samples were heated momentarily in the LightCycler to 94 °C for 1 min and rapidly cooled to 40 °C to create heteroduplexes. The melting curves were obtained by heating from 65 °C to 95 °C at 0.02 °C/s and 25 fluorescent acquisitions per °C. High-resolution melting data were analyzed using the gene scanning module of the LightCycler 480 Software. Allele identification was determined using the combination of rs429358 and rs7412 genotyping results.

4.3.2. Assessment of the *CR1* Density Genetic Polymorphism Using *Hind*III RFLP

The *CR1* density polymorphism on erythrocytes was determined using PCR amplification and *Hind*III restriction enzyme digestion, as described previously [18]. The PCR primers used were 5′–CCTTCAATGGAATGGTGCAT–3′ and 5′–CCCTTGTAAGGCAAGTCTGG–3′. PCR was performed on a MyCycler apparatus using the following conditions: a final volume of 100 μL containing 2 μL of DNA solution (approximately 100–250 ng/μL), a 200-μM concentration of each deoxynucleoside triphosphate, a 0.5-mM concentration of each primer, 2.5 mM MgCl$_2$, and 2.5 U of Taq GOLD DNA polymerase (Perkin Elmer Cetus) in the buffer supplied by the manufacturer. The amplification conditions were as follows: 10 min at 94 °C, followed by 40 cycles of 1 min at 94 °C, 1 min at 61 °C, and 2 min at 72 °C, before being held for 10 min at 72 °C.

For RFLP determination, 30 µL of PCR product and 2 µL of *Hind*III were incubated in a final volume of 50 µL in the buffer supplied by the manufacturer at 37 °C for 2 h, followed by analysis on a 2% ethidium bromide gel.

Using this protocol, *Hind*III digestion did not alter the PCR product (1.8 kb) from individuals who were homozygous for the *CR1* high-density allele (HH). The 1.8-kb band was fully split into two smaller bands of 1.3 and 0.5 kb in samples from individuals who were homozygous for the *CR1* low-density allele (LL).

4.3.3. Determination of the *CR1* Density Genetic Polymorphism by Pyrosequencing

Q981H (Exon 19) Amplification for Pyrosequencing

The PCR primers used were 16aL 5′–GCTACATGCAGGTTGAGACCTTAC–3′ and PCRE111926RE 5′–CTGAGATGTGGCTAGAAAGTAC–3′. PCR was performed on a MyCycler apparatus using the following conditions: 50 µL of final volume containing 1 µL of DNA solution (approximately 100–250 ng/µL), a 200-mM concentration of each deoxynucleoside triphosphate, a 0.5-mM concentration of each primer, 2 mM MgCl$_2$, and 1.25 U of Taq DNA polymerase (Promega, Madison, WI, USA) in the buffer supplied by the manufacturer. The amplification conditions were as follows: 10 min at 94 °C, followed by 40 cycles of 1 min at 94 °C, 1 min at 64 °C, and 1 min at 72 °C, before being held for 10 min at 72 °C. Nested PCR was then performed using the primers PCRE111926RE and PCRssCR1LikeBIOT 5′–AAATCATGTAAAACTCCTCCAGA–3′ biotinylated on its 5′ end to allow immobilization of the PCR product on streptavidin beads and the preparation of single-stranded DNA. One microliter of the first PCR reaction diluted at 1:500 in water was used for the nested PCR. The procedure used for the second PCR procedure was the same as that used in the first PCR. All fragments were subjected to gel electrophoresis on agarose gels containing ethidium bromide, before isolation for pyrosequencing.

Pyrosequencing

Primers were drawn to anneal adjacent to codons Q981H of the *CR1* gene: PCR2Q981H35rev 5′–TGATTCTGGATCCAA–3′. The biotinylated PCR product (40 µL) was immobilized onto 4 µL of streptavidin-coated Sepharose beads (>1.2 nmol binding capacity; Amersham Pharmacia Biotech, Uppsala, Sweden) in 40 µL of binding buffer (10 mM Tris-HCl (pH 7.6), 2 M NaCl, 1 mM EDTA, and 0.1% Tween 20) on a shaker (1400 rpm) at room temperature for 10 min. PCR products immobilized on beads were transferred to a 96-well filter plate (Millipore, Molsheim, France) and vacuum-dried. Single-stranded DNA was obtained by adding 50 µL of denaturation solution (0.2 M NaOH) for 1 min. The immobilized strand was washed twice with 150 µL of washing buffer (10 mM Tris-acetate (pH 7.6)), re-suspended in 45 µL of annealing buffer (20 mM Tris-acetate (pH 7.6) and 2 mM magnesium acetate), and transferred into wells containing 15 pmol of sequencing primer in a volume of 1.5 µL of annealing buffer. The plate was heated at 61 °C for 5 min. Real-time pyrosequencing was performed at 28 °C in an automated 96-well pyrosequencer using PSQ SNP 96, with enzymes and substrate (Pyrosequencing AB, Uppsala, Sweden), with cyclic dispensation of nucleotides. The computer analysis was based on an algorithm that compared the height of the different peaks and the base number of the polymorphic fragment.

4.3.4. Determination of the rate of *CR1* methylation by Pyrosequencing

Amplification of the *CR1* gene (LHR-B segment and LHR-C segment) for Pyrosequencing

The PCR primers used were methF1 (5′–GGAAGTTGATGAGGTATGTATAGTATAA–3′) and methR1biot (5′–AATACCATTTCCAAAAAAAATAAAATCCA–3′). PCR was performed on a MyCycler apparatus using the following conditions: 50 µL of final volume containing 1 µL of DNA solution (approximately 100–150 ng/ µL), a 16-nM concentration of each deoxynucleoside triphosphate, a 0.5-mM concentration of each primer, 1.5 mM $MgCl_2$, and 2 U of AmpliTaq Gold DNA polymerase (Perkin-Elmer, Roissy, France) in the buffer supplied by the manufacturer. The amplification conditions were as follows: 10 min at 94 °C, followed by 40 cycles of 30 s at 94 °C, 30 s at 49 °C, and 30 s at 72 °C, before being held for 7 min at 72 °C. All fragments were subjected to gel electrophoresis on 2% agarose gels containing ethidium bromide, before isolation for pyrosequencing.

Pyrosequencing

Primers were drawn to anneal adjacent to five methylated bases of the LHR-B segment and LHR-C segment of the *CR1* gene (Figure S2): methS0PYRO (5′–TTT-TAT-TTT-TTG-TTT-TTA-GG–3′), methS1PYRO (5′–GGT-TAT-TTA-TTT-GTT-GAA-TGT-ATT-T–3′), and methS2PYRO (5′–ATG-TAT-TTT-TTA-GGG-TAA-TGT-TGT–3′). The biotinylated PCR product (40 µL) was immobilized onto 4 µL of streptavidin-coated Sepharose beads (>1.2 nmol binding capacity; Amersham Pharmacia Biotech, Uppsala, Sweden) in 40 µL of binding buffer (10 mM Tris-HCl (pH 7.6), 2 M NaCl, 1 mM EDTA, and 0.1% Tween 20) on a shaker (1400 rpm) at room temperature for 10 min. PCR products immobilized on beads were transferred to a 96-well filter plate (Millipore, Molsheim, France) and vacuum-dried. Single-stranded DNA was obtained by adding 50 mL of denaturation solution (0.2 M NaOH) for 1 min. The immobilized strand was washed twice with 150 mL of washing buffer (10 mM Tris-acetate (pH 7.6), re-suspended in 40 µL of annealing buffer (20 mM Tris-acetate (pH 7.6) and 2 mM magnesium acetate), and transferred into wells of PSQ 96 Plate Low (40-0010, Qiagen, Courtaboeuf, France) containing 15 pmol of sequencing primer in a volume of 40 µL of annealing buffer. The plate was heated at 81 °C for 2 min. Real-time pyrosequencing was performed at 28 °C in an automated 96-well pyrosequencer using PSQ SNP 96, with enzymes and substrate (Pyrosequencing AB, Uppsala, Sweden), with cyclic dispensation of nucleotides. The computer analysis was based on an algorithm that compared the height of the different peaks and the base number of the polymorphic fragment.

4.3.5. Determination of the *CR1* Length Genetic Polymorphisms Using High-Resolution Melting PCR (HRM-PCR)

The *CR1* length polymorphism was determined at the genetic level using HRM, as described previously [47]. Original primers (CN3: 5′–GGCCTTAGACTTCTCCTGC–3′ and CN3re: 5′–GTTGA CAAATTGGCGGCTTCG–3′) were synthesized by Eurogentec (Seraing, Belgium). PCR was performed in a total volume of 20 µL, using 10 µL of 2× LightCycler 480 High-Resolution Melting Master Kit (Roche, Meylan, France), 1 µL of 300 nM forward primer, 1 µL of 300 nM reverse primer, and 10 ng of DNA. PCR was performed on a 96-well thermal cycler (Veriti; Applied Biosystems, Ontario, Canada) using an amplification protocol of one cycle at 95 °C for 10 min, followed by 45 cycles at 95 °C for 10 s, 62 °C for 15 s, and 72 °C for 20 s. The HRM of the amplicons using the LightCycler 480 System (Roche) displayed the melting-curve profiles corresponding to the four *CR1* length polymorphisms.

4.4. Quantification of sCR1 Using ELISA

Blood samples obtained in dry tubes were centrifuged at 1200× *g* for 10 min. Serum was aliquoted then frozen at −20 °C. An anti-sCR1 ELISA kit (USCN Life Science Inc., Houston, TX, USA) was used according to the manufacturer's instructions. Serum was diluted 1/10, and the minimal dose of detectable sCR1 (sensitivity) was 0.124 ng/mL. The detection range was 0.312–20 ng/mL. All samples and standards were measured in duplicate, and the means were used for statistical analyses.

4.5. Statistical Analyses

Quantitative variables are presented as mean ± standard deviation (m ± SD), and qualitative variables as numbers (percentage). Univariate analysis was performed using comparison of means (Student's *t*, Mann-Whitney *U*, or Kruskal-Wallis tests, or ANOVA), comparison of percentages (chi-squared or Fisher's exact test), or correlation (Pearson correlation coefficient, *r*), as appropriate. Paired tests (Student's paired *t*-test or the Wilcoxon test) were used for comparisons within groups, and unpaired tests for comparisons between groups. Bonferroni's correction was applied to comparisons within groups and between groups for each genotype polymorphism (*Hind*III and Q981H), to control the alpha error due to the risk of inflation from multiple testing. Assuming five comparisons, *p*-values < 0.01 were considered significant for these tests. Agreement between the *Hind*III and Q981H genotypes was measured using the weighted Kappa coefficient. Multivariable analysis was performed to identify factors independently associated with AD. A logistic regression model was constructed using the stepwise method after adjustment for onset/inclusion age and other potential confounders (sex, *APOE*-ε4 genotype, and comorbidities). The interaction between *APOE* and *CR1* was also tested. The threshold for entry into and exit from the model was *p* < 0.20. The goodness-of-fit of the model was tested using the Hosmer-Lemeshow test. The results are presented as odds ratios (ORs) with 95% confidence intervals (CIs). A *p*-value < 0.05 was considered statistically significant. All analyses were performed using the SAS software (version 9.4; SAS Institute, Inc., Cary, NC, USA).

The association between density and methylation was explored using multivariate linear regression. Univariate linear regressions were performed between density and each covariate separately. Manual descending stepwise analysis was performed with methylation-related variables and all variables with a *p*-value strictly less than 0.20. AD was forced during the selection. The selection was complete when the *p*-value of all variables (except AD) was strictly less than 0.05. The overall *p*-values of qualitative variables were calculated using a likelihood ratio test. Mean methylation was calculated as the mean of the five methylation sites whenever the results of at least two methylation sites were available.

The variables selected for the stepwise procedure were density polymorphisms Q981H and *Hind*III, AD, age, *APOE*-ε4, methylation sites 1 to 5, and mean methylation (Table S3). No interaction was introduced into the model (Table S4). The final model included AD, age, density polymorphism *Hind*III, and the second methylation site. A *p*-value < 0.05 was considered statistically significant.

5. Conclusions

Our data (i) confirm the link between the long *CR1* isoform and AD; (ii) show that the long CR1 isoform, despite exhibiting more C3b/C4b binding sites per molecule, is less frequently expressed than the other *CR1* isoforms, probably through a higher methylation level; (iii) show that an acquired decrease in CR1/E in association with a higher level of sCR1 was observed in AD patients compared to controls; (iv) rather than an increase in complement downregulation and immune complexes or deposit removal initially inferred from the presence of the long CR1 isoform, which exhibits one additional C3b/C4b binding site per molecule, a less effective CR1 cleaning ability pattern in AD emerges progressively. This hypothesis opens new avenues for therapeutic research [12,48].

6. Patents

Rachid Mahmoudi, Aymric Kisserli, and Jacques HM Cohen are the inventors of a patent owned by the University of Reims Champagne-Ardenne (URCA) (patent number WO 2015166194).

Supplementary Materials: Supplementary materials can be found at http://www.mdpi.com/1422-0067/19/8/2175/s1. Figure S1: Comparison of the mean number of complement receptor 1 per erythrocyte (CR1/E) and soluble CR1 (sCR1) in AD patients according to AD stage. Box plots of CR1/E are shown. The upper and lower limits of the boxes, and the middle line across the boxes indicate the 75th and 25th percentiles, and the median, respectively. The upper and lower horizontal bars indicate the maximum and minimum values, respectively. A Student's *t*-test was used for normally distributed variables; * *p* = 0.034. (a) Comparison of the mean number of CR1/E in AD patients according to the stage of AD. (b) Comparison of the level of sCR1 in AD patients according to the stage of AD stage; Figure S2: DNA sequence alignment of the high-resolution melting PCR (HRM-PCR) amplicons, and positions of the methylated sites according to the genomic reference sequence (NG 007481.1) corresponding to the long allele of *CR1* (CR1*2). Two areas of the long homologous region B (LHR-B) of *CR1* are amplified at positions 48743 to 48960 and 67300 to 67517 corresponding to the LHR-B segments, and one area is amplified at position 84358 to 84574 corresponding to the LHR-C segment. The positions of the five methylated sites (SM1, SM2, SM3, SM4, and SM5) are framed in color. In blue, SM1 is located at positions 48858, 67415, and 84472, SM3 is located at positions 48919, 67476, and 84533, SM4 is located at positions 48925, 67482, and 84539, and SM5 is located at positions 48939, 67496, and 84553. In orange, SM2 is located at positions 48895 and 67452, but is missing at position 84509; Table S1: Univariate linear regression between CR1/E density and MMSE score as a quantitative variable (values 0–30) and as a qualitative variable in two classes (Mild: 21 to 30 vs. moderate/severe: 0 to 20); Table S2: Univariate linear regression between SCR1 rate and MMSE score, as a quantitative variable (values 0–30) and as a qualitative variable in two classes (Mild: 21 to 30 vs. moderate/severe: 0 to 20); Table S3: Univariate analysis of density and covariates, ordered by *p*-value; Table S4: Tests of interaction.

Author Contributions: Conceptualization, R.M., A.K., and J.H.M.C. Experiments, A.K., V.D., and J.H.M.C. Methodology, R.M., S.F., and J.H.M.C. Investigation, R.M. and J.H.M.C. Statistical analyses, S.F., A.N., and R.M. Writing—original draft preparation, R.M. and J.H.M.C. Writing—review and editing, R.M., T.T., A.K., and J.H.M.C. Recruitment of subjects, L.-A.B., S.B., and V.N. Data management, A.C. All authors read and approved the final version of the manuscript.

Funding: This research was funded by the Reims University Hospitals (grant no: AOL11UF9156) through a local call for projects.

Acknowledgments: The authors thank all the participants in this study, the members of the Plateforme Régionale de Biologie Innovante (PRBI), the staff of the Reims Champagne-Ardenne Resource and Research Memory Center, the staff of the Department of Internal Medicine and Geriatrics, the staff of the Department of Immunology, and the staff of the Department of Research and Innovation of Reims University Hospitals. We also thank Fiona Ecarnot (EA3920, University Hospital Besancon, France) for editorial assistance.

Conflicts of Interest: The authors have no conflicts of interest to declare. The study was approved by the regional ethics committee, and informed consent was obtained from all the participants. The funders had no role in the design of the study; in the collection, analyses, or interpretation of data; in the writing of the manuscript, and in the decision to publish the results.

Abbreviations

AD	Alzheimer's disease
AIDS	Acquired immunodeficiency syndrome
APOE	Apolipoprotein E
CR1	Complement component (C3b/C4b) receptor 1
GWAS	Genome-wide association studies
HRM	High-resolution melting
LHR	Long homologous repeat
SCR	Short consensus repeat
sCR1	Soluble form of CR1
SNP	Single-nucleotide polymorphism

References

1. Harold, D.; Abraham, R.; Hollingworth, P.; Sims, R.; Gerrish, A.; Hamshere, M.L.; Pahwa, J.S.; Moskvina, V.; Dowzell, K.; Williams, A.; et al. Genome-wide association study identifies variants at CLU and PICALM associated with Alzheimer's disease. *Nat. Genet.* **2009**, *41*, 1088–1093. [CrossRef] [PubMed]
2. Hollingworth, P.; Harold, D.; Sims, R.; Gerrish, A.; Lambert, J.C.; Carrasquillo, M.M.; Abraham, R.; Hamshere, M.L.; Pahwa, J.S.; Moskvina, V.; et al. Common variants at ABCA7, MS4A6A/MS4A4E, EPHA1, CD33 and CD2AP are associated with Alzheimer's disease. *Nat. Genet.* **2011**, *43*, 429–435. [CrossRef] [PubMed]
3. Lambert, J.C.; Heath, S.; Even, G.; Campion, D.; Sleegers, K.; Hiltunen, M.; Combarros, O.; Zelenika, D.; Bullido, M.J.; Tavernier, B.; et al. Genome-wide association study identifies variants at CLU and CR1 associated with Alzheimer's disease. *Nat. Genet.* **2009**, *41*, 1094–1099. [CrossRef] [PubMed]
4. Lambert, J.C.; Ibrahim-Verbaas, C.A.; Harold, D.; Naj, A.C.; Sims, R.; Bellenguez, C.; DeStafano, A.L.; Bis, J.C.; Beecham, G.W.; Grenier-Boley, B.; et al. Meta-analysis of 74,046 individuals identifies 11 new susceptibility loci for Alzheimer's disease. *Nat. Genet.* **2013**, *45*, 1452–1458. [CrossRef] [PubMed]
5. Naj, A.C.; Jun, G.; Beecham, G.W.; Wang, L.S.; Vardarajan, B.N.; Buros, J.; Gallins, P.J.; Buxbaum, J.D.; Jarvik, G.P.; Crane, P.K.; et al. Common variants at MS4A4/MS4A6E, CD2AP, CD33 and EPHA1 are associated with late-onset Alzheimer's disease. *Nat. Genet.* **2011**, *43*, 436–441. [CrossRef] [PubMed]

6. Seshadri, S.; Fitzpatrick, A.L.; Ikram, M.A.; DeStefano, A.L.; Gudnason, V.; Boada, M.; Bis, J.C.; Smith, A.V.; Carassquillo, M.M.; Lambert, J.C.; et al. Genome-wide analysis of genetic loci associated with Alzheimer disease. *JAMA* **2010**, *303*, 1832–1840. [CrossRef] [PubMed]

7. Rosenthal, S.L.; Kamboh, M.I. Late-onset Alzheimer's disease genes and the potentially implicated pathways. *Curr. Genet. Med. Rep.* **2014**, *2*, 85–101. [CrossRef] [PubMed]

8. Rogers, J.; Li, R.; Mastroeni, D.; Grover, A.; Leonard, B.; Ahern, G.; Cao, P.; Kolody, H.; Vedders, L.; Kolb, W.P.; et al. Peripheral clearance of amyloid beta peptide by complement C3-dependent adherence to erythrocytes. *Neurobiol. Aging* **2006**, *27*, 1733–1739. [CrossRef] [PubMed]

9. Schifferli, J.A.; Ng, Y.C.; Estreicher, J.; Walport, M.J. The clearance of tetanus toxoid/anti-tetanus toxoid immune complexes from the circulation of humans. Complement- and erythrocyte complement receptor 1-dependent mechanisms. *J. Immunol.* **1988**, *140*, 899–904. [PubMed]

10. Cosio, F.G.; Shen, X.P.; Birmingham, D.J.; Van Aman, M.; Hebert, L.A. Evaluation of the mechanisms responsible for the reduction in erythrocyte complement receptors when immune complexes form in vivo in primates. *J. Immunol.* **1990**, *145*, 4198–4206. [PubMed]

11. Cornacoff, J.B.; Hebert, L.A.; Smead, W.L.; VanAman, M.E.; Birmingham, D.J.; Waxman, F.J. Primate erythrocyte-immune complex-clearing mechanism. *J. Clin. Invest.* **1983**, *71*, 236–247. [CrossRef] [PubMed]

12. Brubaker, W.D.; Crane, A.; Johansson, J.U.; Yen, K.; Garfinkel, K.; Mastroeni, D.; Asok, P.; Bradt, B.; Sabbagh, M.; Wallace, T.L.; et al. Peripheral complement interactions with amyloid β peptides: Erythrocyte clearance mechanisms. *Alzheimers Dement.* **2017**, *13*, 1397–1409. [CrossRef] [PubMed]

13. Johansson, J.U.; Brubaker, W.D.; Javitz, H.; Bergen, A.W.; Nishita, D.; Trigunaite, A.; Crane, A.; Ceballos, J.; Mastroeni, D.; Tenner, A.J.; et al. Peripheral complement interactions with amyloid β peptide in Alzheimer's disease: Polymorphisms, structure, and function of complement receptor 1. *Alzheimers Dement.* **2018**, in press. [CrossRef] [PubMed]

14. Krych-Goldberg, M.; Atkinson, J.P. Structure-function relationships of complement receptor type 1. *Immunol. Rev.* **2001**, *180*, 112–122. [CrossRef] [PubMed]

15. Birmingham, D.J.; Chen, W.; Liang, G.; Schmitt, H.C.; Gavit, K.; Nagaraja, H.N. A CR1 polymorphism associated with constitutive erythrocyte CR1 levels affects binding to C4b but not C3b. *Immunology* **2003**, *108*, 531–538. [CrossRef] [PubMed]

16. Liu, D.; Niu, Z.X. The structure, genetic polymorphisms, expression and biological functions of complement receptor type 1 (CR1/CD35). *Immunopharm. Immunot.* **2009**, *31*, 524–535. [CrossRef] [PubMed]

17. Cohen, J.H.; Lutz, H.U.; Pennaforte, J.L.; Bouchard, A.; Kazatchkine, M.D. Peripheral catabolism of CR1 (the C3b receptor, CD35) on erythrocytes from healthy individuals and patients with systemic lupus erythematosus (SLE). *Clin. Exp. Immunol.* **1992**, *87*, 422–428. [CrossRef] [PubMed]

18. Cornillet, P.; Philbert, F.; Kazatchkine, M.D.; Cohen, J.H. Genomic determination of the CR1 (CD35) density polymorphism on erythrocytes using polymerase chain reaction amplification and HindIII restriction enzyme digestion. *J. Immunol. Methods* **1991**, *136*, 193–197. [CrossRef]

19. He, J.R.; Xi, J.; Ren, Z.F.; Qin, H.; Zhang, Y.; Zeng, Y.X.; Mo, H.Y.; Jia, W.H. Complement receptor 1 expression in peripheral blood mononuclear cells and the association with clinicopathological features and prognosis of nasopharyngeal carcinoma. *Asian Pac. J. Cancer Prev.* **2012**, *13*, 6527–6531. [CrossRef] [PubMed]

20. Mahmoudi, R.; Kisserli, A.; Novella, J.L.; Donvito, B.; Drame, M.; Reveil, B.; Duret, V.; Jolly, D.; Pham, B.N.; Cohen, J.H. Alzheimer's disease is associated with low density of the long CR1 isoform. *Neurobiol. Aging* **2015**, *36*, 1766.e5–1766.e12. [CrossRef] [PubMed]

21. Wilson, J.G.; Murphy, E.E.; Wong, W.W.; Klickstein, L.B.; Weis, J.H.; Fearon, D.T. Identification of a restriction fragment length polymorphism by a CR1 cDNA that correlates with the number of CR1 on erythrocytes. *J. Exp. Med.* **1986**, *164*, 50–59. [CrossRef] [PubMed]

22. Wong, W.W.; Cahill, J.M.; Rosen, M.D.; Kennedy, C.A.; Bonaccio, E.T.; Morris, M.J.; Wilson, J.G.; Klickstein, L.B.; Fearon, D.T. Structure of the human CR1 gene. Molecular basis of the structural and quantitative polymorphisms and identification of a new CR1-like allele. *J. Exp. Med.* **1989**, *169*, 847–863. [CrossRef] [PubMed]

23. Xiang, L.; Rundles, J.R.; Hamilton, D.R.; Wilson, J.G. Quantitative alleles of CR1: Coding sequence analysis and comparison of haplotypes in two ethnic groups. *J. Immunol.* **1999**, *163*, 4939–4945. [PubMed]

24. Herrera, A.H.; Xiang, L.; Martin, S.G.; Lewis, J.; Wilson, J.G. Analysis of complement receptor type 1 (CR1) expression on erythrocytes and of (CR1) allelic markers in Caucasian and African American populations. *Clin. Immunol. Immunopathol.* **1998**, *87*, 176–183. [CrossRef] [PubMed]

25. Pham, B.N.; Kisserli, A.; Donvito, B.; Duret, V.; Reveil, B.; Tabary, T.; Le Pennec, P.Y.; Peyrard, T.; Rouger, P.; Cohen, J.H. Analysis of complement receptor type 1 expression on red blood cells in negative phenotypes of the Knops blood group system, according to CR1 gene allotype polymorphisms. *Transfusion* **2010**, *50*, 1435–1443. [CrossRef] [PubMed]

26. Yoon, S.H.; Fearon, D.T. Characterization of a soluble form of the C3b/C4b receptor (CR1) in human plasma. *J. Immunol.* **1985**, *134*, 3332–3338. [PubMed]

27. Ehlers, M.R.; Riordan, J.F. Membrane proteins with soluble counterparts: Role of proteolysis in the release of transmembrane proteins. *Biochem.* **1991**, *30*, 10065–10074. [CrossRef]

28. Dervillez, X.; Oudin, S.; Libyh, M.T.; Tabary, T.; Reveil, B.; Philbert, F.; Bougy, F.; Pluot, M.; Cohen, J.H. Catabolism of the human erythrocyte C3b/C4b receptor (CR1, CD35): Vesiculation and/or proteolysis? *Immunopharmacology* **1997**, *38*, 129–140. [CrossRef]

29. Hamer, I.; Paccaud, J.P.; Belin, D.; Maeder, C.; Carpentier, J.L. Soluble form of complement C3b/C4b receptor (CR1) results from a proteolytic cleavage in the C-terminal region of CR1 transmembrane domain. *Biochem. J.* **1998**, *329 Pt 1*, 183–190. [CrossRef] [PubMed]

30. Jacquet, M.; Lacroix, M.; Ancelet, S.; Gout, E.; Gaboriaud, C.; Thielens, N.M.; Rossi, V. Deciphering complement receptor type 1 interactions with recognition proteins of the lectin complement pathway. *J. Immunol.* **2013**, *190*, 3721–3731. [CrossRef] [PubMed]

31. Kubiak-Wlekly, A.; Perkowska-Ptasinska, A.; Olejniczak, P.; Rochowiak, A.; Kaczmarek, E.; Durlik, M.; Czekalski, S.; Niemir, Z.I. The comparison of the podocyte expression of synaptopodin, CR1 and neprilysin in human glomerulonephritis: Could the expression of CR1 be clinically relevant? *Int. J. Biomed. Sci.* **2009**, *5*, 28–36. [PubMed]

32. Fonseca, M.I.; Chu, S.; Pierce, A.L.; Brubaker, W.D.; Hauhart, R.E.; Mastroeni, D.; Clarke, E.V.; Rogers, J.; Atkinson, J.P.; Tenner, A.J. Analysis of the putative role of CR1 in Alzheimer's disease: Genetic association, expression and function. *PLoS ONE* **2016**, *11*, e0149792. [CrossRef] [PubMed]

33. Malik, M.; Parikh, I.; Vasquez, J.B.; Smith, C.; Tai, L.; Bu, G.; LaDu, M.J.; Fardo, D.W.; Rebeck, G.W.; Estus, S. Genetics ignite focus on microglial inflammation in Alzheimer's disease. *Mol. Neurodegener.* **2015**, *10*, 52. [CrossRef] [PubMed]

34. Morgan, B.P. Complement in the pathogenesis of Alzheimer's disease. *Semin. Immunopathol.* **2018**, *40*, 113–124. [CrossRef] [PubMed]

35. Brouwers, N.; Van Cauwenberghe, C.; Engelborghs, S.; Lambert, J.C.; Bettens, K.; Le Bastard, N.; Pasquier, F.; Montoya, A.G.; Peeters, K.; Mattheijssens, M.; et al. Alzheimer risk associated with a copy number variation in the complement receptor 1 increasing C3b/C4b binding sites. *Mol. Psychiatry* **2012**, *17*, 223–233. [CrossRef] [PubMed]

36. Hazrati, L.N.; Van Cauwenberghe, C.; Brooks, P.L.; Brouwers, N.; Ghani, M.; Sato, C.; Cruts, M.; Sleegers, K.; St George-Hyslop, P.; Van Broeckhoven, C.; et al. Genetic association of CR1 with Alzheimer's disease: A tentative disease mechanism. *Neurobiol. Aging* **2012**, *33*, 2949.e5–2949.e12. [CrossRef] [PubMed]

37. Kucukkilic, E.; Brookes, K.; Barber, I.; Guetta-Baranes, T.; Morgan, K.; Hollox, E.J. Complement receptor 1 gene (CR1) intragenic duplication and risk of Alzheimer's disease. *Hum. Genet.* **2018**, *137*, 305–314. [CrossRef] [PubMed]

38. Currie, M.S.; Vala, M.; Pisetsky, D.S.; Greenberg, C.S.; Crawford, J.; Cohen, H.J. Correlation between erythrocyte CR1 reduction and other blood proteinase markers in patients with malignant and inflammatory disorders. *Blood* **1990**, *75*, 1699–1704. [PubMed]

39. Ross, G.D.; Yount, W.J.; Walport, M.J.; Winfield, J.B.; Parker, C.J.; Fuller, C.R.; Taylor, R.P.; Myones, B.L.; Lachmann, P.J. Disease-associated loss of erythrocyte complement receptors (CR1, C3b receptors) in patients with systemic lupus erythematosus and other diseases involving autoantibodies and/or complement activation. *J. Immunol.* **1985**, *135*, 2005–2014. [PubMed]

40. Pascual, M.; Lutz, H.U.; Steiger, G.; Stammler, P.; Schifferli, J.A. Release of vesicles enriched in complement receptor 1 from human erythrocytes. *J. Immunol.* **1993**, *151*, 397–404. [PubMed]

41. Mayeux, R.; Stern, Y. Epidemiology of Alzheimer disease. *CSH Perspect. Med.* **2012**, *2*, a006239. [CrossRef] [PubMed]

42. Thomas, B.N.; Donvito, B.; Cockburn, I.; Fandeur, T.; Rowe, J.A.; Cohen, J.H.; Moulds, J.M. A complement receptor-1 polymorphism with high frequency in malaria endemic regions of Asia but not Africa. *Genes. Immun.* **2005**, *6*, 31–36. [CrossRef] [PubMed]

43. American Psychiatric Association. *Diagnostic and Statistical Manual of Mental Disorders*, 4th ed.; American Psychiatric Press: Washington, DC, USA, 1994.

44. McKhann, G.; Drachman, D.; Folstein, M.; Katzman, R.; Price, D.; Stadlan, E.M. Clinical diagnosis of Alzheimer's disease: Report of the NINCDS-ADRDA work group under the auspices of department of health and human services task force on Alzheimer's disease. *Neurology* **1984**, *34*, 939–944. [CrossRef] [PubMed]

45. Cook, J.; Fischer, E.; Boucheix, C.; Mirsrahi, M.; Jouvin, M.H.; Weiss, L.; Jack, R.M.; Kazatchkine, M.D. Mouse monoclonal antibodies to the human C3b receptor. *Mol. Immunol.* **1985**, *22*, 531–539. [CrossRef]

46. Kiss, E.; Csipo, I.; Cohen, J.H.; Reveil, B.; Kavai, M.; Szegedi, G. CR1 density polymorphism and expression on erythrocytes of patients with systemic lupus erythematosus. *Autoimmunity* **1996**, *25*, 53–58. [CrossRef] [PubMed]

47. Kisserli, A.; Tabary, T.; Cohen, J.H.M.; Duret, V.; Mahmoudi, R. High-resolution melting PCR for complement receptor 1 length polymorphism genotyping: An innovative tool for Alzheimer's disease gene susceptibility assessment. *J. Vis. Exp.* **2017**, *125*, 56012. [CrossRef] [PubMed]

48. Crane, A.; Brubaker, W.D.; Johansson, J.U.; Trigunaite, A.; Ceballos, J.; Bradt, B.; Glavis-Bloom, C.; Wallace, T.L.; Tenner, A.J.; Rogers, J. Peripheral complement interactions with amyloid β peptide in Alzheimer's disease: 2. Relationship to amyloid β immunotherapy. *Alzheimers Dement.* **2018**, *14*, 243–252. [CrossRef] [PubMed]

International Journal of
Molecular Sciences

MDPI

Review

Presenilins as Drug Targets for Alzheimer's Disease—Recent Insights from Cell Biology and Electrophysiology as Novel Opportunities in Drug Development

R. Scott Duncan [1], Bob Song [1] and Peter Koulen [1,2,*]

[1] Vision Research Center, Department of Ophthalmology, School of Medicine, University of Missouri-Kansas City, Kansas City, MO 64108, USA; duncanrs@umkc.edu (R.S.D.); bsong1993@gmail.com (B.S.)
[2] Department of Biomedical Science, School of Medicine, University of Missouri-Kansas City, Kansas City, MO 64108, USA
*   Correspondence: koulenp@umkc.edu; Tel.: +1-816-404-1834

Received: 21 April 2018; Accepted: 28 May 2018; Published: 31 May 2018

Abstract: A major cause underlying familial Alzheimer's disease (AD) are mutations in presenilin proteins, presenilin 1 (PS1) and presenilin 2 (PS2). Presenilins are components of the γ-secretase complex which, when mutated, can affect amyloid precursor protein (APP) processing to toxic forms of amyloid beta (Aβ). Consequently, presenilins have been the target of numerous and varied research efforts to develop therapeutic strategies for AD. The presenilin 1 gene harbors the largest number of AD-causing mutations resulting in the late onset familial form of AD. As a result, the majority of efforts for drug development focused on PS1 and Aβ. Soon after the discovery of the major involvement of PS1 and PS2 in γ-secretase activity, it became clear that neuronal signaling, particularly calcium ion (Ca^{2+}) signaling, is regulated by presenilins and impacted by mutations in presenilin genes. Intracellular Ca^{2+} signaling not only controls the activity of neurons, but also gene expression patterns, structural functionality of the cytoskeleton, synaptic connectivity and viability. Here, we will briefly review the role of presenilins in γ-secretase activity, then focus on the regulation of Ca^{2+} signaling, oxidative stress, and cellular viability by presenilins within the context of AD and discuss the relevance of presenilins in AD drug development efforts.

Keywords: γ-secretase; amyloid beta; calcium signaling; drug target discovery; endoplasmic reticulum; inositol 1,4,5-trisphosphate receptor; ion channel; oxidative stress; ryanodine receptor; therapy

---

1. Introduction

Presenilins have long been known to play a role in familial Alzheimer's disease (AD) pathogenesis [1]. With two presenilin genes in vertebrates, homologs of the human genes *PSEN1* and *PSEN2*, the two resulting presenilin proteins, presenilin 1 (PS1) and presenilin 2 (PS2) [1] are constituents of the multi-subunit γ-secretase complex which facilitates proteolytic processing of amyloid precursor protein (APP) [2]. Mutations in APP lead to accumulation of amyloid-beta peptides (Aβ), which can be toxic to neural tissue and contribute to AD pathology in the brain [2] with recent studies indicating that the formation of annular protofibrils by Aβ leads to membrane permeabilization and subsequent dysregulation of ion homeostasis [3]. PS1, specifically, is associated with familial AD in part by influencing Ca^{2+} signaling [4], yet there is still much to be uncovered about presenilins with new studies revealing more about non-canonical (non-γ-secretase–related) functions. Here, we discuss the role of PS1 and PS2 in cellular oxidative stress, in protein degradation/autophagy, and in regulating intracellular endoplasmic reticulum (ER) Ca^{2+} channels (i.e., inositol 1,4,5-trisphosphate receptors (IP$_3$Rs) and ryanodine receptors (RyRs)). Investigating this involvement of presenilins in Ca^{2+} signaling results in unique challenges due to the ubiquitous expression of IP$_3$Rs and RyRs by a wide range of cell types in almost every tissue and organ. This challenge also represents a unique opportunity for drug target discovery and clinical drug development efforts by taking advantage of recently identified mechanisms that place presenilins at the crossroads of oxidative stress, calcium signaling, and neuronal viability. Combining such insights with the newly identified role of presenilin involvement in neuronal calcium signaling represents novel opportunities in drug development for Alzheimer's disease, the focus of the present review.

2. γ-Secretase Activity of Presenilin

The role of presenilins, APP and γ-secretase in AD pathogenesis has been widely studied. Presenilin proteins, PS1 and PS2, are constituents of the γ-secretase complex, which carry out amyloid precursor protein (APP) proteolytic processing [2]. Three new novel PS1 mutations have been uncovered in patients with a vast heterogeneity of clinical phenotypes [5]. Investigation of wild-type γ-secretase with six familial Alzheimer's disease (FAD) mutants in PS1 and five FAD mutants in the Aβ peptide segment of the APP revealed that all mutations were associated with decreased γ-secretase activity and a reduced age of disease onset and death [6]. Furthermore, an increase in the ratio between Aβ expression and γ-secretase activity was an early sign of disease in both sporadic and familial AD [6]. The PS2 K115Efx10 mutation causes PS2 protein truncation, and resembles a PS2 isoform, PS2V, which is found in late onset AD brains [7]. Additionally, PS2V mutants were able to activate γ-secretase activity which, under hypoxic conditions, correlated with an attenuation of the unfolded protein response [7].

Mature PS1 has many distinct conformational states while non-mature PS1 has only one state [8]. Structural studies of PS1 reveal a so-called "gate-plug" structure where the site responsible for endo-proteolytic cleavage is found. Transmembrane 5 and 6 regions (TM5 and TM6) make up the gate while the exon 9 loop region of the protein makes up the plug. A so-called "unplugging mechanism" by endo-proteolysis and subsequent removal of exon 9 loop is associated with the mature PS1, and susceptibility of a gate-plug region to conformational changes may indicate how PS1 mutants initiate disease [8]. Diminished access and inaccurate cleavage of substrate, along with the altered gate-plug activity, may explain why PS1 mutations are correlated with reduced Aβ levels and increase in Aβ$_{42}$:Aβ$_{40}$ ratio [8]. Changes to TM5 and TM6 histidines (H171A and H197A) reduce self-cleavage of PS1 and interaction with additional γ-secretase constituents, leading to reduced Aβ generation [9].

Substitution of histidines with lysine residues in TM5 and TM6 yields structurally normal γ-secretase complexes however with defective enzymatic activity [9].

Saturation of γ-secretase with substrate may mechanistically underlie AD pathogenesis by increasing the Aβ$_{42}$:Aβ$_{40}$ ratio, suggesting that competitive γ-secretase inhibitors may be potential therapeutics for AD [10]. Noncompetitive inhibitors, on the other hand, may worsen AD by promoting APP saturation [10]. Two conserved AXXAXXXG motifs were identified in PS1 and PS2, and their involvement in γ-secretase complex configuration were found to be involved in the alternation between normal and pathological γ-secretase conformations [11].

Small molecule γ-secretase modulators were investigated as potential therapies for AD by reducing Aβ$_{42}$ while not blocking γ-secretase processing of substrates [12]. Using a photo-affinity probe, E2012-BPyne, that specifically labeled the N-terminus of PS1 within the active γ-secretase, but not the full-length PS1 in the active form, γ-secretase displayed several binding sites with separate functions [12].

The subcellular localization of γ-secretase has been investigated as a contributing factor to Aβ production. The protein Retention in Endoplasmic Reticulum 1 (RER1) controls the intracellular trafficking of γ-secretase [13]. While overexpression of RER1 results in decreased localization of γ-secretase to the cell surface and decreased secretion of Aβ secretion, knockdown of RER1 in turn increased both levels of γ-secretase on the cell surface and Aβ secretion [13]. All in all, increased RER1 decreases the mature APP form leading to reduced surface APP accumulation [13].

Mice engineered to express wild type or mutant PS1 in the central nervous system (CNS) and HEK293 cells engineered to express PS paralogs revealed γ-secretase interactions with synaptic vesicle complexes and fusion to cellular membranes as well as H+ transporting lysosomal ATPase complex [14]. The peptidase was mainly co-purified with γ-secretase complexes containing PS2 to control Aβ production [14].

The roles of γ-secretase orthologs from other species have provided clues to non-canonical γ-secretase functions. For example, *Dictyostelium discoideum* γ-secretase orthologs can proteolytically process ectopically expressed human APP to yield Aβ peptides (Aβ_{40} and Aβ_{42}), but γ-secretase-deficient strains cannot generate Aβ peptides [15]. *Dictyostelium* γ-secretase was also found to be important for phagocytosis and cell fate determination. These data suggest that phagocytosis may require an active γ-secretase in mammalian and *Dictyostelium* cells [15].

In AD patients with mutated PS1, Coupland et al. identified a decrease in the DNA methylation of the promoter for the gene encoding microtubule-associated protein tau (MAPT) as a common phenomenon in a specific brain-region of these AD patients [16].

3. Presenilins and Ca^{2+} Signaling

Dysfunction in Ca^{2+} signaling can contribute to age-related central nervous system (CNS) decline [17]. Such damage in brain aging, especially in AD, is thought to be the result of numerous micro-injuries such as oxidative damage in synapses and loss of Ca^{2+} homeostasis leading to increased cytosolic Ca^{2+} concentrations [18]. Long term potentiation is reduced following presynaptic (but not postsynaptic) deletion of presenilins mimicking the depletion of ER Ca^{2+} stores by RyR inhibitors [19]. Presynaptic presenilin deficiency also reduced evoked glutamate release, indicating that presenilins play a role in activity-dependent neurotransmitter release and that presynaptic dysfunction represents an early event in AD development [19].

Neurons expressing mutant PS1 exhibit an increase in calcineurin activity and inhibition or reversal of this elevated calcineurin activity stabilized GluA1 phosphorylation and improved homeostatic synaptic plasticity [20]. Improvement of homeostatic synaptic plasticity leads to attenuation of AD-related cognitive decline and likewise improvement in learning and memory [20]. A novel γ-secretase modulator (compound-1) reduces Aβ expression thus relieving cognitive dysfunction in Tg2576 APP transgenic mice, a common mouse model of AD [21]. In mice embryogenic fibroblast cells, this inhibitor also plays a role in Ca^{2+} signaling by enhancing long-term potentiation (LTP), an indicator of synaptic strength [21].

Presenilins are regulators of intracellular calcium stores. RyRs and IP$_3$Rs, major intracellular Ca^{2+} channels residing in the ER, are regulated by PS proteins. Furthermore, the expression of ER resident Ca^{2+} channels is increased in neurons expressing mutant PS1 [20]. The presenilin–ryanodine receptor (PS–RyR) interaction, where PS1 and PS2 N-termini bind the cytoplasmic face of RyR, regulates channel activity [22] similar to the actions of other AD related proteins binding to the RyR [23]. Investigation of the expression patterns of PS1 and PS2 identified an overall decrease in PS1 level with increase in PS2 level in older mice [24].

A PS1 N-terminal fragment (NTF), which lacks four cysteine residues, decreased total RyR-mediated Ca^{2+} release, while a PS2 NTF, which contains four cysteine residues, had no effect [25]. These cysteines were mutated, allowing conversion of PS1 NTF function to PS2 NTF-like function and vice versa, likely based on differential RyR binding [25]. Inactivation of presenilin in the hippocampus has no effect on ER Ca^{2+} concentration, but in the absence of presenilin, RyR levels and function were decreased in the hippocampus [26]. This suggests a connection between presenilin and Ca^{2+} homeostasis via RyR, further supporting the idea that loss of Ca^{2+} homeostasis is an early pathologic injury in AD [26].

The effect of Aβ plaque proximity to disruptions in hippocampal pyramidal neuron Ca^{2+} signaling was investigated. No significant correlation between Aβ plaque proximity to cells with altered Ca^{2+} signaling was found [27]. These data suggest that early disruptions in pyramidal cell Ca^{2+} signaling occur through Aβ plaque-independent mechanisms [27]. Neuronal presenilins in *Drosophila* have no role on resting Ca^{2+} channels but cause deficits in intracellular Ca^{2+} stores [28]. In addition, calmodulin null mutations suppress presenilin-induced deficits in Ca^{2+} stores [28].

Lee et al., 2015, studied the notion that the mechanism by which PS1 deletion impacts AD was through lysosomal acidification [29]. Their studies revealed that an increased pH in the lysosomes of PS1 knockout (PS1KO) cells caused abnormal Ca^{2+} efflux from lysosomes, resulting in increased cytosolic Ca^{2+} concentrations [29]. Normalizing lysosomal pH restored Ca^{2+} homeostasis, but restored Ca^{2+} homeostasis in turn by itself did not result in adequate acidification of lysosomes or reverse proteolytic and autophagic effects. This led the authors to conclude that an instable lysosomal vesicular ATPase (vATPase) subunit in PS1-deficient cells causes the deficits in lysosomal autophagy [29].

4. Presenilins and Oxidative Stress

Oxidative stress is a contributing factor to Alzheimer Disease pathogenesis, with several theories supporting a connection between oxidative stress and the accumulation of Aβ [30]. As monomeric Aβ facilitates glutathione release from astrocytes, it potentially contributes to protection from oxidative stress, a function that is reduced with $Aβ_{42}$ aggregation and subsequent depletion of monomeric $Aβ_{42}$ [31]. Presenilins are involved in neuroprotection against oxidative stress [30]. PS1 was determined to be important for neurotrophic factor-mediated neuroprotection against excitotoxicity and oxidative stress and was not dependent on the role of PS1 in γ-secretase activity, as γ-secretase inhibitors lacked any effect on trypsin-induced neuroprotection [32]. This mechanism seems to stem from PS1 mutants being unable to use trypsin to subsequently rescue neurons from excitotoxicity by activating extracellular signal-regulated kinase 1/2 (ERK1/2) [32]. As expected, PS mutants inhibited neuronal protection against toxic insults [32]. Exposure of neurons to low concentrations (0.25 ppm) of ozone lead to significant increases in $Aβ_{42}$ in mitochondrial fractions, reduction in $Aβ_{40}$, overexpression of PS2, and reductions in ADAM10 expression [30], suggesting that $Aβ_{42}$ accumulation may be involved in mitochondrial dysfunction and subsequent oxidative stress [30]. Sarasija et al. also studied Ca^{2+} transfer, but instead investigated a presenilin analog SEL-12 which regulates ER Ca^{2+} release, demonstrating that mutations in SEL-12 causes mitochondrial fragmentation and dysfunction [33]. This role in mitochondrial damage did not require γ-secretase activity and amyloid plaques [33].

The effect of certain diabetes drugs on Aβ production and oxidative stress has been investigated. Administration of the insulin sensitizer, metformin, increases APP and presenilin expression via NF-κB activation [34]. In contrast, insulin administration antagonized the effects of metformin by decreasing Aβ levels and reducing oxidative stress and mitochondrial dysfunction [34]. Interestingly, monomeric $Aβ_{42}$ is capable of activating the phosphatidylinositol-3-kinase pathway and thereby generates neuroprotection via insulin-like growth factor-1 and other receptors [35]. This raises the interesting notion that part of Aβ toxicity is the result of a depletion of $Aβ_{42}$ subsequent to $Aβ_{42}$ oligomerization and polymerization [35].

The relationship between mitochondrial function and chaperone-mediated RyR degradation in cardiomyocytes (as well as fibroblast number) was studied in AD patients with PS1 mutations [35]. Fibroblasts with the AD mutation had elevated $Aβ_{42}$, reduced ATP levels, reduced mitochondrial respiration, and impaired mitochondrial respiratory capacity [36].

Copper (Cu^{2+}) is important for enzymatic antioxidant activity, namely as a cofactor in the antioxidant enzyme superoxide dismutase (SOD) [37]. While PS1 and PS2 play roles in Cu^{2+} uptake, presenilin knockdown in *Drosophila* reduces Cu^{2+} levels and consequently decreases SOD [37]. These presenilin knockdown *Drosophila* were sensitive to SOD-inducing chemical paraquat, supporting the role of presenilin on SOD activity [37]. Interestingly, in Zebrafish, a truncated PS2 isoform, PSV2, is induced in spontaneous AD under hypoxic conditions and conditions of high cholesterol [38].

PSV2 normally increases γ-secretase activity [38]. Zebrafish possess another presenilin isoform, PS1IV, an isoform similar to PS2V in humans [38]. It is associated with changes in cytokine gene expression, such as IL1β and CCR5, and in addition, the absence of PS1IV under hypoxic conditions is associated with changes in vascular development, protein synthesis, Ca^{2+} homeostasis, and cell proliferation [38].

Drosophila presenilin interacts with the enzymes thiol-specific antioxidant (TSA) and proliferation-associated gene (PAG), both involved in cellular antioxidant activity, and thereby affects Notch signaling [39]. Transgenic presenilin expression in precursor cells of wing and sensory organ caused a Notch loss-of-function phenotype [38]. In fact, co-expression of presenilin with proteins resulted in a more severe and penetrant Notch loss-of-function phenotype than PS expression alone [39]. Such signaling mechanisms involved in inflammation appear to be of particular importance given the role inflammation has in AD development in the presence of high Aβ levels [40] and that other pathogenic signaling mechanism such as tau protein cleavage and of the formation of neurofibrillary tangles respond to intervention with antioxidants [41].

Pedrozo et al. induced chaperone-mediated autophagy (CMA) in cardiomyocytes with geldanamycin, which prevented the loss of RyR2 degradation, suggesting that presenilins were involved in this process [42]. Presenilins, therefore, are involved in CMA and can target oxidatively damaged RyR2 [42].

5. The Role of Presenilins in Proteasome Function and Autophagy

Presenilin has many roles including, but not limited to, RyR regulation and interaction with other regulatory pathways. Hwang et al. demonstrated that PS2 mutations can lead to NF-κB mediated amyloidosis [43]. Presenilins have two roles: proteolysis-dependent activity in the γ-secretase complex and activities in cellular signaling that are independent of proteolytic activity [44]. The coupling of ubiquitin conjugation to endoplasmic reticulum degradation (CUE) ubiquitin binding domain of PS1 coordinates polyubiquitination at lysine 63 [45].

Recent studies determined the effect of presenilins in the autophagy/lysosome system and found that presenilin deficit led to a reduction in lysosomal Ca^{2+} stores regardless of lysosome accumulation, and prevention of the organization of two-pore channels 1 and 2 (TPC1 and TPC2) [46]. This indicates that modifications in lysosomal Ca^{2+} due to presenilin deficiency can lead to interference of autophagy [46]. In addition, genetic deletion or knockdown of presenilins can lead to a buildup of autophagosomes independent of γ-secretase activity [47]. Ablation of *Dictyostelium* presenilins lead to PS1-mediated restoration of the terminal differentiation of multiple cell types independent of its proteolytic effect [44]. Presenilin loss in *Dictyostelium* leads to elevated cAMP concentrations and elevated Ca^{2+} release, indicating that presenilins indeed regulate signaling pathways [44].

The impact of loss of PS1 activity on lysosomal alkalization and subsequent impairment of autophagosomal function was determined, but investigations were unable to identify presenilin involvement in controlling autophagy [48]. Studies of mice brains lacking PS, however, revealed a function for PS in regulating lysosomal formation [48].

Tequila and mammalian analog *Prss12* gene expression is reduced by presenilins in brains of *Drosophila melanogaster* larvae and in mouse embryonic fibroblasts [49]. A mature γ-secretase complex was found to be essential for inhibiting neurotrypsin expression and reduction of agrin cleavage, but PS1 processing of γ-secretase substrates was not required for this activity [49]. Silencing of the *Drosophila* ortholog of presenilins (dPsn) lowered the heart rate, while dPsn overexpression increased it [50]. dPSN silencing also increased dIP_3R expression and decreased dSERCA expression, while dPsn overexpression lowered dRyR expression [50]. All in all, changes in presenilin expression resulted in cardiac dysfunction via aberrant Ca^{2+} signaling and disrupted Wnt signaling [50] (summarized in Table 1).

Table 1. Presenilin function within cells.

Presenilin Function	Protein/Signaling Targets	References
γ-secretase complex activity	APP	[2,6,7,10–15]
Ca^{2+} signaling	IP$_3$R, RyR (mammalian); regulation of dIP$_3$R, dSERCA and dRyR expression (*Drosophila melanogaster*); SEL-12 (*Caenorhabditis elegans*)	[4,17–22,25,26,28,29,33,46]
Oxidative stress	trypsin-mediated ERK1/2 activation, mitochondrial proteins, thiol-specific antioxidant (TSA) and proliferation-associated gene (PAG)	[18,30,32,39,42]
Proteolysis	Trypsin, CREB activity	[32,49]
Lysosome/Autophagy	vATPase regulation, chaperone-mediated autophagy, two-pore calcium channel expression, lysosomal proteolysis, lysosomal acidification	[29,42,46–48]
Cellular signaling	Notch, inflammatory signaling	[38,39]
Cu^{2+} uptake	reduced Cu^{2+} uptake, reduced SOD expression	[37]
Cellular differentiation/development	Proteolytic agrin cleavage	[15,44,49]

6. Functions of Presenilins Outside of AD

Besides its well-documented role in AD, presenilins also play many roles in other diseases (see Table 2). This results in both a more differentiated view of the involvement of PS and potentially opens up new avenues for drug targeting and drug discovery. The role of a gene, which interacts with PTEN-induced putative kinase in mitochondrial homeostasis and during early-onset Parkinson disease, called presenilin-associated rhomboid-like (*PARL*), was investigated [51]. Single nucleotide polymorphisms in PARL represented a rare cause of Parkinson disease [51].

Presenilin is also involved in variants of cancer, as PS1 was amplified in gastric cancer and correlated with a poor survival and increased metastasis [52]. This mechanism may be explained by the E-cadherin cleavage and β-catenin release by PS1, thus allowing β-catenin nuclear translocation and transcriptional activations to promote gastric cancer progression [52]. Fusion transcripts between large tumor suppressor 1 (*LATS1*) and PS1 genes were unable to phosphorylate yes-associated protein and subsequently inhibit the growth of malignant mesothelioma cells [53].

PS1 is also involved in the development of the skin disorder hidradenitis suppurativa or acne inversa. Defective Notch signaling due to loss of function mutations of PS-1 and other γ-secretase subunits likely contributes to the pathogenesis of hidradenitis suppurativa affecting integral membrane proteins such as Notch, E-cadherin, or CD44 [54]. A Mutation of PS2 was identified as a genetic cause for familial comedones syndrome, which has clinical phenotypes similar to hidradentis suppurativa [55].

While a clear link has been demonstrated between development of Alzheimer's disease and increasing age, links have also been found between PS function and normal aging. A preclinical model for aging was used to identify changes in cerebellar and forebrain PS expression that correlate with performance in motor function, memory, and learning in aged rats, where PS1 was decreased while PS2 was increased [24]. Puig et al. identified the roles of mutant APP and PS1 in the enteric nervous system [56]. They found that APP/PS1 mice had normal gastrointestinal function, but they had higher luminal IgA and APP, indicating elevated proinflammatory factors and immune cell activation [56].

Presenilins also play a role in cardiac function. Chaperone-mediated autophagy (CMA), a process involved in the degradation of soluble proteins in the cytosol, occurs by lysosome associated membrane protein type 2A- (LAMP-2A)-facilitated degradation [42]. LAMP2 mutations can lead to Danon disease, characterized by hypertrophic cardiomyopathy [42]. Pedrozo et al. discovered that RyR2 is degraded by CMA, suggesting that oxidative damage targets RyR2 for turnover via presenilins and CMA [42]. Li et al. discovered that silencing the *Drosophila* ortholog of presenilins (dPsn) reduced heart rates and generated an age-dependent rise in end-diastolic vertical dimensions; conversely, dPsn overexpression

Int. J. Mol. Sci. **2018**, *19*, 1621

led to higher heart rates [50]. Silencing of dPsn elevated the expression levels of the *Drosophila* ortholog of IP$_3$R and reduced expression of the *Drosophila* ortholog of SERCA while overexpression of dPsn led to reduced expression of the *Drosophila* ortholog of the RyR [50], offering a mechanism for how cardiac dysfunction occurs via changes in PS expression. Overall, presenilin changes lead to cardiac dysfunction secondary to abnormal Ca^{2+} channel activity and disrupted Wnt signaling [50]. Presenilins also play a role in embryogenesis. Donoviel et al. generated PS1/PS2 double null mice and noticed embryonic lethality [57]. In addition, embryos deficient in both presenilins demonstrated developmental dysregulation such as absence of segmentation, defects in ventral neural tube formation, delays in the closure of the anterior neuropore, and irregular heart development [57].

Table 2. Presenilin involvement in diseases and conditions.

Disease/Condition	System/Organ	References
Normal neuronal function (cognition, memory)	Brain, intestine	[19,21,24,26,28,32,42,43]
Alzheimer's disease	Brain	[1,4–6,16,17,19]
Parkinson's disease	Brain	[51]
Familial comedones	Skin	[54,55]
Cancer	gastrointestinal	[52,53]
Cardiac dysfunction (embryonic development)	heart	[42,50,57]

7. Conclusions

Overall, the involvement of PS as part of the γ-secretase complex and in other roles in both excitable and non-excitable cells, but especially in immune cells such as T-cells and macrophages (Table 1), opens up a wide range of possible roles for PS as targets for AD drug target discovery and drug development (Table 2).

Author Contributions: P.K. conceived and designed the review; R.S.D., B.S. and P.K. wrote the paper.

Acknowledgments: This publication was supported in part by grants from the National Eye Institute (EY014227, EY022774 and EY027005), the National Institute on Aging (AG022550 and AG027956), the National Center for Research Resources and National Institute of General Medical Sciences (RR027093) of the National Institutes of Health (P.K.). The content is solely the responsibility of the authors and does not necessarily represent the official views of the National Institutes of Health. Additional support by the Felix and Carmen Sabates Missouri Endowed Chair in Vision Research and a Challenge Grant from Research to Prevent Blindness (P.K.) is gratefully acknowledged.

Conflicts of Interest: The authors declare no conflict of interest.

Abbreviations

Aβ	amyloid beta
AD	Alzheimer's disease
APP	amyloid precursor protein
CMA	chaperone-mediated autophagy
CNS	central nervous system
ER	endoplasmic reticulum
ERK1/2	extracellular signal-regulated kinase 1/2
FAD	familial Alzheimer's disease
IP$_3$R	inositol 1,4,5-trisphosphate receptor
LAMP	by lysosome associated membrane protein
LATS1	large tumor suppressor 1
MAPT	microtubule-associated protein tau
NTF	N-terminal fragment
PS1	presenilin 1
PS2	presenilin 2

RyR   ryanodine receptor
SOD   superoxide dismutase
TPC   two-pore channels

References

1. Clark, R.F.; Hutton, M.; Fuldner, R.A.; Froelich, S.; Karran, E.; Talbot, C.; Crook, R.; Lendon, C.; Prihar, G.; He, C.; et al. The structure of presenilin 1(S182) gene and identification of six novel mutation in early onset AD families. *Nat. Genet.* **1995**, *11*, 219–222. [CrossRef] [PubMed]
2. Xia, W.; Zhang, J.; Kholodenko, D.; Citron, M.; Podlisny, M.B.; Teplow, D.B.; Haass, C.; Seubert, P.; Koo, E.H.; Selkoe, D.J. Enhanced production and oligomerization of the 42-residue amyloid beta-protein by Chinese hamster ovary cells stably expressing mutant presenilins. *J. Biol. Chem.* **1997**, *272*, 7977–7982. [CrossRef] [PubMed]
3. Lasagna-Reeves, C.A.; Glabe, C.G.; Kayed, R. Amyloid-β annular protofibrils evade fibrillar fate in Alzheimer disease brain. *J. Biol. Chem.* **2011**, *286*, 22122–22130. [CrossRef] [PubMed]
4. Mattson, M.P.; Guo, Q.; Furukawa, K.; Pedersen, W.A. Presenilins, the endoplasmic reticulum, and neuronal apoptosis in Alzheimer's disease. *J. Neurochem.* **1998**, *70*, 1–14. [CrossRef] [PubMed]
5. Roeber, S.; Müller-Sarnowski, F.; Kress, J.; Edbauer, D.; Kuhlmann, T.; Tüttelmann, F.; Schindler, C.; Winter, P.; Arzberger, T.; Müller, U.; et al. Three novel presenilin 1 mutations marking the wide spectrum of age at onset and clinical patterns in familial Alzheimer's disease. *J. Neural Transm.* **2015**, *122*, 1715–1719. [CrossRef] [PubMed]
6. Svedružić, Ž.M.; Popović, K.; Šendula-Jengić, V. Decrease in catalytic capacity of γ-secretase can facilitate pathogenesis in sporadic and Familial Alzheimer's disease. *Mol. Cell. Neurosci.* **2015**, *67*, 55–65. [CrossRef] [PubMed]
7. Moussavi Nik, S.H.; Newman, M.; Wilson, L.; Ebrahimie, E.; Wells, S.; Musgrave, I.; Verdile, G.; Martins, R.N.; Lardelli, M. Alzheimer's disease-related peptide PS2V plays ancient, conserved roles in suppression of the unfolded protein response under hypoxia and stimulation of γ-secretase activity. *Hum. Mol. Genet.* **2015**, *24*, 3662–3678. [CrossRef] [PubMed]
8. Somavarapu, A.K.; Kepp, K.P. The dynamic mechanism of presenilin-1 function: Sensitive gate dynamics and loop unplugging control protein access. *Neurobiol. Dis.* **2016**, *89*, 147–156. [CrossRef] [PubMed]
9. Pardossi-Piquard, R.; Yang, S.P.; Kanemoto, S.; Gu, Y.; Chen, F.; Böhm, C.; Sevalle, J.; Li, T.; Wong, P.C.; Checler, F.; et al. APH1 polar transmembrane residues regulate the assembly and activity of presenilin complexes. *J. Biol. Chem.* **2009**, *284*, 16298–16307. [CrossRef] [PubMed]
10. Svedružić, Z.M.; Popović, K.; Smoljan, I.; Sendula-Jengić, V. Modulation of γ-secretase activity by multiple enzyme-substrate interactions: Implications in pathogenesis of Alzheimer's disease. *PLoS ONE* **2012**, *7*, e32293. [CrossRef] [PubMed]
11. Marinangeli, C.; Tasiaux, B.; Opsomer, R.; Hage, S.; Sodero, A.O.; Dewachter, I.; Octave, J.N.; Smith, S.O.; Constantinescu, S.N.; Kienlen-Campard, P. Presenilin transmembrane domain 8 conserved AXXXAXXXG motifs are required for the activity of the γ-secretase complex. *J. Biol. Chem.* **2015**, *290*, 7169–7184. [CrossRef] [PubMed]
12. Pozdnyakov, N.; Murrey, H.E.; Crump, C.J.; Pettersson, M.; Ballard, T.E.; Am Ende, C.W.; Ahn, K.; Li, Y.M.; Bales, K.R.; Johnson, D.S. γ-Secretase modulator (GSM) photoaffinity probes reveal distinct allosteric binding sites on presenilin. *J. Biol. Chem.* **2013**, *288*, 9710–9720. [CrossRef] [PubMed]
13. Park, H.J.; Shabashvili, D.; Nekorchuk, M.D.; Shyqyriu, E.; Jung, J.I.; Ladd, T.B.; Moore, B.D.; Felsenstein, K.M.; Golde, T.E.; Kim, S.H. Retention in endoplasmic reticulum 1 (RER1) modulates amyloid-β (Aβ) production by altering trafficking of γ-secretase and amyloid precursor protein (APP). *J. Biol. Chem.* **2012**, *287*, 40629–40640. [CrossRef] [PubMed]
14. Jeon, A.H.; Böhm, C.; Chen, F.; Huo, H.; Ruan, X.; Ren, C.H.; Ho, K.; Qamar, S.; Mathews, P.M.; Fraser, P.E.; et al. Interactome analyses of mature γ-secretase complexes reveal distinct molecular environments of presenilin (PS) paralogs and preferential binding of signal peptide peptidase to PS2. *J. Biol. Chem.* **2013**, *288*, 15352–15366. [CrossRef] [PubMed]
15. McMains, V.C.; Myre, M.; Kreppel, L.; Kimmel, A.R. Dictyostelium possesses highly diverged presenilin/ gamma-secretase that regulates growth and cell-fate specification and can accurately process human APP: A system for functional studies of the presenilin/gamma-secretase complex. *Dis. Models Mech.* **2010**, *3*, 581–594. [CrossRef] [PubMed]

16. Coupland, K.G.; Kim, W.S.; Halliday, G.M.; Hallupp, M.; Dobson-Stone, C.; Kwok, J.B. Effect of PSEN1 mutations on MAPT methylation in early-onset Alzheimer's disease. *Curr. Alzheimer Res.* **2015**, *12*, 745–751. [CrossRef] [PubMed]

17. Mattson, M.P.; Rydel, R.E.; Lieberburg, I.; Smith-Swintosky, V.L. Altered calcium signaling and neuronal injury: Stroke and Alzheimer's disease as examples. *Ann. N. Y. Acad. Sci.* **1993**, *679*, 1–21. [CrossRef] [PubMed]

18. Beal, M.F. Aging, energy, and oxidative stress in neurodegenerative diseases. *Ann. Neurol.* **1995**, *38*, 357–366. [CrossRef] [PubMed]

19. Zhang, C.; Wu, B.; Beglopoulos, V.; Wines-Samuelson, M.; Zhang, D.; Dragatsis, I.; Südhof, T.C.; Chen, J. Presenilins are essential for regulating neurotransmitter release. *Nature* **2009**, *460*, 632–636. [CrossRef] [PubMed]

20. Kim, S.; Violette, C.J.; Ziff, E.B. Reduction of increased calcineurin activity rescues impaired homeostatic synaptic plasticity in presenilin 1 M146V mutant. *Neurobiol. Aging* **2015**, *36*, 3239–3246. [CrossRef] [PubMed]

21. Hayama, T.; Murakami, K.; Watanabe, T.; Maeda, R.; Kamata, M.; Kondo, S. Single administration of a novel γ-secretase modulator ameliorates cognitive dysfunction in aged C57BL/6J mice. *Brain Res.* **2016**, *1633*, 52–61. [CrossRef] [PubMed]

22. Payne, A.J.; Kaja, S.; Koulen, P. Regulation of ryanodine receptor-mediated calcium signaling by presenilins. *Recept. Clin. Investig.* **2015**, *2*, e449.

23. Grillo, M.A.; Grillo, S.L.; Gerdes, B.C.; Kraus, J.G.; Koulen, P. Control of Neuronal Ryanodine Receptor-Mediated Calcium Signaling by Calsenilin. *Mol. Neurobiol.* **2018**, 1–10. [CrossRef] [PubMed]

24. Kaja, S.; Sumien, N.; Shah, V.V.; Puthawala, I.; Maynard, A.N.; Khullar, N.; Payne, A.J.; Forster, M.J.; Koulen, P. Loss of Spatial Memory, Learning, and Motor Function During Normal Aging Is Accompanied by Changes in Brain Presenilin 1 and 2 Expression Levels. *Mol. Neurobiol.* **2015**, *52*, 545–554. [CrossRef] [PubMed]

25. Payne, A.J.; Gerdes, B.C.; Naumchuk, Y.; McCalley, A.E.; Kaja, S.; Koulen, P. Presenilins regulate the cellular activity of ryanodine receptors differentially through isotype-specific N-terminal cysteines. *Exp. Neurol.* **2013**, *250*, 143–150. [CrossRef] [PubMed]

26. Wu, B.; Yamaguchi, H.; Lai, F.A.; Shen, J. Presenilins regulate calcium homeostasis and presynaptic function via ryanodine receptors in hippocampal neurons. *Proc. Natl. Acad. Sci. USA* **2013**, *110*, 15091–15096. [CrossRef] [PubMed]

27. Briggs, C.A.; Schneider, C.; Richardson, J.C.; Stutzmann, G.E. β amyloid peptide plaques fail to alter evoked neuronal calcium signals in APP/PS1 Alzheimer's disease mice. *Neurobiol. Aging* **2013**, *34*, 1632–1643. [CrossRef] [PubMed]

28. Michno, K.; Knight, D.; Campusano, J.M.; van de Hoef, D.; Boulianne, G.L. Intracellular calcium deficits in Drosophila cholinergic neurons expressing wild type or FAD-mutant presenilin. *PLoS ONE* **2009**, *4*, e6904. [CrossRef]

29. Lee, J.H.; McBrayer, M.K.; Wolfe, D.M.; Haslett, L.J.; Kumar, A.; Sato, Y.; Lie, P.P.; Mohan, P.; Coffey, E.E.; Kompella, U.; et al. Presenilin 1 Maintains Lysosomal Ca^{2+} Homeostasis via TRPML1 by Regulating vATPase-Mediated Lysosome Acidification. *Cell Rep.* **2015**, *12*, 1430–1444. [CrossRef] [PubMed]

30. Hernández-Zimbrón, L.F.; Rivas-Arancibia, S. Oxidative stress caused by ozone exposure induces β-amyloid 1–42 overproduction and mitochondrial accumulation by activating the amyloidogenic pathway. *Neuroscience* **2015**, *304*, 340–348. [CrossRef] [PubMed]

31. Ye, B.; Shen, H.; Zhang, J.; Zhu, Y.G.; Ransom, B.R.; Chen, X.C.; Ye, Z.C. Dual pathways mediate β-amyloid stimulated glutathione release from astrocytes. *Glia* **2015**, *63*, 2208–2219. [CrossRef] [PubMed]

32. Nikolakopoulou, A.M.; Georgakopoulos, A.; Robakis, N.K. Presenilin 1 promotes trypsin-induced neuroprotection via the PAR2/ERK signaling pathway. Effects of presenilin 1 FAD mutations. *Neurobiol. Aging* **2016**, *42*, 41–49. [CrossRef] [PubMed]

33. Sarasija, S.; Norman, K.R. A γ-Secretase Independent Role for Presenilin in Calcium Homeostasis Impacts Mitochondrial Function and Morphology in *Caenorhabditis elegans*. *Genetics* **2015**, *201*, 1453–1466. [CrossRef] [PubMed]

34. Picone, P.; Nuzzo, D.; Caruana, L.; Messina, E.; Barera, A.; Vasto, S.; Di Carlo, M. Metformin increases APP expression and processing via oxidative stress, mitochondrial dysfunction and NF-κB activation: Use of insulin to attenuate metformin's effect. *Biochim. Biophys. Acta* **2015**, *1853*, 1046–1059. [CrossRef] [PubMed]

35. Giuffrida, M.L.; Caraci, F.; Pignataro, B.; Cataldo, S.; De Bona, P.; Bruno, V.; Molinaro, G.; Pappalardo, G.; Messina, A.; Palmigiano, A.; et al. Beta-amyloid monomers are neuroprotective. *J. Neurosci.* **2009**, *29*, 10582–10587. [CrossRef] [PubMed]

36. Gray, N.E.; Quinn, J.F. Alterations in mitochondrial number and function in Alzheimer's disease fibroblasts. *Metab. Brain Dis.* **2015**, *30*, 1275–1278. [CrossRef] [PubMed]

37. Southon, A.; Greenough, M.A.; Ganio, G.; Bush, A.I.; Burke, R.; Camakaris, J. Presenilin promotes dietary copper uptake. *PLoS ONE* **2013**, *8*, e62811. [CrossRef] [PubMed]

38. Ebrahimie, E.; Moussavi Nik, S.H.; Newman, M.; Van Der Hoek, M.; Lardelli, M. The Zebrafish Equivalent of Alzheimer's Disease-Associated PRESENILIN Isoform PS2V Regulates Inflammatory and Other Responses to Hypoxic Stress. *J. Alzheimers Dis.* **2016**, *52*, 581–608. [CrossRef] [PubMed]

39. Wangler, M.F.; Reiter, L.T.; Zimm, G.; Trimble-Morgan, J.; Wu, J.; Bier, E. Antioxidant proteins TSA and PAG interact synergistically with Presenilin to modulate Notch signaling in Drosophila. *Protein Cell* **2011**, *2*, 554–563. [CrossRef] [PubMed]

40. Armstrong, R.A. β-amyloid (Aβ) deposition in cognitively normal brain, dementia with Lewy bodies, and Alzheimer's disease: A study using principal components analysis. *Folia Neuropathol.* **2012**, *50*, 130–139. [PubMed]

41. Means, J.C.; Gerdes, B.C.; Koulen, P. Distinct Mechanisms Underlying Resveratrol-Mediated Protection from Types of Cellular Stress in C6 Glioma Cells. *Int. J. Mol. Sci.* **2017**, *18*, 1521. [CrossRef] [PubMed]

42. Pedrozo, Z.; Torrealba, N.; Fernández, C.; Gatica, D.; Toro, B.; Quiroga, C.; Rodriguez, A.E.; Sanchez, G.; Gillette, T.G.; Hill, J.A.; et al. Cardiomyocyte ryanodine receptor degradation by chaperone-mediated autophagy. *Cardiovasc. Res.* **2013**, *98*, 277–285. [CrossRef] [PubMed]

43. Hwang, C.J.; Park, M.H.; Choi, M.K.; Choi, J.S.; Oh, K.W.; Hwang, D.Y.; Han, S.B.; Hong, J.T. Acceleration of amyloidogenesis and memory impairment by estrogen deficiency through NF-κB dependent beta-secretase activation in presenilin 2 mutant mice. *Brain Behav. Immun.* **2016**, *53*, 113–122. [CrossRef] [PubMed]

44. Ludtmann, M.H.; Otto, G.P.; Schilde, C.; Chen, Z.H.; Allan, C.Y.; Brace, S.; Beesley, P.W.; Kimmel, A.R.; Fisher, P.; Killick, R.; et al. An ancestral non-proteolytic role for presenilin proteins in multicellular development of the social amoeba Dictyostelium discoideum. *J. Cell Sci.* **2014**, *127*, 1576–1584. [CrossRef] [PubMed]

45. Duggan, S.P.; Yan, R.; McCarthy, J.V. A ubiquitin-binding CUE domain in presenilin-1 enables interaction with K63-linked polyubiquitin chains. *FEBS Lett.* **2015**, *589*, 1001–1008. [CrossRef] [PubMed]

46. Neely Kayala, K.M.; Dickinson, G.D.; Minassian, A.; Walls, K.C.; Green, K.N.; Laferla, F.M. Presenilin-null cells have altered two-pore calcium channel expression and lysosomal calcium: Implications for lysosomal function. *Brain Res.* **2012**, *1489*, 8–16. [CrossRef] [PubMed]

47. Neely, K.M.; Green, K.N.; LaFerla, F.M. Presenilin is necessary for efficient proteolysis through the autophagy-lysosome system in a γ-secretase-independent manner. *J. Neurosci.* **2011**, *31*, 2781–2791. [CrossRef] [PubMed]

48. Zhang, X.; Garbett, K.; Veeraraghavalu, K.; Wilburn, B.; Gilmore, R.; Mirnics, K.; Sisodia, S.S. A role for presenilins in autophagy revisited: Normal acidification of lysosomes in cells lacking PSEN1 and PSEN2. *J. Neurosci.* **2012**, *32*, 8633–8648. [CrossRef] [PubMed]

49. Almenar-Queralt, A.; Kim, S.N.; Benner, C.; Herrera, C.M.; Kang, D.E.; Garcia-Bassets, I.; Goldstein, L.S. Presenilins regulate neurotrypsin gene expression and neurotrypsin-dependent agrin cleavage via cyclic AMP response element-binding protein (CREB) modulation. *J. Biol. Chem.* **2013**, *288*, 35222–35236. [CrossRef] [PubMed]

50. Li, A.; Zhou, C.; Moore, J.; Zhang, P.; Tsai, T.H.; Lee, H.C.; Romano, D.M.; McKee, M.L.; Schoenfeld, D.A.; Serra, M.J.; et al. Changes in the expression of the Alzheimer's disease-associated presenilin gene in drosophila heart leads to cardiac dysfunction. *Curr. Alzheimer Res.* **2011**, *8*, 313–322. [CrossRef] [PubMed]

51. Wüst, R.; Maurer, B.; Hauser, K.; Woitalla, D.; Sharma, M.; Krüger, R. Mutation analyses and association studies to assess the role of the presenilin-associated rhomboid-like gene in Parkinson's disease. *Neurobiol. Aging* **2016**, *39*, 217.

52. Li, P.; Lin, X.; Zhang, J.R.; Li, Y.; Lu, J.; Huang, F.C.; Zheng, C.H.; Xie, J.W.; Wang, J.B.; Huang, C.M. The expression of presenilin 1 enhances carcinogenesis and metastasis in gastric cancer. *Oncotarget* **2016**, *7*, 10650–10662. [CrossRef] [PubMed]

53. Miyanaga, A.; Masuda, M.; Tsuta, K.; Kawasaki, K.; Nakamura, Y.; Sakuma, T.; Asamura, H.; Gemma, A.; Yamada, T. Hippo pathway gene mutations in malignant mesothelioma: Revealed by RNA and targeted exon sequencing. *J. Thorac. Oncol.* **2015**, *10*, 844–851. [CrossRef] [PubMed]

54. Prens, E.; Deckers, I. Pathophysiology of hidradenitis suppurativa: An update. *J. Am. Acad. Dermatol.* **2015**, *73*, S8–S11. [CrossRef] [PubMed]

55. Panmontha, W.; Rerknimitr, P.; Yeetong, P.; Srichomthong, C.; Suphapeetiporn, K.; Shotelersuk, V. A Frameshift Mutation in PEN-2 Causes Familial Comedones Syndrome. *Dermatology* **2015**, *231*, 77–81. [CrossRef] [PubMed]

56. Puig, K.L.; Lutz, B.M.; Urquhart, S.A.; Rebel, A.A.; Zhou, X.; Manocha, G.D.; Sens, M.; Tuteja, A.K.; Foster, N.L.; Combs, C.K. Overexpression of mutant amyloid-β protein precursor and presenilin 1 modulates enteric nervous system. *J. Alzheimers Dis.* **2015**, *44*, 1263–1278. [PubMed]

57. Donoviel, D.B.; Hadjantonakis, A.K.; Ikeda, M.; Zheng, H.; Hyslop, P.S.; Bernstein, A. Mice lacking both presenilin genes exhibit early embryonic patterning defects. *Genes Dev.* **1999**, *13*, 2801–2810. [CrossRef] [PubMed]

International Journal of
Molecular Sciences

MDPI

Article

Multiple Layers of *CDK5R1* Regulation in Alzheimer's Disease Implicate Long Non-Coding RNAs

Marco Spreafico [1], Barbara Grillo [1], Francesco Rusconi [1], Elena Battaglioli [1,2] and Marco Venturin [1,*]

1   Dipartimento di Biotecnologie Mediche e Medicina Traslazionale, Università degli Studi di Milano, Via Fratelli Cervi 93, 20090 Segrate, Italy; marco.spreafico@unimi.it (M.S.); barbara.grillo@unimi.it (B.G.); francesco.rusconi@unimi.it (F.R.); elena.battaglioli@unimi.it (E.B.)
2   Istituto di Neuroscienze, CNR, Via Vanvitelli 32, 20129 Milano, Italy
*   Correspondence: marco.venturin@unimi.it; Tel.: +39-02-503-30443; Fax: +39-02-503-30365

Received: 17 May 2018; Accepted: 3 July 2018; Published: 11 July 2018

Abstract: Cyclin-dependent kinase 5 regulatory subunit 1 (*CDK5R1*) gene encodes for p35, the main activator of Cyclin-dependent kinase 5 (CDK5). The active p35/CDK5 complex is involved in numerous aspects of brain development and function, and its deregulation is closely associated to Alzheimer's disease (AD) onset and progression. We recently showed that miR-15/107 family can negatively regulate *CDK5R1* expression modifying mRNA stability. Interestingly, miRNAs belonging to miR-15/107 family are downregulated in AD brain while *CDK5R1* is upregulated. Long non-coding RNAs (lncRNAs) are emerging as master regulators of gene expression, including miRNAs, and their dysregulation has been implicated in the pathogenesis of AD. Here, we evaluated the existence of an additional layer of *CDK5R1* expression regulation provided by lncRNAs. In particular, we focused on three lncRNAs potentially regulating *CDK5R1* expression levels, based on existing data: NEAT1, HOTAIR, and MALAT1. We demonstrated that NEAT1 and HOTAIR negatively regulate *CDK5R1* mRNA levels, while MALAT1 has a positive effect. We also showed that all three lncRNAs positively control miR-15/107 family of miRNAs. Moreover, we evaluated the expression of NEAT1, HOTAIR, and MALAT1 in AD and control brain tissues. Interestingly, NEAT1 displayed increased expression levels in temporal cortex and hippocampus of AD patients. Interestingly, we observed a strong positive correlation between *CDK5R1* and NEAT1 expression levels in brain tissues, suggesting a possible neuroprotective role of NEAT1 in AD to compensate for increased *CDK5R1* levels. Overall, our work provides evidence of another level of *CDK5R1* expression regulation mediated by lncRNAs and points to NEAT1 as a biomarker, as well as a potential pharmacological target for AD therapy.

Keywords: *CDK5R1*; lncRNAs; Alzheimer's disease; miR-15/107; NEAT1; HOTAIR; MALAT1

1. Introduction

Alzheimer's Disease (AD) is the most common neurodegenerative disorder, causing a severe and permanent impairment of both cognitive and behavioral functions. It accounts for about 70% of the 50 million people suffering from dementia worldwide and it is currently estimated that, with global population aging, the prevalence of AD will triple by 2050 [1], with a significant economic and social burden on both patients' families and society.

AD is characterized by a plethora of pathological features, including neuronal loss, dendritic hypotrophy and synaptic alteration, microglial malfunction, cerebrovascular amyloid angiopathy, inflammation, and mitochondrial dysfunction [2,3]. However, the most distinctive features are the presence of extracellular senile plaques, formed by fibrillary β-amyloid (Aβ) and neurofibrillary tangles

(NFTs), composed of hyperphosphorylated Tau [4]. Abnormal kinase activity is believed to play a major role in AD pathogenesis [5]. In particular, deregulation of Cyclin-dependent kinase 5 (CDK5), a proline-directed serine/threonine kinase involved in several developmental and physiological processes in the central nervous system (CNS) [6,7], has been suggested to play a pivotal role in the onset of the two main pathological hallmarks of AD by inducing Aβ peptide production and mediating Tau protein hyperphosphorylation [8].

CDK5 requires the p35 regulatory subunit to become active and its kinase activity is strictly dependent on the amount of its activator. p35 is encoded by the cyclin-dependent kinase 5 regulatory subunit 1 (*CDK5R1*) gene, which displays a large and highly conserved 3′-UTR, suggestive of an important role of post-transcriptional regulation in the control of its expression. Indeed, we previously demonstrated that *CDK5R1* expression is regulated at the post-transcriptional level by neuronal ELAV (nELAV) RNA-binding proteins [9,10] and by heterogeneous nuclear ribonucleoproteins A2/B1 (hnRNP A2/B1) [10]. In addition, we recently found that the miR-15/107 family of microRNAs is also involved in negatively regulating *CDK5R1* expression. More interestingly, this group of microRNAs turned out to be downregulated in the hippocampus and cerebral cortex of AD patients while *CDK5R1* mRNA levels were upregulated in AD hippocampus [11].

An additional layer of complexity to the regulation of *CDK5R1* expression that can be relevant for AD pathogenesis might be provided by long non-coding RNAs (lncRNAs). LncRNAs are a highly heterogeneous class of RNA molecules of more than 200 bases in length with no protein-coding capacity. They are involved in the control of gene expression at multiple levels, from nuclear architecture to transcription regulation, mRNA splicing and maturation to mRNA localization and stability, and protein translation and stability to regulation of miRNA activity [12]. Owing to this versatility, lncRNAs are now considered as master regulators of gene expression [13]. In particular, lncRNAs have been shown to post-transcriptionally regulate the levels of several target genes by the formation of lncRNA/miRNA/target gene axes, and the dysregulation of the crosstalk between the two types of ncRNAs has been found to be a crucial contributor to disease pathogenesis [14].

The role of lncRNAs in malignancies and their significance as both diagnostic and prognostic markers has been extensively studied and is well established [15], but an involvement of lncRNAs in the pathogenesis of neurodegenerative diseases is now clearly emerging. In particular, different lncRNAs have been found dysregulated in Alzheimer's disease and involved in AD pathogenesis by promoting β-amyloid production, including BACE1-AS, 17A, and NDM29 [16]. For example, the expression of BACE1-AS, the antisense transcript of the β-secretase encoding gene *BACE1*, is upregulated in AD brains specimens. BACE1-AS was reported to increase the stability of *BACE1* mRNA and to prevent the binding of miRNA 485-5p, therefore positively regulating BACE1 protein levels and promoting Aβ42 synthesis [16,17].

In the present work, we focused on three different lncRNAs which had the potential for regulating *CDK5R1* expression levels and deserved to be analyzed in AD brain tissues, namely NEAT1, HOTAIR, and MALAT1. NEAT1 (nuclear enriched abundant transcript 1) is a lncRNA that regulates gene expression by binding to the promoter of active chromatin sites [18,19]. Moreover, NEAT1 is known to act as a scaffold for paraspeckles [20], representing specific subnuclear bodies that are involved in gene expression regulation by sequestration and retention of specific RNAs and proteins [21]. Relevantly, NEAT1 levels were found to be deregulated in different neurodegenerative diseases [22]. MALAT1 (metastasis-associated lung adenocarcinoma transcript 1), also known as NEAT2 (nuclear-enriched abundant transcript 2), is predominantly localized to nuclear speckles, where it regulates alternative splicing by modulating the phosphorylation status of SR family of splicing factors [23]. MALAT1 has been linked to several human tumors, in most cases being overexpressed in malignant tissues [24]. Both NEAT1 and MALAT1 have been demonstrated to regulate the expression of members of the miR-15/107 group of miRNAs [25,26], which are known *CDK5R1* negative regulators [11]. HOTAIR (HOX antisense intergenic RNA) is transcribed from the antisense strand of the HOXC locus and represses expression of the downstream HOXD locus together with several genes on other chromosomes. HOTAIR is involved in the control of cell apoptosis, growth, metastasis, angiogenesis, DNA repair and, like MALAT1, it has been shown to be upregulated in different types of cancer [27]. Interestingly, HOTAIR can also serve as a scaffold for Lysine-specific histone demethylase 1A (LSD1) complex and polycomb repressive complex 2 (PRC2) [28]. Since the expression of *CDK5R1* is repressed by LSD1 [29], HOTAIR can potentially impact *CDK5R1* levels.

Here, we demonstrated that NEAT1 and HOTAIR negatively regulate *CDK5R1* mRNA levels, while MALAT1 has a positive effect on *CDK5R1* expression. We also showed that all three lncRNAs positively control the levels of miR-15/107 family of microRNAs. Moreover, we evaluated the expression of NEAT1, HOTAIR, and MALAT1 in AD and control brain tissues. Interestingly, NEAT1 displayed increased expression levels in temporal cortex and hippocampus of AD patients, compared to controls. In addition, we observed a strong positive correlation between *CDK5R1* and NEAT1 expression levels in brain tissues, suggesting a novel molecular marker of AD pathogenesis, warranting further studies. Overall, our work provides evidence of another level of *CDK5R1* expression regulation mediated by long non-coding RNAs, which can also impact on Alzheimer's disease research.

2. Results

2.1. NEAT1, HOTAIR, and MALAT1 Long Non-Coding RNAs Differently Regulate CDK5R1 Expression

In order to test the hypothesis that lncRNAs might be involved in the regulation of *CDK5R1*, we analyzed the effect of NEAT1, HOTAIR, and MALAT1 downregulation on *CDK5R1* expression. We transfected HeLa cells with 10 nM of specific 2'OMe-PS antisense oligonucleotides (ASO) to specifically knockdown the three lncRNAs. Total RNA was extracted 24 h after transfection and the levels of lncRNAs and *CDK5R1* mRNA were assessed by qRT-PCR. The analysis showed that NEAT1, HOTAIR, and MALAT1 levels were reduced by 61%, 71%, and 78% respectively, compared to the control oligonucleotide (Figure 1A). Remarkably, increased *CDK5R1* transcript levels were observed after NEAT1 and HOTAIR silencing, meaning that these two lncRNAs negatively regulate *CDK5R1* expression (Figure 1B). On the contrary, *CDK5R1* mRNA levels were significantly decreased after MALAT1 silencing compared to controls, indicating a positive action of this lncRNA on *CDK5R1* expression (Figure 1B).

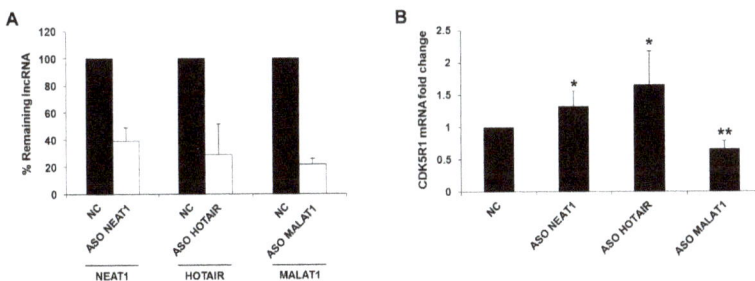

Figure 1. Effect of NEAT1, HOTAIR, and MALAT1 silencing on *CDK5R1* mRNA levels. (**A**) NEAT1, HOTAIR, and MALAT1 levels 24 h after transfection with specific ASOs. The levels of each lncRNA were reduced by at least 60%, compared to a control ASO (NC)-transfected cells. (**B**) Increased *CDK5R1* transcript levels were observed after NEAT1 and HOTAIR silencing, compared to the normal control. On the contrary, *CDK5R1* mRNA levels were significantly decreased after MALAT1 silencing. $n = 5$, mean ± s.d., * $p < 0.05$, ** $p < 0.01$, Student's *t*-test.

2.2. NEAT1, HOTAIR, and MALAT1 Upregulate miR-15/107 Expression

Since we previously demonstrated that *CDK5R1* expression is negatively regulated by the miR-15/107 group of microRNAs [11], we also verified by qRT-PCR on the RNA previously extracted from HeLa cells if NEAT1, HOTAIR, and MALAT1 silencing was able to affect miR-15/107 expression.

The levels of all the analyzed miR-15/107 family members were reduced after NEAT1, HOTAIR, and MALAT1 silencing, compared to the control treatment (Figure 2), being HOTAIR the most efficient with a reduction of miRNA targets of about 50%. NEAT1 and MALAT1 led to a less pronounced but significant reduction of all miRNAs, with the exception of miR-15b after NEAT1 knock-down, whose reduction did not reach the statistical significance (Figure 2).

These data suggest that HOTAIR and NEAT1 might negatively regulate *CDK5R1* expression through a positive action on miR-15/107 levels. On the contrary, the positive effect of MALAT1 on *CDK5R1* mRNA cannot be explained by the action of these miRNAs, and a different mechanism must be involved in MALAT1-mediated positive effect on *CDK5R1* expression.

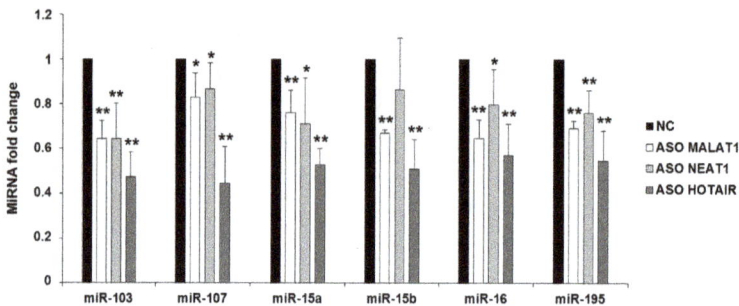

Figure 2. Effect of NEAT1, HOTAIR, and MALAT1 silencing on miR-15/107 levels. Decreased levels of all miR-15/107 miRNAs were detected after the knock-down of the three lncRNAs. $n = 5$, mean $\pm$ s.d., * $p < 0.05$, ** $p < 0.005$, Student's *t*-test.

CDK5R1 also represents a target of EGR1 transcription factor whose expression is induced by ERK/MAPK pathway activation [30]. Since MALAT1 has been described to be a positive modulator of ERK/MAPK pathway [31], we evaluated the expression of *EGR1* following MALAT1 silencing. Consistently, qRT-PCR analysis showed that the levels of *EGR1* are strongly reduced (83%) in cells treated with MALAT1 specific antisense oligonucleotide, compared to normal control (Figure 3). These results suggest that the positive regulation exerted by MALAT1 on *CDK5R1* expression can be due to MALAT1-mediated enhancement of *EGR1* levels, likely overcoming the concurrent downregulation of miR-15/107 miRNAs.

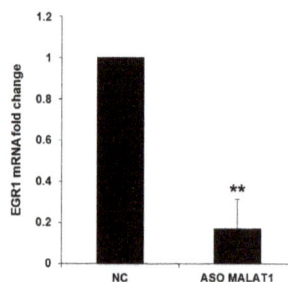

Figure 3. Effect of MALAT1 silencing on *EGR1* mRNA levels. *EGR1* mRNA levels were significantly decreased after MALAT1 silencing, compared to the normal control (NC). $n = 3$, mean $\pm$ s.d., ** $p < 0.005$, Student's *t*-test.

2.3. NEAT1 is Upregulated in AD Temporal Cortex and Hippocampus

We recently showed that miR-15/107 miRNAs level is reduced in the hippocampus and the temporal cortex, but not in the cerebellum, of AD brains. Furthermore, we showed that increased *CDK5R1* mRNA levels are displayed by AD hippocampus tissue, compared to controls [11]. These data are consistent with the hypothesis that an increase of *CDK5R1* expression, and consequent enhanced CDK5 activity, caused by downregulation of the miR-15/107 family has a role in the pathogenesis of AD.

To verify whether NEAT1, HOTAIR, and MALAT1 expression is also altered in Alzheimer's disease, we quantified their levels by qRT-PCR in the temporal cortex, hippocampus, and cerebellum of the same AD patients and age-matched healthy controls which were analyzed in our previous work.

Remarkably, we found that NEAT1 was significantly overexpressed in temporal cortex and hippocampus and downregulated in cerebellum of AD patients, compared to control individuals (Figure 4A). Comparing NEAT1 distribution among the different brain areas of control individuals we observed similar expression levels, while NEAT1 was significantly higher in temporal cortex and hippocampus compared to cerebellum in AD patients (Figure 4B). On the contrary, MALAT1 expression showed no difference between AD patients and controls (Figure 5A), even though higher levels were detected in cerebellum, compared to temporal cortex and hippocampus, in both groups (Figure 5B). Finally, HOTAIR was expressed at very low levels in hippocampus and cerebellum and was not detectable in temporal cortex. Particularly, HOTAIR was downregulated in cerebellum in AD patients, compared to controls. No difference in HOTAIR expression between hippocampus and cerebellum was observed in both groups.

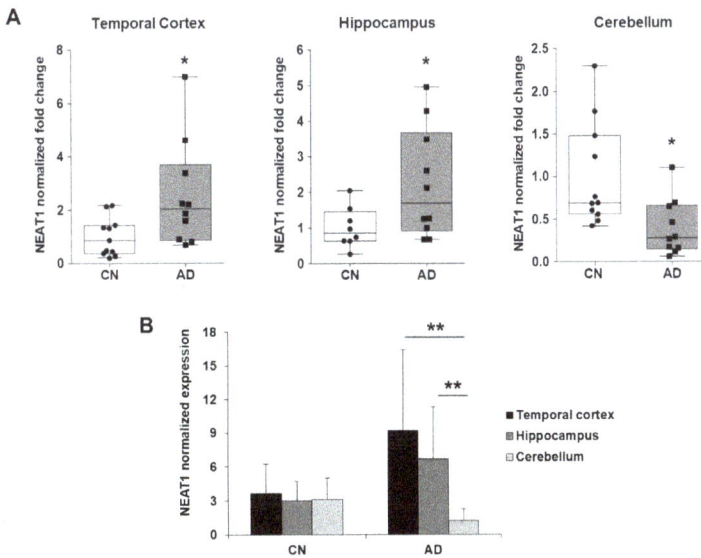

Figure 4. Comparison between the levels of NEAT1 expression in AD and control brain tissues. (**A**) Dot-Box-plots of the levels of NEAT1 expression in three different brain areas (temporal cortex, hippocampus, and cerebellum) of AD patients ($n = 10$) and controls ($n = 8$–11). Dark horizontal lines represent the median, with the box representing the 25th and 75th percentiles, the whiskers the 5th and 95th percentiles. The average of control values was set to 1 and all values were calculated relatively. NEAT1 levels are significantly upregulated in temporal cortex and hippocampus and downregulated in cerebellum of AD patients, compared to control individuals. (**B**) Higher NEAT1 expression levels were observed in temporal cortex and hippocampus, compared to cerebellum, in AD patients, but not in control individuals. * $p < 0.05$, ** $p < 0.01$, Student's *t*-test.

A

B

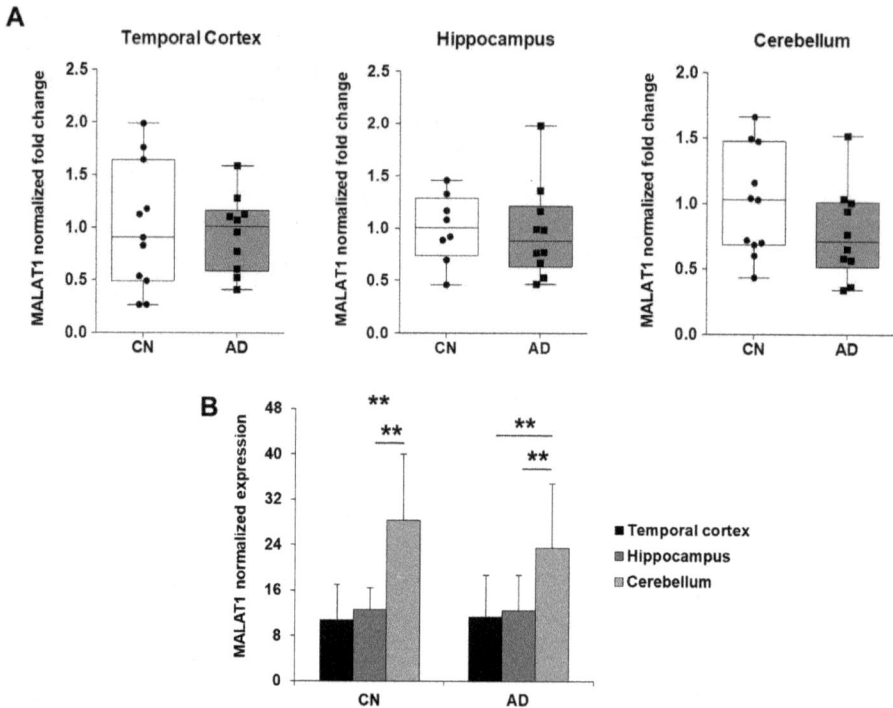

Figure 5. Comparison between the levels of MALAT1 expression in AD and control brain tissues. (**A**) Dot-Box-plots of the levels of MALAT1 expression in three different brain areas (temporal cortex, hippocampus, and cerebellum) of AD patients ($n = 10$) and controls ($n = 8–11$). Dark horizontal lines represent the median, with the box representing the 25th and 75th percentiles, the whiskers the 5th and 95th percentiles. The average of control values was set to 1 and all values were calculated relatively. We observed no difference in MALAT1 levels between AD patients and control individuals in any analyzed tissues. (**B**) Higher MALAT1 expression levels were observed in cerebellum, compared to temporal cortex and hippocampus, in both AD patients and controls individuals. ** $p < 0.01$, Student's *t*-test.

2.4. NEAT1 and CDK5R1 Overexpression as a Biomarker of AD

In order to verify the existence of a correlation between NEAT1 and *CDK5R1* expression in AD and control brain tissues, we performed a Pearson's correlation analysis between the normalized expression levels of NEAT1 and those previously obtained for *CDK5R1* [11].

The analysis showed that a significant positive correlation between *CDK5R1* and NEAT1 levels was only displayed by AD patients' postmortem specimens of hippocampi and temporal cortices with very significant values of Pearson's *r* (Figure 6). In other words, in AD, NEAT1 increases along with *CDK5R1*, indicating a peculiar functional relationship (in vitro defined as a negative NEAT1 control over *CDK5R1*) which is specific for AD and that can be either a protective response-related mechanism aimed at limiting (inefficiently) *CDK5R1* upregulation or part of the disease pathogenesis. Notably, in the cerebellum, a brain area that is almost unaffected by the disease and in which *CDK5R1* does not appear to be upregulated, the correlation between NEAT1 levels and those of *CDK5R1* is still evident only in AD. This observation suggests that in the cerebellum a protective, NEAT1-associated mechanism might efficiently control *CDK5R1* levels.

Another interesting observation is the opposite correlation between the expression of NEAT1 and miR-15/107 miRNAs in AD brains and controls. Indeed, while in controls we observed high expression of miR-15/107 and low expression of NEAT1, in AD patients, to higher NEAT1 levels correspond very low miR-15/107 levels, particularly in temporal cortex and hippocampus (Figure 7). In conclusion, a picture emerges in which not only NEAT1 is unable to increase its levels in a sufficient manner to counteract *CDK5R1* increase in AD brains, but also, it loses the ability to positively regulate miR-15/107, validated negative regulators of *CDK5R1*. In this way, converging pathological mechanisms based on a failure of lncRNA NEAT1 and miR-15/107 homeostatic role towards *CDK5R1* expression could result in *CDK5R1* upregulation.

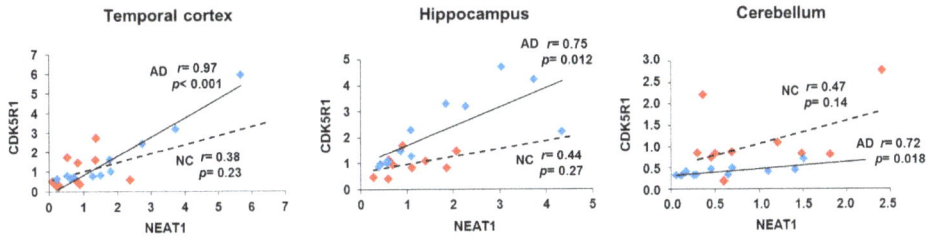

Figure 6. Correlation analysis between NEAT1 and *CDK5R1* expression in temporal cortex, hippocampus and cerebellum samples of AD patients (blue diamonds) and controls (red diamonds). *r* = Pearson's correlation coefficient, solid line = linear regression line of AD patients, dashed line = linear regression line of normal controls (NC).

3. Discussion

Cyclin-dependent kinase 5 (CDK5) has a major role in CNS development and functioning and its deregulation can contribute to different pathological events implicated in the pathogenesis of Alzheimer's disease [8]. Monomeric CDK5 itself does not display kinase activity and requires, in order to be active, the association with its regulatory subunits, p35 or p39, although p35, encoded by the *CDK5R1* gene, is considered the most important CDK5 activator [32]. Multiple layers of regulation govern *CDK5R1* expression and ensure p35 levels and CDK5 activity to be tightly controlled. They include transcriptional activation by EGR1 transcription factor and repression by LSD1 demethylase [29,30], as well as well various post-transcriptional mechanisms which involve the binding to the long and evolutionary conserved *CDK5R1* 3′-UTR of both RNA-binding proteins and microRNAs [9–11].

In this work, we took into account another class of non-coding RNAs, long non-coding RNAs (lncRNAs), as potential regulators of *CDK5R1* expression. In particular, our attention was focused on three lncRNAs, NEAT1, HOTAIR, and MALAT1. Our results showed that these three lncRNAs are able to influence *CDK5R1* expression. In particular, NEAT1 and HOTAIR exert a negative regulatory effect on *CDK5R1* levels, while MALAT1 has an opposite, positive action. In addition, all these lncRNAs were proven to positively regulate the miRNAs belonging to the miR-15/107 family.

We hypothesize that the negative regulatory effect of NEAT1 on *CDK5R1* expression might depend on its capacity to exert a positive control on miR-15/107 levels. Interestingly, we also found that NEAT1 is significantly overexpressed in temporal cortex and hippocampus of AD patients, compared to control individuals, suggesting that NEAT1 upregulation can be considered a biomarker of the disease. Recent studies have linked altered expression and function of long non-coding RNAs to the pathogenesis of neurodegenerative diseases (reviewed in [22]). In particular, different lncRNAs have been found to be dysregulated in Alzheimer's disease (e.g., BACE1-AS and NDM29) and to be involved in AD pathogenesis by promoting β-amyloid production. In this work we show that, in vitro, NEAT1 negatively regulates *CDK5R1* expression. In line, NEAT1 upregulation in AD patients would predict a corresponding downregulation of *CDK5R1*. Notably, this was not the case. Indeed, in AD brains the expression of both *CDK5R1* and NEAT1 is increased compared to healthy controls.

A possibility is that the negative control of NEAT1 over *CDK5R1* levels is not efficient either because the ratio between *CDK5R1* and NEAT1 typical of controls is increased in AD brains (Figure 6), or because NEAT1 loses its positive control towards miR-15/107 (Figure 7). As a result, we can infer that the critical NEAT1 level that would be necessary to counteract *CDK5R1* expression is not reached in AD temporal cortex and hippocampus. For these reasons, NEAT1 overexpression as a pathomechanism in Alzheimer's disease is unlikely, although our data do not allow to fully reject this hypothesis. Moreover, several lines of evidence suggest that NEAT1 and paraspeckles may have a neuroprotective role in neurodegenerative diseases. An increase in paraspeckles formation and NEAT1 levels has been detected in spinal motor neurons of early phase amyotrophic lateral sclerosis (ALS) patients compared to control individuals [33] and compromised paraspeckles formation has been proposed as a pathogenic factor in FUSopathies [34]. Moreover, NEAT1 levels are also increased in the brains of patients affected by frontotemporal lobar degeneration (FTLD) [35]. Importantly, Sunwoo and colleagues [36] showed that NEAT1 is overexpressed in Huntington's disease patients and plays a protective role against cell injury. These data suggest that NEAT1 may contribute to neuronal survival in the degenerating brain. Analogously, our work showed that NEAT1 is also overexpressed in AD patients. In this context, putative beneficial effects of NEAT1 are still unknown. However, enhanced amounts of *CDK5R1* are predicted to cause CDK5 hyperactivation, which is a typical hallmark of the disease [8]. It is worth noting that CDK5 can phosphorylate p53, which is also known to be upregulated in AD [37], thereby inducing its stabilization and transcriptional activation, contributing to neuronal cell death [38]. Remarkably, p53 was recently demonstrated to activate NEAT1 expression [39]. These findings provide a possible molecular link between *CDK5R1* and NEAT1 upregulation in AD brains, albeit they do not indicate the reason why *CDK5R1* escapes NEAT1 control in AD condition.

The negative action exerted by HOTAIR on *CDK5R1* expression is likely mediated by different converging mechanisms. On the one hand, HOTAIR can negatively regulate *CDK5R1* at the post-transcriptional level via the same miR-15/107 miRNA-mediated mechanism as NEAT1, on the other hand it could regulate *CDK5R1* also at the transcriptional level participating to recruiting and regulating the LSD1 and PRC2 repressing complexes [28]. Interestingly, HOTAIR also represses the transcription of *BDNF* [27], which normally induces the ERK-mediated expression of *CDK5R1* [29].

On the contrary, our silencing experiments suggest that MALAT1 positively affect *CDK5R1* expression. Since reduction of miR-15/107 levels after MALAT1 silencing would predict an increase in the amount of *CDK5R1* mRNA, as expected for their inhibitory action, there must be other predominant regulatory mechanisms leading to *CDK5R1* upregulation by MALAT1. As we have also shown that MALAT1 silencing causes a strong reduction in the levels of EGR1, which is the main activator of *CDK5R1* transcription, we thus speculate that MALAT1 can enhance *CDK5R1* expression mainly by upregulating EGR1 transcription factor through activation of ERK/MAPK signaling pathway [31].

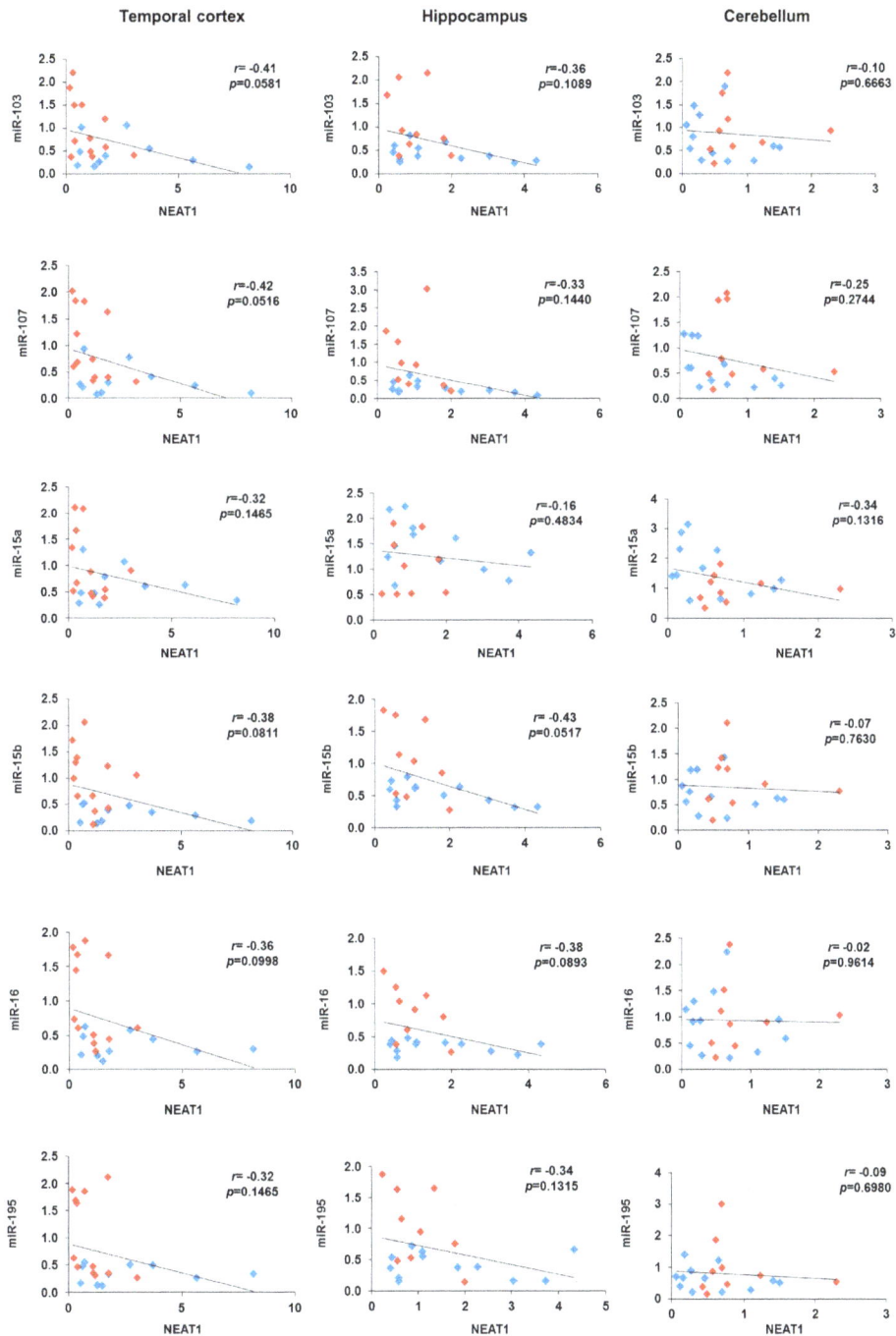

Figure 7. Correlation analysis between miR-103, miR-107, miR-15a, miR-15b, miR-16, and miR-195 and NEAT1 expression in temporal cortex, hippocampus, and cerebellum samples of AD patients (blue diamonds) and controls (red diamonds). *r* = Pearson's correlation coefficient.

Moreover, the activation of this pathway is known to play a critical role in promoting neurite outgrowth [31]. Given that the p35/CDK5 complex is also essential for neurite outgrowth during neuronal differentiation [40], our evidence raises the interesting hypothesis that MALAT1 induces axonal elongation via *CDK5R1*/p35 upregulation.

Mounting evidence suggests that lncRNAs can function as miRNA sponges, by sequestering the mature miRNA molecules and preventing the binding to their target mRNAs [41]. However, this mechanism is predicted to increase—or leave unchanged—the levels of the sequestered miRNAs when the lncRNA acting as sponge is silenced [14,42]. Since we observed that the silencing of NEAT1, HOTAIR, and MALAT1 lead to a reduction of miR-15/107 miRNAs, we hypothesize that this effect could be mediated by a positive regulatory action of these lncRNAs on transcription factors that promote the expression of this family of miRNAs or, alternatively, by their interaction with the microprocessor to enhance pri-miRNA processing, as already demonstrated for NEAT1 [43].

Overall, our data suggest that lncRNAs can provide a further layer and a higher degree of complexity to the control of *CDK5R1* expression. In addition, we show that NEAT1 is upregulated in AD brain, possibly as a part of a protective mechanism against neuronal death, and can be considered a marker of the disease and represents a potential pharmacological target for therapeutic intervention in AD.

4. Materials and Methods

4.1. Cell Cultures

HeLa cells (code CCL-2, ATCC, Manassas, VA, USA) were cultured in DMEM high glucose (Euroclone, Pero, Italy) medium with 10% fetal bovine serum (FBS) (Euroclone), 100 U/mL penicillin-streptomycin (Euroclone) and 0.01 mM L-glutamine (Euroclone). Cultures were maintained at 37 °C in a 5% CO_2 incubator.

4.2. Brain Tissues

Post-mortem frozen brain tissue samples of AD patients and age- and sex-matched non-demented individuals were obtained from MRC London Neurodegenerative Diseases Brain Bank (King's College London), Newcastle Brain Tissue Resource (Newcastle University), and South West Dementia Brain Bank (University of Bristol) and are described in [10]. The approval of the Ethics Committee of the University of Milan was obtained for the use of post-mortem tissues for research purposes (Project identification code: RV_RIC_AT16MVENT_M, 15 June 2016).

4.3. Antisense Oligonucleotides Transfection

2′-*O*-methyl phosphorothioate antisense oligonucleotides (2′OMe-PS ASO) were designed as described by [44] (NEAT1 1473, HOTAIR 1259, MALAT1 5326, NC1) and purchased from Consorzio Futuro in Ricerca, Università degli Studi di Ferrara (Ferrara, Italy).

ASOs were used at 10 nM concentration. 200×10^3 HeLa cells were seeded in 6-well plates in order to extract total RNA. The cells were transfected 24 h after seeding with 2′OMe-PS ASOs, using Lipofectamine 2000 (Thermo Fisher Scientific, Waltham, MA, USA) transfection reagent according to the manufacturer's instructions. Cell extracts were prepared for analysis 24 h after the transfection.

4.4. Real-Time PCR

Total RNA from transfected/nontransfected cells and from brain tissues (100 mg of each sample tissue) was isolated using TRIzol reagent (Thermo Fisher Scientific), according to the manufacturer's instructions. Concentration and purity of RNA were measured using the Nanodrop spectrophotometer (ThermoFisher Scientific). All RNA samples had an A260/280 value of 1.8–2.1.

For the measurement of *CDK5R1*, *EGR1* mRNA, NEAT1, HOTAIR, and MALAT1 RNA, a DNase reaction was performed on 1 µg of total RNA using RQ1 RNase-Free Dnase (Promega,

Madison, WI, USA) and then cDNA was synthetized in 20 µL reactions using the High Capacity cDNA Reverse Transcription Kit (Thermo Fisher Scientific), according to the manufacturer's instructions. SYBR Green Real-Time PCR was performed using the GoTaq qPCR Master Mix (Promega) and the following primers: *CDK5R1* fw: TGAGCGGGTCTAGTGGAAAG; *CDK5R1* rev: AGCAGCAGACAAGGGGGTAG; *EGR1* fw: GAGCACCTGACCGCAGAGTC; *EGR1* rev: GTGTTGCCACTGTTGGGTGC; *HOTAIR* fw: GGCAAGACGGGCACTCACAG; *HOTAIR* rev: CTGGGCGTTCATGTGGCGAG; *MALAT1* fw: AGGGAAAGCGAGTGGTTGGT; *MALAT1* rev: GAAATCGGCCTACGTCCCCA; *NEAT1* fw: CGGAGGTGAGGGGTGGTCTG; *NEAT1* rev: GCAGTCCCCGCCTGTCAAAC; *EIF4A2* fw: GGTCAGGGTCAAGTCGTGTT; *EIF4A2* rev: CCCCCTCTGCCAATTCTGTG; *CYC1* fw: TAGAGTTTGACGATGGCACCC; *CYC1* rev: CGTTTTCGATGGTCGTGCTC; *SYP* fw: CTTCGCCATCTTCGCCTTTG; *SYP* rev: TACACTTGGTGCAGCCTGAAG; *ENO2* fw: CTGAAGCCATCCAAGCGTGC; *ENO2* rev: CCCACCACCAGGTCAGCAAT. 20 µL PCR reactions were prepared with 2× SYBR Green mix containing 1.6 µL of reverse transcriptase product and 0.4 µL of each primer (10 µM). The PCR mixtures were incubated at 95 °C for 3 min, followed by 40 cycles of 95 °C for 10 s, 60 °C for 20 s, and 72 °C for 10 s. The calculation of gene expression levels was based on the ΔΔCt method in transfection experiments and on the ΔCt method for gene expression analysis in brain tissues. The geometric mean of the expression values of *EIF4A2* and *CYC1* housekeeping genes was used as internal control in transfection experiments, while gene expression levels in brain tissues were normalized on the geometric mean of the same housekeeping genes and the neuronal markers *SYP* and *ENO2*.

For the measurement of miRNAs, a two-step Taq-Man real-time PCR assay was performed using primers and probes obtained from Thermo Fisher Scientific. The reverse transcriptase reaction was performed using the TaqMan MicroRNA Reverse Transcription kit (Thermo Fisher Scientific), according to the manufacturer's instructions. cDNA was synthesized from 50 ng of total RNA in 15 µL reactions, using the stem-loop primer for miR-15a (ID000389), miR-15b (ID000390), miR-16 (ID000391), miR-103 (ID000439), miR-107 (ID000443), miR-195 (ID000494), and U6 snRNA (ID001973). The PCR reaction (20 µL) contained 1.3 µL of reverse transcriptase product, 10 µL of Taq-Man Universal PCR Master Mix (Thermo Fisher Scientific) and 1 µL of the appropriate TaqMan MicroRNA Assay (20×) containing primers and probes for the miR of interest. The PCR mixtures were incubated at 95 °C for 10 min, followed by 40 cycles of 95 °C for 15 s and 60 °C for 60 s. The expression of miRs was based on the ΔΔCt methods, using U6 snRNA as an endogenous control. All PCRs were performed in triplicate using an iQ5 Real-Time PCR detection system (Bio-Rad, Hercules, CA, USA).

4.5. Statistical Analysis

Each experiment was carried out at least three times. Histograms represent the mean values and bars indicate the standard deviation of the mean. The box plots show median, 25th and 75th percentile values and whiskers to the minimum and maximum value. The statistical significance of the results was determined using Student's *t*-test, with data considered significant when $p < 0.05$. The degree of linear relationship between *CDK5R1* gene and NEAT1 expression levels was calculated using Pearson's correlation coefficient (*r* value). The *p* value was calculated from an extra sum-of-squares *F* test.

Author Contributions: Conceptualization, M.S., F.R., E.B., and M.V.; Formal analysis, M.S., B.G., F.R., E.B., and M.V.; Funding acquisition, E.B., F.R., and M.V.; Investigation, M.S. and B.G.; Methodology, M.S., B.G., and M.V.; Project administration, M.V.; Resources, E.B. and M.V.; Supervision, M.V.; Validation, M.V.; Visualization, M.S. and M.V.; Writing—original draft, M.S. and M.V.; Writing—review & editing, M.S., B.G., F.R., E.B., and M.V.

Acknowledgments: The authors would like to thank the MRC London Neurodegenerative Diseases Brain Bank, the Newcastle Brain Tissue Resource, and the South West Dementia Brain Bank (SWDBB) for providing brain tissue for this study. The SWDBB is supported by BRACE (Bristol Research into Alzheimer's and Care of the Elderly), Brains for Dementia Research and the Medical Research Council. This work was supported by the Academic Grant "Bando BIOMETRA—Fondi Incentivo alla Ricerca 2015" (Università degli studi di Milano, project no. RV_RIC_AT16MVENT_M) to M.V., the Ministero dell'Istruzione, dell'Università e della Ricerca (award no. Epigenomics Flagship Project), Telethon Foundation project (award no. GGP14074) and Fondazione Cariplo (award no. 2016-0204) to E.B., and Fondazione Cariplo (award no. 2014-0972) to F.R.

Conflicts of Interest: The authors declare no conflict of interest.

References

1. Hebert, L.E.; Weuve, J.; Scherr, P.A.; Evans, D.A. Alzheimer disease in the United States (2010–2050) estimated using the 2010 census. *Neurology* **2013**, *80*, 1778–1783. [CrossRef] [PubMed]
2. Serrano-Pozo, A.; Frosch, M.P.; Masliah, E.; Hyman, B.T. Neuropathological alterations in Alzheimer disease. *Cold Spring Harb. Perspect. Med.* **2011**, *1*, a006189. [CrossRef] [PubMed]
3. Desler, C.; Lillenes, M.S.; Tønjum, T.; Rasmussen, L.J. The role of mitochondrial dysfunction in the progression of Alzheimer's disease. *Curr. Med. Chem.* **2017**, *24*. [CrossRef] [PubMed]
4. Jiang, T.; Chang, R.C.; Rosenmann, H.; Yu, J.T. Advances in Alzheimer's disease: From bench to bedside. *Biomed. Res. Int.* **2015**, *2015*, 202676. [CrossRef] [PubMed]
5. Wang, J.Z.; Grundke-Iqbal, I.; Iqbal, K. Kinases and phosphatases and tau sites involved in Alzheimer neurofibrillary degeneration. *Eur. J. Neurosci.* **2007**, *25*, 59–68. [CrossRef] [PubMed]
6. Fischer, A.; Sananbenesi, F.; Schrick, C.; Spiess, J.; Radulovic, J. Cyclin-dependent kinase 5 is required for associative learning. *J. Neurosci.* **2002**, *22*, 3700–3707. [CrossRef] [PubMed]
7. McLinden, K.A.; Trunova, S.; Giniger, E. At the fulcrum in health and disease: Cdk5 and the balancing acts of neuronal structure and physiology. *Brain Disord. Ther.* **2012**, *2012* (Suppl. 1), 001. [CrossRef] [PubMed]
8. Liu, S.L.; Wang, C.; Jiang, T.; Tan, L.; Xing, A.; Yu, J.T. The Role of Cdk5 in Alzheimer's Disease. *Mol. Neurobiol.* **2016**, *53*, 4328–4342. [CrossRef] [PubMed]
9. Moncini, S.; Bevilacqua, A.; Venturin, M.; Fallini, C.; Ratti, A.; Nicolin, A.; Riva, P. The 3' untranslated region of human Cyclin-Dependent Kinase 5 Regulatory subunit 1 contains regulatory elements affecting transcript stability. *BMC Mol. Biol.* **2007**, *8*, 111. [CrossRef] [PubMed]
10. Zuccotti, P.; Colombrita, C.; Moncini, S.; Barbieri, A.; Lunghi, M.; Gelfi, C.; de Palma, S.; Nicolin, A.; Ratti, A.; Venturin, M.; et al. hnRNPA2/B1 and nELAV proteins bind to a specific U-rich element in CDK5R1 3'-UTR and oppositely regulate its expression. *Biochim. Biophys. Acta* **2014**, *1839*, 506–516. [CrossRef] [PubMed]
11. Moncini, S.; Lunghi, M.; Valmadre, A.; Grasso, M.; del Vescovo, V.; Riva, P.; Denti, M.A.; Venturin, M. The miR-15/107 Family of microRNA Genes Regulates CDK5R1/p35 with Implications for Alzheimer's Disease Pathogenesis. *Mol. Neurobiol.* **2017**, *54*, 4329–4342. [CrossRef] [PubMed]
12. Khorkova, O.; Hsiao, J.; Wahlestedt, C. Basic biology and therapeutic implications of lncRNA. *Adv. Drug Deliv. Rev.* **2015**, *87*, 15–24. [CrossRef] [PubMed]
13. Nie, L.; Wu, H.J.; Hsu, J.M.; Chang, S.S.; Labaff, A.M.; Li, C.W.; Wang, Y.; Hsu, J.L.; Hung, M.C. Long non-coding RNAs: Versatile master regulators of gene expression and crucial players in cancer. *Am. J. Transl. Res.* **2012**, *4*, 127–150. [PubMed]
14. Bayoumi, A.S.; Sayed, A.; Broskova, Z.; Teoh, J.P.; Wilson, J.; Su, H.; Tang, Y.L.; Kim, I.M. Crosstalk between Long Noncoding RNAs and MicroRNAs in Health and Disease. *Int. J. Mol. Sci.* **2016**, *17*, 356. [CrossRef] [PubMed]
15. Li, Y.; Wang, X. Role of long noncoding RNAs in malignant disease (Review). *Mol. Med. Rep.* **2016**, *13*, 1463–1469. [CrossRef] [PubMed]
16. Faghihi, M.A.; Modarresi, F.; Khalil, A.M.; Wood, D.E.; Sahagan, B.G.; Morgan, T.E.; Finch, C.E.; St Laurent, G., 3rd; Kenny, P.J.; Wahlestedt, C. Expression of a noncoding RNA is elevated in Alzheimer's disease and drives rapid feed-forward regulation of beta-secretase. *Nat. Med.* **2008**, *14*, 723–730. [CrossRef] [PubMed]
17. Faghihi, M.A.; Zhang, M.; Huang, J.; Modarresi, F.; Van der Brug, M.P.; Nalls, M.A.; Cookson, M.R.; St-Laurent, G., 3rd; Wahlestedt, C. Evidence for natural antisense transcript-mediated inhibition of microRNA function. *Genome Biol.* **2010**, *11*, R56. [CrossRef] [PubMed]
18. Chakravarty, D.; Sboner, A.; Nair, S.S.; Giannopoulou, E.; Li, R.; Hennig, S.; Mosquera, J.M.; Pauwels, J.; Park, K.; Kossai, M.; et al. The oestrogen receptor alpha-regulated lncRNA NEAT1 is a critical modulator of prostate cancer. *Nat. Commun.* **2014**, *5*, 5383. [CrossRef] [PubMed]
19. West, J.A.; Davis, C.P.; Sunwoo, H.; Simon, M.D.; Sadreyev, R.I.; Wang, P.I.; Tolstorukov, M.Y.; Kingston, R.E. The long noncoding RNAs NEAT1 and MALAT1 bind active chromatin sites. *Mol. Cell* **2014**, *55*, 791–802. [CrossRef] [PubMed]
20. Fox, A.H.; Lamond, A.I. Paraspeckles. *Cold Spring Harb. Perspect. Biol.* **2010**, *2*, a000687. [CrossRef] [PubMed]

21. Hirose, T.; Virnicchi, G.; Tanigawa, A.; Naganuma, T.; Li, R.; Kimura, H.; Yokoi, T.; Nakagawa, S.; Bénard, M.; Fox, A.H.; et al. NEAT1 long noncoding RNA regulates transcription via protein sequestration within subnuclear bodies. *Mol. Biol. Cell* **2014**, *25*, 169–183. [CrossRef] [PubMed]

22. Riva, P.; Ratti, A.; Venturin, M. The Long Non-Coding RNAs in Neurodegenerative Diseases: Novel Mechanisms of Pathogenesis. *Curr. Alzheimer Res.* **2016**, *13*, 1219–1231. [CrossRef] [PubMed]

23. Tripathi, V.; Ellis, J.D.; Shen, Z.; Song, D.Y.; Pan, Q.; Watt, A.T.; Freier, S.M.; Bennett, C.F.; Sharma, A.; Bubulya, P.A.; et al. The nuclear-retained noncoding RNA MALAT1 regulates alternative splicing by modulating SR splicing factor phosphorylation. *Mol. Cell* **2010**, *39*, 925–938. [CrossRef] [PubMed]

24. Gutschner, T.; Hämmerle, M.; Diederichs, S. MALAT1—A paradigm for long noncoding RNA function in cancer. *J. Mol. Med. (Berl.)* **2013**, *91*, 791–801. [CrossRef] [PubMed]

25. Wang, P.; Wu, T.; Zhou, H.; Jin, Q.; He, G.; Yu, H.; Xuan, L.; Wang, X.; Tian, L.; Sun, Y.; et al. Long noncoding RNA NEAT1 promotes laryngeal squamous cell cancer through regulating miR-107/CDK6 pathway. *J. Exp. Clin. Cancer Res.* **2016**, *35*, 22. [CrossRef] [PubMed]

26. Zhang, H.; Wang, G.; Yin, R.; Qiu, M.; Xu, L. Comprehensive Identification of MicroRNAs Regulated by Long Non-coding RNA MALAT1. *Zhongguo Fei Ai Za Zhi* **2016**, *19*, 247–251. [CrossRef] [PubMed]

27. Yu, X.; Li, Z. Long non-coding RNA HOTAIR: A novel oncogene (Review). *Mol. Med. Rep.* **2015**, *12*, 5611–5618. [CrossRef] [PubMed]

28. Tsai, M.C.; Manor, O.; Wan, Y.; Mosammaparast, N.; Wang, J.K.; Lan, F.; Shi, Y.; Segal, E.; Chang, H.Y. Long noncoding RNA as modular scaffold of histone modification complexes. *Science* **2010**, *329*, 689–693. [CrossRef] [PubMed]

29. Toffolo, E.; Rusconi, F.; Paganini, L.; Tortorici, M.; Pilotto, S.; Heise, C.; Verpelli, C.; Tedeschi, G.; Maffioli, E.; Sala, C.; et al. Phosphorylation of neuronal Lysine-Specific Demethylase 1 LSD1/KDM1A impairs transcriptional repression by regulating interaction with CoREST and histone deacetylases HDAC1/2. *J. Neurochem.* **2014**, *128*, 603–616. [CrossRef] [PubMed]

30. Harada, T.; Morooka, T.; Ogawa, S.; Nishida, E. ERK induces p35, a neuron-specific activator of Cdk5, through induction of Egr1. *Nat. Cell Biol.* **2001**, *3*, 453–459. [CrossRef] [PubMed]

31. Chen, L.; Feng, P.; Zhu, X.; He, S.; Duan, J.; Zhou, D. Long non-coding RNA Malat1 promotes neurite outgrowth through activation of ERK/MAPK signalling pathway in N2a cells. *J. Cell. Mol. Med.* **2016**, *20*, 2102–2110. [CrossRef] [PubMed]

32. Ko, J.; Humbert, S.; Bronson, R.T.; Takahashi, S.; Kulkarni, A.B.; Li, E.; Tsai, L.H. p35 and p39 are essential for cyclin-dependent kinase 5 function during neurodevelopment. *J. Neurosci.* **2001**, *21*, 6758–6771. [CrossRef] [PubMed]

33. Nishimoto, Y.; Nakagawa, S.; Hirose, T.; Okano, H.J.; Takao, M.; Shibata, S.; Suyama, S.; Kuwako, K.; Imai, T.; Murayama, S.; et al. The long non-coding RNA nuclear-enriched abundant transcript 1_2 induces paraspeckle formation in the motor neuron during the early phase of amyotrophic lateral sclerosis. *Mol. Brain* **2013**, *6*, 31. [CrossRef] [PubMed]

34. Shelkovnikova, T.A.; Robinson, H.K.; Troakes, C.; Ninkina, N.; Buchman, V.L. Compromised paraspeckle formation as a pathogenic factor in FUSopathies. *Hum. Mol. Genet.* **2014**, *23*, 2298–2312. [CrossRef] [PubMed]

35. Tollervey, J.R.; Curk, T.; Rogelj, B.; Briese, M.; Cereda, M.; Kayikci, M.; König, J.; Hortobágyi, T.; Nishimura, A.L.; Zupunski, V.; et al. Characterizing the RNA targets and position-dependent splicing regulation by TDP-43. *Nat. Neurosci.* **2011**, *14*, 452–458. [CrossRef] [PubMed]

36. Sunwoo, J.S.; Lee, S.T.; Im, W.; Lee, M.; Byun, J.I.; Jung, K.H.; Park, K.I.; Jung, K.Y.; Lee, S.K.; Chu, K.; et al. Altered Expression of the Long Noncoding RNA NEAT1 in Huntington's Disease. *Mol. Neurobiol.* **2017**, *54*, 1577–1586. [CrossRef] [PubMed]

37. Hooper, C.; Meimaridou, E.; Tavassoli, M.; Melino, G.; Lovestone, S.; Killick, R. p53 is upregulated in Alzheimer's disease and induces tau phosphorylation in HEK293a cells. *Neurosci. Lett.* **2007**, *418*, 34–37. [CrossRef] [PubMed]

38. Lee, J.H.; Kim, H.S.; Lee, S.J.; Kim, K.T. Stabilization and activation of p53 induced by Cdk5 contributes to neuronal cell death. *J. Cell Sci.* **2007**, *120*, 2259–2271. [CrossRef] [PubMed]

39. Mello, S.S.; Sinow, C.; Raj, N.; Mazur, P.K.; Bieging-Rolett, K.; Broz, D.K.; Imam, J.F.C.; Vogel, H.; Wood, L.D.; Sage, J.; et al. Neat1 is a p53-inducible lincRNA essential for transformation suppression. *Genes Dev.* **2017**, *31*, 1095–1108. [CrossRef] [PubMed]

40. Nikolic, M.; Dudek, H.; Kwon, Y.T.; Ramos, Y.F.; Tsai, L.H. The cdk5/p35 kinase is essential for neurite outgrowth during neuronal differentiation. *Genes Dev.* **1996**, *10*, 816–825. [CrossRef] [PubMed]

41. Rinn, J.L.; Chang, H.Y. Genome regulation by long noncoding RNAs. *Annu. Rev. Biochem.* **2012**, *81*, 145–166. [CrossRef] [PubMed]

42. Yoon, J.H.; Abdelmohsen, K.; Gorospe, M. Functional interactions among microRNAs and long noncoding RNAs. *Semin. Cell Dev. Biol.* **2014**, *34*, 9–14. [CrossRef] [PubMed]

43. Jiang, L.; Shao, C.; Wu, Q.J.; Chen, G.; Zhou, J.; Yang, B.; Li, H.; Gou, L.T.; Zhang, Y.; Wang, Y.; et al. NEAT1 scaffolds RNA-binding proteins and the Microprocessor to globally enhance pri-miRNA processing. *Nat. Struct. Mol. Biol.* **2017**, *24*, 816–824. [CrossRef] [PubMed]

44. Lennox, K.A.; Behlke, M.A. Cellular localization of long non-coding RNAs affects silencing by RNAi more than by antisense oligonucleotides. *Nucleic Acids Res.* **2016**, *44*, 863–877. [CrossRef] [PubMed]

International Journal of
Molecular Sciences

MDPI

Review

Cellular Receptors of Amyloid β Oligomers (AβOs) in Alzheimer's Disease

Barbara Mroczko [1,2,3,*], Magdalena Groblewska [3], Ala Litman-Zawadzka [1], Johannes Kornhuber [4] and Piotr Lewczuk [1,4]

1   Department of Neurodegeneration Diagnostics, Medical University of Białystok, 15-089 Białystok, Poland; ala.litman-zawadzka@umb.edu.pl (A.L.-Z.); piotr.lewczuk@uk-erlangen.de (P.L.)
2   Department of Biochemical Diagnostics, Medical University of Białystok, 15-089 Białystok, Poland
3   Department of Biochemical Diagnostics, University Hospital in Białystok, 15-276 Białystok, Poland; magdalena.groblewska@umb.edu.pl
4   Department of Psychiatry and Psychotherapy, Universitätsklinikum Erlangen, and Friedrich-Alexander Universität Erlangen-Nürnberg, 91054 Erlangen, Germany; johannes.kornhuber@uk-erlangen.de
*   Correspondence: mroczko@umb.edu.pl; Tel.: +48-85-686-5168; Fax: +48-85-686-5169

Received: 23 May 2018; Accepted: 22 June 2018; Published: 26 June 2018

Abstract: It is estimated that Alzheimer's disease (AD) affects tens of millions of people, comprising not only suffering patients, but also their relatives and caregivers. AD is one of age-related neurodegenerative diseases (NDs) characterized by progressive synaptic damage and neuronal loss, which result in gradual cognitive impairment leading to dementia. The cause of AD remains still unresolved, despite being studied for more than a century. The hallmark pathological features of this disease are senile plaques within patients' brain composed of amyloid beta (Aβ) and neurofibrillary tangles (NFTs) of Tau protein. However, the roles of Aβ and Tau in AD pathology are being questioned and other causes of AD are postulated. One of the most interesting theories proposed is the causative role of amyloid β oligomers (AβOs) aggregation in the pathogenesis of AD. Moreover, binding of AβOs to cell membranes is probably mediated by certain proteins on the neuronal cell surface acting as AβO receptors. The aim of our paper is to describe alternative hypotheses of AD etiology, including genetic alterations and the role of misfolded proteins, especially Aβ oligomers, in Alzheimer's disease. Furthermore, in this review we present various putative cellular AβO receptors related to toxic activity of oligomers.

Keywords: amyloid-β oligomer; protein aggregation; AβO receptors; Alzheimer's disease; neurodegeneration

1. Introduction

Alzheimer's disease (AD) affects tens of millions of people worldwide and estimated number of AD patients would increase to over 130 million by 2050 [1]. AD is a big socio-economical problem because of comprising not only suffering patients, but also their relatives and caregivers. Additionally, current therapeutic strategies provide only palliative, not disease-modifying, agents.

AD belongs to a large group of neurodegenerative diseases (NDs), which include also Parkinson's disease (PD) and PD-related disorders, prion disease, motor neuron diseases (MND), Huntington's disease (HD), spinocerebellar ataxia (SCA), spinal muscular atrophy (SMA), and others [2]. Characterized features of NDs are the progressive degeneration and/or death of neuron cells. In AD, this neuronal loss is accompanied by progressive synaptic damage, which results in gradual cognitive impairment and, finally, dementia.

The main histopathological hallmarks of AD are the extracellular plaques within brain tissue consisted of variant forms of amyloid β (Aβ) and neurofibrillary tangles (NFTs) of many forms of phosphorylated Tau proteins (pTau), localized intraneuronally [3]. Primarily, these pathological alterations are seen within medial temporal lobe, whereas in later stages of AD they progress subsequently to brain regions associated with neocortex [4,5]. Formation of Aβ plaques and neuronal cell damage are preceded by reduced synaptic transmission and loss of dendritic spines, which lead to synaptic dysfunction in AD brain. It is estimated that these changes may anticipate the first cognitive decline symptoms even for two decades [6]. Furthermore, declined levels of Aβ42 in cerebrospinal fluid (CSF) and the presence of Aβ plaques in neuroimaging may head other AD-related alterations by many years [7].

2. Postulated Hypotheses of AD Etiology

2.1. Risk Factors of AD

The risk factors of AD include increasing age, genetic, and vascular factors, smoking, obesity and diabetes [8]. The presence of genetic mutations as the etiological factors of AD has been identified in 1–5% cases [8]. Although most of sporadic AD cases are unrelated to any autosomal-dominant inheritance, certain genetic changes may be linked with a significant risk of AD development. The mutations in presenilin1 (*PS1*), presenilin2 (*PS2*), and amyloid precursor protein (*APP*) genes are associated with the familial Alzheimer's disease (FAD) (reviewed by Hardy [9]), while the presence of apolipoprotein E (*APOE*) ε4 genotype links to sporadic form of AD [10,11].

2.2. Amyloid Hypothesis

The precise etiology of AD remains unknown, despite over century passing from the first report of its symptoms by Alois Alzheimer [12]. Scientific efforts to elucidate AD etiopathogenesis lead to several different, partly complementary hypotheses. The relevant role of Aβ42 aggregation in AD pathogenesis has been disputed for over 25 years. Originally, the imbalance between production and clearance of Aβ42 in the very early stages of AD have been assumed as a causative and initiating factor of this disease. This dyshomeostasis may be the result of the mutations either in APP genes or in genes encoding presenilin, the substrate and enzyme of the reaction that generates Aβ42, respectively. It leads to the presence of Aβ deposits and the damage of the nerve tissue (reviewed by Selkoe [13]). Amyloid hypothesis may by supported by the observation that progressive Aβ deposition is present already in early, preclinical stages of AD and, finally, in all AD patients.

2.3. Isoform APOE4

The relationship between impaired amyloid deposition/clearance and genetic risk factors in AD were highlighted. As it was mentioned above, the best known genetic risk factor of AD is the ε4 allele of the APOE [14,15], which is associated with sporadic, late-onset AD. APOE is a polymorphic lipoprotein, with three major gene alleles: *APOE-ε2*, *APOE-ε3*, and *APOE-ε4*. It was shown that these three APOE isoforms bind Aβ differentially and modulate its fibrillogenesis [16–18]. The isoform APOE4 is unable to stimulate degradation of Aβ effectively, which results in a decreased brain clearance of Aβ and leads to the accumulation of amyloid deposits in the brain [19]. Moreover, there are more vascular and plaque deposits of Aβ observed in APOE4 carriers than in humans expressing only APOE3 [20]. This observation was also confirmed in genetically engineered mice [10]. Additionally, a quantitative evaluation in transgenic mice bearing human APP and APOE genes has shown decreased Aβ clearance in *APOE4* carriers in comparison with E3 and E2 mice, which was paralleled by the degree of Aβ deposition [21].

2.4. Mutations in Presenilins PS1 and PS2 or APP Genes

In the familial form of AD, most cases are related to mutations in genes encoding one of three proteins: presenilins PS1 and PS2 or APP (i.e., the proteases and their substrate for generation of Aβ, respectively) [22]. APP is an essential membrane protein expressed mainly by the synapses and involved in their formation [23] as well as in neural plasticity [24] and iron export [25]. This protein is the precursor molecule that proteolytic processing generates various peptide fragments, including polypeptides of Aβ with 37 to 49 amino acid residues and molecular weight of approximately 4 kDa [26]. Proteolytic cleavage of APP is completed by enzymes of secretase family: α-, β-, and γ-secretase. Whereas APP processing by α- and β-secretase leads to removal of almost entire extracellular domain and produces membrane-anchored C-terminal fragments (reviewed by Zheng [27]), γ-secretase processing of APP results in generation of Aβ fragment [28]. γ-Secretase is a large, multi-subunit enzyme whose catalytic subunit is presenilin, a multi-pass transmembrane protein. The amyloidogenic processing of APP [29] and γ-secretase activity [30] have been associated with lipid rafts within cellular membrane. The role of cholesterol in lipid raft maintenance has been cited as a likely explanation for observations that high cholesterol and APOE-ε4 genotype are the major risk factors for Alzheimer's disease [31].

Most mutations in the presenilin and APP genes enhance the production of Aβ42 [32] and early-onset deposition of this peptide [33–35]. Especially, the mutations in the region of APP molecule corresponding to the Aβ sequence lead to the production of more self-aggregating forms of amyloid [13]. Similarly, different presenilin mutations result in decreased ability of processing of APP by γ-secretase, and consequently increase the relative production of longer, more hydrophobic and more self-aggregating peptides of Aβ [13]. Peptides Aβ42, Aβ43, and longer express high potential of self-aggregation, whereas Aβ40 may rather be anti-amyloidogenic [36]. However, some of the pathogenic presenilin mutations only alter the ratio between Aβ42 and the other peptides of Aβ, especially Aβ40, but do not increase Aβ42 levels [37,38].

2.5. Down's Syndrome

A gene for *APP* is located on chromosome 21. In subjects affected with Down's syndrome due to the trisomy of this chromosome and possessing three copies of *APP* gene, AD is most likely to develop within the first 40 years of life [39,40]. This duplication of the wild-type *APP* gene leads to early-onset Aβ deposition, which occurs already in the teenagers, is then followed by microgliosis, astrocytosis, and accumulation of NFTs typical for AD. On the contrary, inheritance of a missense mutation in *APP*, that decreases the production and aggregation of Aβ, protects against AD and age-related cognitive decline [41].

2.6. Deposition of Misfolded Tau Protein

It was proposed that amyloid cascade is not the only pathway to AD (discussed by [42–44]). Although accumulation of Aβ in AD brain is followed by progressive deposition Tau protein, another hypothesis assumes that the abnormalities in the Tau protein initiate the cascade of events in AD [45]. In normal conditions, Tau is a soluble protein that is responsible for the association and stabilization of microtubules. In nerve cells, Tau is typically found in axons, but in the tauopathies, such as AD, progressive supranuclear palsy (PSP), corticobasal degeneration (CBD), or inherited frontotemporal dementia and Parkinsonism linked to chromosome 17 (FTDP-17), this protein is present in an abnormal filamentous form and redistributed to the cell body and neurites [46]. Hyperphosphorylated forms of Tau aggregate in NFTs within the neurons [47]. This results in the disintegration of the microtubule and the destruction of neuronal transport [48], leading to impaired communication between neurons and, finally, their death [49]. This hypothesis may be partially confirmed in a model of AD created by Jack et al. [7], where Tau pathology in brain precedes Aβ deposition in time, but only on at a sub-threshold biomarker detection level. Although some human neuropathological studies suggest that NFTs may occur prior to presence of amyloid plaques (for review see: [13]), it is possible, that such studies might not have searched systematically for diffuse plaques or soluble, oligomeric forms of Aβ in the brain. Moreover, genetic studies prove that Aβ-elevating *APP* mutations lead to downstream alteration and aggregation of wild-type Tau, whereas Tau mutations do not lead to Aβ deposition and amyloid-related dementia. Some researchers suggest that Aβ can trigger AD-type Tau alterations, whereas Tau expression seems to permit certain downstream neuronal consequences of progressive Aβ build-up to arise [50]. This triggering feature is particularly addressed to soluble AβO [51].

2.7. Neuroinflammation

The immune system participates also in AD pathogenesis. It was demonstrated that AD patients express decreased levels of naturally-produced antibodies against Aβ when compared with healthy individuals [52]. The inflammatory reaction, oxidative stress and dyshomeostasis of metals metabolism also play an important role in AD pathogenesis [53,54]. It appears that insoluble Aβ deposits are recognized as foreign material and trigger activation of inflammatory response cascade [55,56]. Additionally, inflammation within AD brain may be partially linked with APOE4's role as an aberrant immunomodulatory factor. The function of macrophages and microglia is regulated by APOE and may vary depending on isoform of this lipoprotein. Especially, APOε4 is associated with an enhanced inflammatory response compared to macrophages not expressing this allele [57]. It was shown that microglia derived from homozygotic mice possessing both alleles *APOε4* demonstrated a pro-inflammatory phenotype, altered cell morphology, increased NOS2 mRNA levels and NO production, as well as higher pro-inflammatory cytokine production compared to microglia derived from *APOε3* mice [58]. This effect was gene dose-dependent and increased with the number of *APOε4* gene alleles. Although the immune aspects of AD draw increasing attention of researchers, the aim of this paper was to concentrate on other aspects of AD, such as soluble Aβ oligomers (AβOs) toxic activity and their putative cellular receptors.

2.8. Soluble AβOs Toxicity

Although initiated by Aβ, progression of AD is subsequently complicated and accelerated by other pathological processes, such as Tau pathology or inflammation. It is known that Aβ peptide may be present in various distinct states, including oligomeric forms of Aβ. These oligomeric species are antigenically distinct from monomeric and fibrillar conformations of Aβ peptide [59,60]. Currently, it is supposed that soluble AβOs, but not fibrillar Aβ42 within neuritic plaques, may be the toxic factors acting on a very early stage of AD, perhaps even initiating pathological cascade.

Mechanisms of AβO toxicity include synapse loss, the strongest pathological correlate of cognitive deficits in AD (Figure 1). AβO-induced decrease in synapse density is observed already in the earliest stages of AD [61], and the degree of synapse loss is greatest in close proximity of amyloid plaques [62]. The evidence for toxic AβO activity comes from observation that the loss of synapses in AD transgenic animals is correlated with the degree of colocalization of Aβ soluble oligomers with synaptic puncta [63].

Another mechanism of AβO toxicity is oligomers-induced disruption of synaptic transmission. It was demonstrated that soluble AβO could inhibit long-term potentiation (LTP) in mouse hippocampal tissue samples, suggesting that this form of Aβ might be the species triggering loss of synapses and memory impairment in AD [64]. It was also shown in mice model of AD that transgenic animals overexpressing mutant form of human APP exhibited lower density of presynaptic terminals, as well as severe impairments in synaptic transmission in the hippocampus for months before the presence of amyloid plaques [65]. This toxic activity of oligomers was confirmed in the study of Shankar et al., who demonstrated that soluble AβOs isolated from AD patients' brains decreased number of synapses in animal models of AD, leading to enhanced long-term synaptic depression (LTD) and LTP in regions of brain which are responsible for memory [51].

What is interesting, it seems that intracellular soluble AβOs may be transmitted between neurons using synaptic connections, reaching even distant areas of the brain [66]. It was confirmed in AD mice models, where intracerebral injections of brain extracts from AD or Aβ aggregates induced amyloidogenesis [67]. Although various forms of AβOs may disseminate between neurons in this way, they do not spread between glial cells [68]. It is thought that the self-replication of Aβ and Tau aggregates and their spreading in a prion-like manner may contribute to the progressive nature of AD.

Moreover, AβOs can be transferred not only within the brain, from one region to another, but it is supposed that oligomers might be transmitted between people [69].

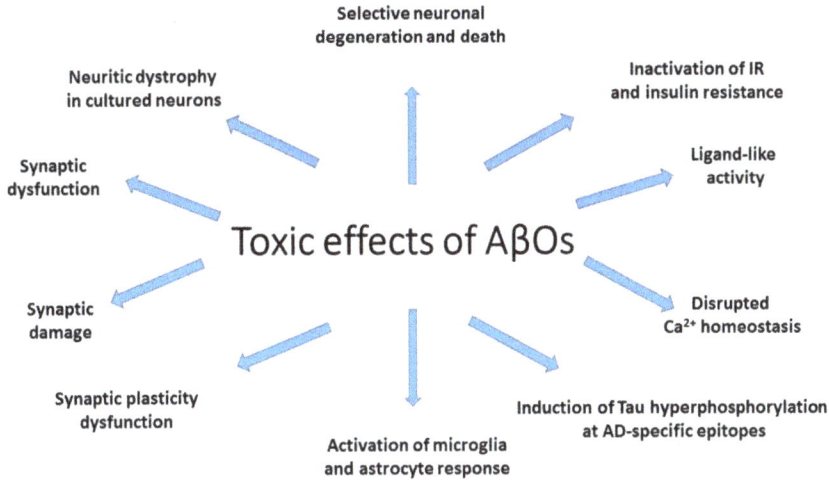

Figure 1. Toxic activity of amyloid β oligomers (AβOs).

A limitation of AβOs' hypothesis may be the fact that oligomers are not homogenous species. Implications for the phenotypic diversity of Aβ in AD include clinical and neuropathological heterogeneity of AD, various distribution and significant differences in the expression of Aβ species in brain tissue, as well as different concentrations of Aβ peptides in CSF of AD patients [69]. AβOs' molecular weight, morphology and conformation are highly variegated. There are different Aβ forms, from small, dimeric molecules, through trimers, and low molecular weight (LMW) and high molecular weight (HMW) oligomers up to protofibrils and fibrils, which range from relatively small molecules about 4 kDa to assemblies of 100 kDa. Therefore, the precise role of different oligomer species still remains to be elucidated [70,71].

3. Cellular Receptors Related to AβOs Activity

Although it is known that extracellular AβOs are able to bind to the surface of neurons, resulting in synaptic dysfunction and neurodegeneration, precise mechanism remains uncertain. It was suggested that AβOs may damage neuronal membranes directly, forming pores, which leads to the ionic dyshomeostasis, especially an increase in intracellular Ca^{2+} levels [72]. It was also proposed that AβOs at high concentration may interact directly with negatively charged phospholipid bilayers, leading to changes in their conductance in non-specific manner [73]. However, it is unclear how such membrane-disrupting activity could explain the selectivity of AβOs for the central nervous system (CNS), especially for the synapses, in AD.

Some other potential mechanisms of AβOs and their targets, including the abnormal activation of signaling pathways, are under extensive investigation [74]. As it was mentioned above, extracellular AβOs accrue generally at synapses, especially at synaptic spines, but detailed specificity of AβOs to particular cells was uncertain. It was shown that in hippocampal cultures Aβ oligomers were bound mostly to neurons, whereas in cortical and cerebellar cultures this binding occurred in a lesser degree [75,76]. It was also demonstrated that both AβOs prepared in vitro and those extracted from AD brain were the ligands targeting cultured mouse hippocampal cells. Soluble AβO isolated from AD patients bound to hippocampal dendrites in cultured mouse neurons with high, "ligand-like" specificity [77]. This specific targeting of neuronal cells is in line with rapid disruption of hippocampal

LTP and LTD induced by oligomers [78,79] and with selective neuronal degeneration induced by soluble AβOs seen in brain slice preparations [80].

Several studies postulate various possible receptors involved in the toxicity of AβOs. Binding of AβOs to cell membranes is probably mediated by particular cell surface proteins that act as toxin receptors. This hypothesis could explain various mechanisms of AβOs' activity, resulting in synaptic dysfunction and neurodegeneration [64].

It is possible that these receptor proteins might be expressed only on certain cells, converting action of AβOs into harmful responses. Moreover, such receptors for extracellular AβO are rather localized at neuronal synapses and should have a high affinity for AβO. These receptors should be also more selective for AβOs than for monomeric or fibrillar Aβ, because monomers of Aβ are present ubiquitously in all individuals, while their levels do not substantially change with disease. Furthermore, putative AβO receptors should have the ability to transduce extracellular triggering factors into certain intracellular changes. It may be achieved either directly or by connection with other active molecules.

There are over 20 candidates for AβO receptors, including glutamate, adrenergic, acetylcholine receptors and others. Unfortunately, no single candidate receptor protein has been shown yet to be responsible for all features of AβO activity. Moreover, the heterogeneity of AβOs results in their diverse affinity as ligands when binding to various putative oligomers' receptors [81].

3.1. Glutamate Receptor NMDAR

Synaptotoxic activity of AβOs includes inappropriate increase of extracellular glutamate concentration and activation of glutamate receptors, which results in rapid impairment of synaptic plasticity [82]. The N-methyl-D-aspartate (NMDA) receptor is a glutamate receptor with ion channel activity. It plays a role in controlling of synaptic plasticity and synapse formation, which are responsible for memory function, learning and formation of neuronal networks in CNS [83]. It was suggested that oligomeric Aβ toxicity may involve NMDAR activation, although it remains controversial whether AβOs trigger loss or gain of its function.

Some studies indicate that AβOs initiate impairment of NMDAR activity by removal from the cell surface and triggering of synaptic depression signaling pathways [84–86]. By modulation of NMDAR-dependent signaling pathway, AβOs induce reversible synapse loss causing the decrease in spine density [85,87]. Both in vivo and in vitro studies demonstrated that Aβ can disrupt induction of LTP depending on this type receptor [88]. Moreover, activity of this receptor is required for AβOs-induced synaptic depression [87].

On the contrary, other authors demonstrated that AβOs cause an increase of NMDAR receptor function. AβOs induce neuronal oxidative stress through an NMDAR-dependent mechanism. This activity of AβOs is blocked by memantine, an uncompetitive NMDAR antagonist and the drug used to relief AD symptoms [89]. Moreover, it was reported in animal models of AD, that chronic treatment with memantine reduced Aβ deposition in the brain, both insoluble Aβ fibrils and soluble AβOs. Memantine not only inhibited the formation of different types of Aβ aggregates in a concentration-dependent manner, but also led to disaggregation of Aβ42 fibrils [90]. Interestingly, specific antibody to the extracellular domain of the NR1 subunit of NMDARs led also to reduction of AβOs binding to neurons and completely blocked the formation of reactive oxygen species (ROS) [89].

Dysregulation of Ca^{2+} signaling and membrane disturbance, which is thought as a ubiquitous mechanism of soluble AβOs neurotoxicity [91], may also be mediated by activation of NMDAR [92]. AβOs disrupt NMDAR-mediated postsynaptic Ca^{2+} signaling in response to presynaptic stimulation by enhancing the accessibility of extracellular glutamate as well as directly disturbing the NMDARs [93]. This excessive activation of NMDAR leads to disproportionate inflow of Ca^{2+} to neurons and may cause excitotoxicity, a pathological mechanism recognized in some NDs, including AD. It is thought that this aberrant regulation of intracellular Ca^{2+} signaling is an early event in AD, prior to the presence of clinical symptoms. Dysregulation of Ca^{2+} signaling is also believed to be a crucial factor contributing to AD pathogenesis (for review see: [94]).

Moreover, AβOs interfere specifically with several proteins involved in calcium-related signaling pathways, such as calcineurin, which is Ca^{2+}-dependent phosphatase, and Ca^{2+}/calmodulin-dependent kinase II (CaMKII) [88]. The dynamic balance between these enzymes is presumed to be important for synaptic plasticity. It was demonstrated that LMW AβOs may inhibit CaMKII activity and thus disrupt the equilibrium between above mentioned enzymes [95]. In addition, activation of NMDAR by soluble AβOs involves Ca^{2+}-mediated mitochondrial dysfunction as well as decreased CaMKII levels at synapses. This results in dramatic loss of synaptic proteins such as postsynaptic density-95 (PSD-95), dynamin-1 and synaptophysin [96].

NMDAR may also mediate the toxic impact of AβOs on glucose metabolism in neurons. AMP-activated kinase (AMPK) is a key enzyme in energy sensing and metabolic reprogramming under cellular energy restriction, which is associated with some peripheral metabolic diseases, including diabetes. An impaired AMPK function has been linked recently to certain neurological disorders, such as AD [97]. The intracellular ATP levels and AMPK activity were decreased in cultured hippocampal neurons already after short-term exposure to AβOs. This AβO-dependent reduction in AMPK activity is also mediated by glutamate receptors NMDARs, which results in removal of glucose transporters (GLUTs) from the surfaces of hippocampal neurons [97].

3.2. Glutamate Receptor AMPAR

The α-amino-3-hydroxy-5-methyl-4-isoxazolepropionic acid receptor (AMPAR) is also a glutamate ionotropic transmembrane receptor that mediates synaptic transmission in CNS. AMPAR is classified as a non-NMDA-type receptor. AMPARs are tetrameric receptors composed of four subunits, labelled as GluA1, GluA2, GluA3, and GluA4. Subunits GluA1 and GluA2 play an important role in synaptic plasticity and LTP [98]. The permeability of AMPAR to Ca^{2+} is related to the GluA2 subunit [99]. Phosphorylation of AMPARs influences ion channel localization and its conductance. Subunit GluA1 has four known phosphorylation sites, but serine 845 (S845) is a residue that plays an essential role in the trafficking of AMPARs toward extrasynaptic sites [100].

Aβ oligomers may cause synaptic dysfunction also by inducing calcineurin-dependent internalization of AMPAR [101]. It was shown in cortical cultures that soluble oligomers of Aβ, but not monomers, mediate the internalization of the AMPAR subunits GluA1/GluA2 by endocytosis [102]. Short-term exposure of hippocampal neurons to AβOs led to noticeable removal of AMPARs from postsynaptic surface and to impaired insertion of this receptor during synaptic potentiation [103]. It is in accordance with the finding that acute exposure of cultured neurons to soluble AβOs induced AMPAR ubiquitination associated with the removal of this receptor from the plasma membrane [104]. Aβ oligomers reduce basal levels of S845 phosphorylation and surface expression of AMPARs affecting AMPAR subunit composition contributing to early synapse dysfunction in a transgenic mouse model of AD [105].

Binding of AβOs to neurons occurs in dendritic spines expressing AMPARs, preferentially GluA2, which is calcium impermeable [106]. Furthermore, pharmacological inhibition of AMPARs leads to reduced AβOs binding. It was demonstrated that the process of rapid internalization of AβOs with surface AMPAR subunits is mediated by calcineurin, whereas inhibition of this phosphatase and AMPARs prevents AβOs-induced synaptic disruption and spine loss [106].

Whereas the role of AMPARs in hippocampal pyramidal neurons containing GluA1 and GluA2 subunits (GluA1/2) has been extensively examined, the importance of AMPAR type having GluA2 and GluA3 (GluA2/3) for synapse physiology was not clear. It was recently revealed that activation of GluA3 AMPARs may constitute novel type of plasticity at synapses [107]. Animal studies shown that in basal conditions GluA2/3 AMPARs are in low-conductance state, shifting to a high-conductance GluA2/3 channels with increased intracellular cyclic AMP (cAMP) levels, which led to synaptic potentiation [107]. It was also indicated that some forms of LTP, such as vestibulo-cerebellar motor learning, may rather require GluA3-AMPARs activation by increasing single-channel conductance mediated by cAMP signaling [108].

Furthermore, the presence of GluA3-containing AMPARs may be also relevant for synaptic and cognitive deficits mediated by AβOs. It was shown in experiments in AD mouse models that all the effects on synapses and memory mediated by soluble oligomeric clusters of Aβ required presence of AMPA receptor subunit GluA3 [109]. Moreover, AβOs blocked synaptic LTP only in neurons expressing this subunit, whereas GluA3-deficient hippocampal neurons were resistant to toxic AβO activity, such as synaptic depression and spine loss. What is important, mice lacking GluA3 subunit did not express memory impairments [109].

Interestingly, abnormal Tau phosphorylation may contribute to AβOs-induced signaling deficits of AMPAR [110]. AβOs led to abnormal Tau distribution in dendritic spines in cultured rodent hippocampal neurons. Aberrant Tau localization was dependent on the phosphorylation of this protein and resulted in early cognitive deficits and synaptic loss [110].

3.3. Metabotropic Glutamate Receptor 5 mGluR5

Glutamate is one of the main excitatory neurotransmitters in human CNS and glutamatergic neurotransmission is involved in most aspects of normal human brain function [111]. This neurotransmitter signals through ligand-gated ion channels, such as AMPAR, or through metabotropic glutamate receptors (mGluRs), a family of several G protein-coupled receptors. Two principal signal transduction pathways involving mGluRs are known: cAMP and phosphatidylinositol signal pathways [112].

Metabotropic glutamate receptor 5 (mGluR5) belongs to group I of metabotropic glutamate receptors and activates phospholipase C. This type of receptor has been implicated in a diverse variety of physiological neuronal functions. Moreover, mGluR5 acts postsynaptically as a co-receptor for AβO [113]. Soluble extracellular AβOs bind to lipid-anchored cellular prion protein (PrP^C) with high affinity and specificity [114,115]. The coexpression of mGluR5 allows PrP^C-bound AβO for activation of intracellular Fyn kinase, what results in the disruption of synapses [113]. Complexes of AβO with PrP^C generate mGluR5-mediated influx of Ca^{2+} in neurons. This influx may be also driven by human AD brain extracts. Aβ peptides also disturb intracellular Ca^{2+} homeostasis. It was demonstrated that Aβ42 oligomers, but not monomers, significantly altered Ca^{2+} release from intracellular stores, which involved mGluR5 and required network activity [116]. In addition, dendritic spine loss is also mediated by AβO-PrP^C-mGluR5 complexes signaling pathway [113].

3.4. Cellular Prion Protein PrP^C

It seems that significant part of AβO toxicity in AD may be mediated after initial interaction with PrP^C on the neuronal surface. In the normal brain, the expression of PrP^C is controlled by a feedback loop with amyloid intracellular domain (AICD). PrP^C inhibits the activity of β-secretase (β-site APP cleaving enzyme-1, BACE1) [117] as well as AICD production [118], whereas AICD upregulates PrP^C expression, which maintains the inhibitory effect of PrP^C on BACE1.

This reaction is disrupted in AD, resulting in the binding of increased levels of AβOs to PrP^C and disturbed regulation of BACE1 activity. Moreover, PrP^C inhibits formation of fibrillar aggregates of Aβ, trapping this peptide in an oligomeric state [119]. Only recently it was demonstrated that PrP^C specifically inhibits elongation of Aβ fibrils by binding to the ends of growing polymers [120]. It was shown that this inhibitory effect requires the globular C-terminal domain of PrP^C, which suggests that PrP^C might recognize specific structure that is common to the ends of both oligomeric and fibrillar form of Aβ [120]. This interaction could probably contribute to the neurotoxicity of AβOs.

As it was mentioned above, cellular prion protein PrP^C was identified as AβO co-receptor, although the infectious form PrP^{Sc} conformation is not required [115]. PrP^C binds Aβ42-oligomers with high affinity and high selectivity. Purified recombinant PrP^C interacted directly with AβOs, whereas the binding of synthetic AβOs to neurons decreased in PrP^C-null mice.

Moreover, PrP^C mediates impairment of synaptic plasticity by AβOs [115]. The effects of interaction between PrP^C and AβOs on LTP were compared between wild-type and PrP^C-null mice [115]. It was shown that soluble AβOs reduced LTP in the wild-type mice, but not in the PrP^C-null mice. It may indicate that PrP^C is required to mediate these toxic effects of AβOs. It was also demonstrated that binding onto PrP^C induces intracellular Ca^{2+} increase in neurons via the complex PrP^C-mGluR5, with harmful effects on synaptic transmission [121].

Although additional receptors may contribute to mediation of AβO action, recent investigations indicate that PrP^C supposedly plays a primary role (reviewed by Del Rio [122]). PrP^C is a glycosylphosphatidylinositol (GPI)-anchored protein. Thus, the mediation of the signal transduction requires the formation of complexes between PrP^C and certain transmembrane proteins, such as acetylcholine and glutamate receptors [113,123–126]. The complex PrP^C-mGluR5 plays an important role in AβO binding and activity of oligomers in neurons. The signal transduction downstream of AβO-PrP^C complexes involves mGluR5, as well as kinases Fyn and Pyk2 [113]. Additionally, after AβOs to binding PrP^C and activation of Fyn tyrosine kinase, NMDARs are phosphorylated, which in turn results in altered surface expression, dysregulation of receptor function, excitotoxicity, and dendritic spine retraction [113]. This mechanism is consistent with previous discovery that Fyn is essential for AβO-induced synaptotoxicity [64,127].

Interestingly, it was shown that PrP^c also appears to be relevant in α-synucleopathies, such as PD, participating in α-synuclein binding and brain spreading [122].

3.5. β$_2$-Adrenergic Receptors

The β$_2$-adrenergic receptors (β2ARs) are expressed in the brain, especially in regions involved in AD pathogenesis, i.e. hippocampus and cortex [128]. β2ARs play an important role in cognitive functioning. The activation of β2ARs is essential for normal learning and memory [129,130]. Stimulation of β2ARs promotes synaptic LTP in dentate gyrus and hippocampus [131–136]. The role of β2AR in memory formation may be confirmed by enhanced expression of β2ARs in dendritic spines [137,138]. β2AR roles in brain are associated with the AMPA-type glutamate receptor [137].

It was demonstrated that β2ARs activation enhances neurogenesis in APP/PS1 mice, a mouse model of AD [139]. Stimulation of these receptors attenuated memory deficits and reduced Aβ accumulation in mouse brain. Moreover, activation of β2ARs enabled the recovery of memory deficits in APP/PS1 mice, enhanced neurogenesis in the dentate gyrus, restored dendritic branches, and spine density in the hippocampus as well as increased the levels of synapse-associated proteins such as synaptophysin, synapsin 1, and PSD-95 [139]. These findings suggest that activation of β2AR protects synapses in this animal model of AD.

Alterations in β2ARs function have been linked to AD, although the results were not consistent. Decreased levels of β2ARs in certain regions of post-mortem human AD brain, such as locus coeruleus and hippocampus, were demonstrated [140–142]. Activation of β2ARs resulted in enhanced γ-secretase activity and intensified amyloid plaque formation [143], whereas use of β2AR antagonists conversely attenuated production of Aβ induced by acute stress [144].

On the other side, the administration of ICI, a selective β2AR antagonist, enhanced neuropathological changes, such as increased Aβ plaque burden, as well as accumulation of phosphorylated Tau in a mouse model of AD [145]. Moreover, blockade of β2AR led to cognitive deficits in mice. These results suggest that selective pharmacologic inhibition of β2ARs may have negative effects on AD-like pathology in this animal model of AD. It should be highlighted that the link between β2ARs and AD is likely highly complex.

It was shown that human AβOs, when applied to slices of rodent brain, are able to induce the degradation of β2ARs [146,147]. β2AR levels in hippocampal slices were decreased significantly after exposition to AβOs. Although HMW soluble oligomers of Aβ extracted from AD brain had faint or none cytotoxic activity, they dissociated in alkaline environment to smaller, LMW oligomers (approximately 8–70 kDa). Postincubation LMW were much more bioactive. They induced impaired hippocampal LTP, activated brain microglia and led to decrease in the neuronal levels of β2ARs in mice in vivo [148].

3.6. Acetylcholine Receptor α7nAChR

It was shown that Aβ42 binds with high, picomolar affinity to α7 nicotinic acetylcholine receptor (α7nAChR), a neuronal pentameric cation channel [149]. This binding is accompanied with the loss of cholinergic neurons in the brain, resulting in receptor internalization and intracellular accumulation of Aβ [150]. Furthermore, formation of the α7nAChR-Aβ42 complex was suppressed by shorter chains of Aβ (12–28), indicating that this sequence region contains the binding epitope of amyloid [149].

It was also demonstrated in immunohistochemical studies on human sporadic AD brains that α7nAChR is present in neuritic plaques and co-localizes with Aβ in individual cortical neurons [149]. Moreover, the presence of intracellular AβOs was shown in human cholinergic basal forebrain neurons, suggesting the role of amyloid oligomers in cholinergic deficiency [151]. It was confirmed in a triple-transgenic mouse model of AD, where loss of the α7nAChRs was restricted to brain regions that accumulate Aβ intraneuronally [152].

The loss of α7nAChR enhances AβOs accumulation in a mouse model of AD, exacerbating early-stage cognitive decline and septo-hippocampal pathology [153]. In α7nAChR-null mice crossed with those transgenic for mutant human APP, a neurodegeneration in hippocampus and cognitive decline were found already in early, pre-plaque stage of AD. These changes were associated with the appearance of a small, dodecameric form of AβO [153]. Presented findings suggest that α7nAChR plays a protective role for AβOs toxicity. What is more, restoring LTP impaired by AβOs is possible by using a selective neuronal nicotinic receptor partial agonist SSR180711, which completely rescued both early and late LTP impaired by Aβ42 oligomers [154].

3.7. Insulin Receptor

A pathophysiological connection between AD and diabetes was confirmed in numerous studies (reviewed by de Felice [155]). An increasing body of evidence indicates that AD may be called a "brain-specific form of diabetes" or "type 3 diabetes" [156,157]. Both diseases are characterized by key pathological features such as insulin resistance, inflammation, and altered metabolism. Diabetic pathophysiology includes reduction in brain insulin signaling, decreased levels of brain insulin, and elevated levels of glucose.

Toxic activity of AβOs may be also linked with impaired insulin signaling and brain insulin resistance, which lead to elevated Aβ production and reduced AβO clearance, resulting in oligomers' deposits in the brain and neuronal damage [158]. Furthermore, it was shown that signal transduction by neuronal insulin receptors (IRs) is highly sensitive to soluble AβO. AβOs themselves can influence IRs and decrease brain insulin signaling [158]. In addition, AβOs bind to neuronal IRs and affect its insulin-induced autophosphorylation, preventing activation of specific kinases required for LTP [159]. In cultures of mature hippocampal neurons, soluble oligomers caused a rapid, substantial loss of surface IRs, especially on dendrites bound by AβOs [106].

3.8. p75 Neurotrophin Receptor p75NTR

It was suggested that AβOs may induce neuronal death via nerve growth factor (NGF) receptor by alteration of NGF-mediated signaling in cultured cells [75,160]. NGF mediates cell loss through low-affinity receptor for nerve growth factor, also called p75 neurotrophin receptor (p75NTR), which belongs to the tumor necrosis factor (TNF) receptor superfamily [74]. Precisely, toxic effects of Aβ mediated by p75NTR depend on a death domain in the cytoplasmic part of this receptor molecule [161,162]. Synapse targeting of AβOs involves activation of p75NTR. AβOs, together with PrPC, bind at the membrane receptors, forming annular amyloid pores and ion channels to induce aberrant cytoskeletal changes in dendritic spines [96].

In the mouse hippocampus, the expression of p75NTR induced by AβOs involves insulin-like growth factor 1 receptor (IGF-1R) signaling [163]. Significantly elevated hippocampal expression of membrane-associated p75NTR protein was shown in transgenic AD mice and was associated with the age-dependent increase of Aβ42 levels. Moreover, it was demonstrated that microinjections of AβOs induced p75NTR expression in the hippocampus through phosphorylation of IGF-1R, whereas co-administration of IGF-1R inhibitor blocked AβOs-induced overexpression of p75NTR [163].

Conflicting evidence exists regarding the role of p75NTR in AD, especially against toxicity of AβOs. Although an important role of p75NTR in Aβ metabolism and Aβ-mediated neurodegeneration in AD brains was shown, this protein also promotes the differentiation and survival of vertebrate neurons [164]. Furthermore, a conflicting role of p75NTR in the cytotoxic function of Aβ depends on the different state of this peptide. In fact, the neurotoxicity of the two forms of Aβ, insoluble fibrillar or soluble oligomeric form, occurs with different mechanisms. Primarily, it was proved that the expression of p75NTR is required for cell death by fibrillar form of Aβ [165]. Interestingly, the toxicity of fibrillar Aβ species is strictly dependent on p75NTR, whereas neurotoxicity of soluble AβOs is independent of p75NTR and is even decreased by the presence of this receptor. Moreover, the expression of p75NTR protects against the neurotoxicity of oligomers [166]. This protective effect results from an active function of the juxtamembrane sequence of the cytoplasmic region of p75NTR and is mediated by phosphatidylinositide 3-kinase (PI3K) activity [166]. These results suggest that p75NTR might have diverse functions in cell death and survival.

3.9. Immunoglobulin and Immunoglobulin-Like Receptors

Human leukocyte immunoglobulin-like receptor B2 (LilrB2) belongs to the subfamily B class of leukocyte immunoglobulin-like receptors (LIR) expressed on immune cells. LilrB2 inhibits stimulation of an immune response, controls inflammatory responses and cytotoxicity, and limits autoreactivity of immune system. LilrB2 binds to major histocompatibility complex (MHC) class I molecules on antigen-presenting cells. It was indicated that MHC class I molecules have additional functions in CNS [167]. Furthermore, numerous MHC class I antigens and their binding partners are found to be expressed in CNS neurons and might be involved in activity-dependent synaptic plasticity [167]. LilrB2 also participates in the process of synaptic plasticity and neurite growth in CNS [168].

Murine homolog of LilrB2, paired immunoglobulin-like receptor B (PirB), is an immune inhibitory receptor, primarily identified in mouse immune cells [169]. Expression of PirB is also observed in subsets of neurons throughout mouse brain. In addition, PirB participates in the inhibition of axonal regeneration [170,171]. It was also suggested that PirB plays an important role in age-related hippocampal aging, synaptic loss and neurotransmitter release, which causes cognitive dysfunction associated with AD [172].

Importantly, murine PirB and its human orthologue LilrB2 are thought to be nanomolar affinity receptors for Aβ oligomers [168]. The interaction between AβOs and PirB/LilrB2 are mediated by the first two extracellular immunoglobulin domains of the receptors [168]. PirB regulates synaptic plasticity, affecting hippocampal LTP, which contributes to Aβ-induced deficits of memory in a mouse model of AD [168]. Moreover, high PirB expression is required for the harmful effect of AβOs on hippocampal formation [168]. In double transgenic APP/PS1 mice, ocular dominance plasticity (ODP) was defective during the very early period of synaptic plasticity development [173]. This observation is in contrast with enhanced ODP during the critical period and in adult mice lacking PirB [174]. It suggests that impaired ODP is one of the earliest Aβ-induced deficits in a mouse model of AD. While Aβ42 oligomers robustly bound to LilrB2-expressing heterologous cells, only a minimal binding of monomeric Aβ42 to LilrB2 was observed [168], which suggests selectivity of AβOs reaction with this receptor. Although similar levels of LilrB2 were detected either in human AD brains or in specimens from non-demented adults, downstream signaling was altered in AD specimens.

It was suggested that FcγRIIb (Fragment crystallizable gamma receptor II b) may also play a role as a AβOs receptor, mediating neurodegeneration and toxic activity of oligomers [175]. FcγRIIb belongs to family of Fc-gamma receptors (FcγR) which have a binding specificity for the Fc (Fragment, crystallizable) region of immunoglobulin gamma (IgG) [176]. They are present on the surface of B lymphocytes, dendritic cells, natural killer cells, macrophages, granulocytes, mast cells, and other cells of the immune system. Additionally, all of the Fcγ receptors (FcγR) belong to the immunoglobulin superfamily and differ in their affinities for IgG due to variegated molecular structure of different IgG subclasses.

It was demonstrated that FcγRIIb is an important factor contributing to the AβOs' neurotoxicity and memory impairment. This protein was significantly upregulated in the hippocampus of AD brains and neuronal cells exposed to synthetic Aβ [175]. Soluble Aβ oligomers interacted with FcγRIIb both in vitro and in AD brains, whereas inhibition of that interaction blocked neurotoxicity of synthetic AβO. Moreover, in mouse model of AD, genetic depletion of FcγRIIb rescued memory impairments and prevented AβO-induced inhibition of LTP, which supports an idea that this receptor could play an essential role in Aβ-mediated neuronal dysfunction [175].

3.10. Triggering Receptor Expressed on Myeloid Cells 2 TREM2

Triggering receptor expressed on myeloid cells 2 (TREM2) is a transmembrane-glycoprotein receptor that is present on the surface of immune cells of myeloid origin [177]. As a lipid-sensing activating receptor, TREM2 binds to phospholipids, apolipoproteins, and lipoproteins through its immunoglobulin-like domain [178]. Moreover, TREM2 interacts with TYRO protein tyrosine kinase-binding protein, also known as DNAX-activating protein of 12 kDa (DAP12), which is an adapter protein for this receptor. In the brain, this interaction triggers the phagocytosis of apoptotic neurons and Aβ peptide in microglia with no inflammatory effects [179].

It was shown that certain coding variants in *TREM2* gene are associated with increased risk for AD [180], which suggests that immune cell dysfunction may also play a role in AD pathogenesis. Normal proteolytic maturation of full-length TREM2 at the plasma membrane is disturbed in mutations of *TREM2* gene, resulting in impaired phagocytosis, which may contribute to the pathogenesis of AD [179].

In normal conditions, TREM2 directly binds to AβOs with nanomolar affinity. Only recently, it was demonstrated that in AD-associated TREM2 mutations this binding is reduced [181]. Moreover, the degradation of Aβ in primary microglial culture and mouse brain was impaired in TREM2 deficiency, resulting in microglial depolarization, induction of K^+ current into cells as well as increased cytokine expression and secretion, cells migration, proliferation, apoptosis, and morphological changes are dependent on TREM2 [182]. Additionally, TREM2-DAP12 interaction was enhanced by AβOs, which demonstrates that TREM2 may act as a microglial AβO receptor that mediates physiological and AD-related pathological effects [181].

3.11. Tyrosine Kinase Ephrin Receptors Eph4A and EphB2

It was suggested that the tyrosine kinase Eph receptors may also play a role in AβOs-induced synaptotoxicity [183]. Eph receptors were named after the cell line from which the cDNA was first isolated, erythropoietin-producing hepatocellular carcinoma. Based on the affinities for binding ligands and similarity of extracellular domain sequences, they are divided into two functionally different groups: EphA and EphB. There are nine EphA receptors (1–9), which bind to ephrin-A ligands (ephrin-A 1–5), proteins anchored to the cell membrane by GPI motif, whereas five EphB receptors (1–5) bind to ephrin-B ligands (ephrin-B 1–3) with a transmembrane domain and a short cytoplasmic region [184].

Eph receptors and their ligands play a key role in the physiological functioning, development and maturation of nervous system [183]. Since Eph ligands and receptors are both membrane-bound proteins, the Eph/ephrin binding and activation of their intracellular signalling pathways may occur via direct cell-to-cell interactions only. In particular, their presence both in pre- and postsynaptic regions is necessary for the development and stabilization of synapses, although EphB and EphA play opposite roles [185].

It was shown, that EphB promotes morphogenesis and growth of dendritic spines, whereas their development is aberrant in the absence of these receptors in hippocampal neurons [186]. Moreover, the formation of synapses is induced by activation of EphB2 receptor via interaction with NMDAR [187].

EphB receptors are also important factors in the pathophysiology of AD and other neuropathologies [188]. It was demonstrated that EphB2 levels in the membrane of hippocampal neurons were decreased after short term treatment of AβOs [189], which could be a result of NMDAR activation [190]. It was also shown that AβOs binding to the fibronectin (FN) type III repeat domain of EphB2 triggers to endocytosis of this receptor and its degradation in the proteasome [191]. The results of EphB2 degradation are the impairment of NMDAR functioning and cognitive deficits. What is interesting, these interaction sites of the EphB2 FN domain with AβOs may be blocked by a small, 10 amino acids length peptide Pep63, which rescued memory deficits in mouse model of AD [192]. These results suggest that inhibition of EphB2-AβOs interactions may be a promising strategy for AD treatment. Furthermore, it was shown induction of the EphB2 expression in the dentate gyrus prevented the cognitive deficits and LTP impairments in mice model of AD [106]. On the other hand, the decrease of AMPAR and NMDAR levels induced by AβOs may be prevented by overexpression of EphB2. This protective effect could be directly related to PDZ-binding motif of EphB2 [191,193,194].

EphA receptors bind membrane-bound ephrinA family ligands residing on adjacent cells, leading to contact-dependent bidirectional signalling [195]. EphA4 receptor plays also an important role in the regulation of synapses functioning in the nervous system and in the repairment after injury, preventing axonal regeneration as well as in the angiogenesis and formation of vessels within central nervous system [183].

EphA4 mediates dendritic spine remodeling and contributes to homeostatic plasticity through the regulation of AMPAR levels [195–197]. Activation of EphA4 receptor induces a decrease in the strength of the excitatory synapse as well as reduction of spine length and density in hippocampal slices [195]. This receptor is also associated with the loss of dendritic spines, their retraction and growth cone collapse [195]. It was confirmed in EphA4-knockout mice, which expressed disorganized, longer, and more numerous spines than wild-type mice [183].

Moreover, EphA receptors seems to be key player in the pathophysiology of AD and other neuropathologies, such as motor neuron degeneration in amyotrophic lateral sclerosis (ALS) [198,199]. It was shown in AD brains, that hippocampal distribution of EphA4 was co-localized with neuritic plaques already at early stages (Braak stage II), which suggests that EphA4 may contribute to synaptic dysfunction [200]. Additionally, it was demonstrated that levels of EphA4 mRNA in synaptoneurosomes from AD patients were twofold higher than in non-demented controls [201].

Furthermore, AβOs aberrantly activate EphA4 leading to dendritic spine elimination, whereas blockade or absence of this receptor in hippocampal neurons prevents synaptic loss [202,203]. This AβOs-EphA4 axis involves c-Abl tyrosine kinase activation by AβOs in dendritic spines of cultured hippocampal neurons, which is required for AβOs-induced synaptic loss [183,203].

3.12. Receptor for Advanced Glycation Endproducts RAGE

RAGE is a small, 35 kDa transmembrane protein, which belongs to the immunoglobulin superfamily and plays a role in innate immunity. RAGE is composed of three extracellular Ig-like domains (Vd, C1d, C2d), with a single transmembrane domain, and a short cytoplasmic tail [204]. This receptor is described as a "pattern recognition" receptor because of its ability to recognize common structural motifs. RAGE is able to bind multiple ligands, such as advanced glycation endproducts (AGE), glycans and glycoproteins, as well as chromatin protein high mobility group box 1 protein (HMGB-1), calprotectin and S100B [205]. After stimulation, RAGE activates certain pro-inflammatory genes, which mediate Aβ-induced oxidative stress.

Moreover, activation of RAGE results in continuous instigation of nuclear factor kappa-light-chain-enhancer of activated B cells (NF-κB) [206,207]. On the other hand, RAGE itself is upregulated by NF-κB, thus creating RAGE/NF-κB axis. This forms a positive feed-back loop leading to chronic inflammation, altered micro- and macrovasculature, and tissue damage, pathological events observed also in AD.

Expression of RAGE is increased in the AD brain. Moreover, enhanced levels of RAGE ligands were observed in a range of inflammatory diseases, atherosclerosis, diabetes and cancer as well as in AD, which suggests a causative role of this receptor in inflammatory chronic state [207]. RAGE was also identified as one of the cell-surface binding sites for Aβ peptide at the plasma membrane of neurons, microglial cells, and endothelial cells of the vessel wall [206].

Furthermore, RAGE participates in the clearance of Aβ. The level of Aβ peptides as well as other substances in the brain is not only a result of specific equilibrium between their synthesis and degradation, but also depends on the transport into the brain from blood and efflux from the brain into blood through blood-brain barrier (BBB). Both in normal aging and in AD, the rate of CSF reabsorption into the blood, known as bulk flow, is also impaired. The main receptors for the transport of Aβ across BBB are RAGE and low-density lipoprotein receptor related protein-1 (LRP1). RAGE is responsible for Aβ influx, whereas LRP1 is the main receptor controlling the efflux across the BBB to the plasma [208]. Soluble form of LRP1 sequesters 70–90% of plasma Aβ peptides in normal conditions. In AD, this Aβ clearance is disturbed [206]. Both in AD mouse models and in AD patients, the brain endothelial expression of RAGE is elevated, whereas plasma levels of sLRP1 and its Aβ-binding capacity are decreased, leading to increase in free Aβ fraction in plasma [209].

Interestingly, distinct regions of RAGE are induced by different Aβ conformations in AD-related apoptosis [204]. It was demonstrated that anti-RAGE antibodies significantly improved survival of cortical rat neurons and RAGE-expressing cells exposed to either AβOs or aggregated Aβ. Moreover, the use of site-specific antibodies against domain Vd of this receptor prevented AβOs-induced neurotoxicity, whereas blockade of the apoptosis induced by aggregated Aβ required neutralization of C1d domain of RAGE [204].

3.13. Megalin Receptor

Megalin, also known as glycoprotein 330 (gp330) or low-density lipoprotein-related protein 2 (LRP2), is a large, approximately 600 kDa protein, which is a multiligand binding receptor expressed in the plasma membrane of epithelial cells [210]. In CNS, megalin is present in choroid plexus epithelium and ependymal cells covering the brain ventricles. LRP2 is a member of a family of receptors with structural similarities to the low-density lipoprotein receptor (LDLR). Megalin mediates endocytosis of its ligands, which results in the degradation in lysosomes or transcytosis [211]. This receptor has been shown to interact with various ligands, vitamin-binding proteins, carrier proteins, lipoproteins, hormones and hormone precursors, as well as drugs and toxins [212]. Moreover, this receptor has also functions in cellular communication and signal transduction, with PSD-95 as an interaction partner [213].

LRP2 is also an endocytic receptor for apolipoprotein J (ApoJ)/clusterin involved in rapid receptor-mediated uptake or bidirectional exchanges of soluble Aβ across BBB [214]. ApoJ has been revealed to be the major protein binding Aβ in CSF. Megalin mediates cellular uptake and transport of ApoJ alone and ApoJ complexed with Aβ-40, the most abundant amyloid isoform found in Aβ deposits of the blood vessels, from the periphery into the brain at the cerebral vascular endothelium and choroid epithelium [215]. This interaction of ApoJ-Aβ complex with megalin is thought to be another, besides RAGE and LRP1, mechanism preventing pathological accumulation of Aβ [216]. It was shown that Aβ alone did not bind directly to LRP-2, whereas complexes of Aβ-40 with apoJ were able to react with megalin. Moreover, ApoJ/Aβ binding interaction was blocked polyclonal anti-megalin antibodies, which supports the role of LRP-2 as a mediator of the clearance of ApoJ/Aβ complex from CSF and in the regulation of Aβ accumulation [216].

3.14. Nuclear Receptors

Nuclear receptors (NRs) constitute a class of proteins that mediate certain, relatively small, molecules pathways, thus controlling the development, homeostasis, and various metabolic processes. There are currently 48 nuclear receptors known in the human genome, most of them have identified specific ligands [217]. NRs are involved in the synthesis and metabolism of steroid and thyroid hormones as well as and various other lipid-soluble signals, including retinoic acid, oxysterols, vitamin D, cholesterol, lipids, and bile acids or thyroid hormone [217].

Moreover, many of NRs, but not all, directly bind to signalling molecules. These molecules are small and have lipophilic character, therefore they can easily enter the target cell. Thus, unlike described above membrane-bound receptors, NRs are intracellular proteins which are capable of direct binding to DNA, thus controlling the expression of adjacent genes, which is their unique property that differentiates them from other classes of receptors. Because of this ability, NRs are classified as transcription factors [218].

The nuclear receptor superfamily may be divided according to their amino acid sequence similarities in six subfamilies, which are thyroid hormone receptor-like, retinoid x receptor-like, estrogen receptor-like, nerve growth factor IB-like, steroidogenic factor-like and germ cell nuclear factor-like [217]. Moreover, some NRs require heterodimerization with retinoid X receptor (RXR) [217]. Furthermore, NRs-ligands interactions are characterized by certain redundancy: ligands are nonselective for particular receptors, which also share their transcriptional targets, serving as transcriptional inducers of one another. However, several NRs remain with unknown ligands and are described as "orphan receptors" [219].

All NRs are conservative and similar in their general structure, which includes a ligand-binding/dimerization domain (LBD) and DNA-binding/weak dimerization domain (DBD) as well as at least one N-terminal ligand-independent transactivation region, referred to as AF-1 for activation function 1 (or the A/B domain), and a ligand-dependent transcription region AF-2. To bind DNA, AF-2 may form complexes with co-regulatory proteins that can act as co-activators or co-repressors. AF-2 co-activators regulate histone acetyltransferase activity, whereas its co-repressors control histone deacetylase activity [218].

Certain NRs are also linked with AD pathology as well as with AβOs toxicity. One of these receptors is vitamin D receptor (VDR) that is broadly expressed in brain and regulates many genes. VDR mediates action of Vitamin D (1,25-$(OH)_2D_3$), an important neurosteroid, which plays key role in the brain functioning, such as calcium signaling, cell proliferation and differentiation. Vitamin D is also a neurotrophic factor that regulates neurotransmission and synaptic plasticity. It was revealed recently, that Vitamin D treatment results in significant increase of LRP1 expression both in-vivo and in-vitro studies [220]. Moreover, it was suggested that VDR deficiency/inhibition can be a potential risk factor for AD [221]. It was shown that Vitamin D may be also involved in Aβ clearance. 1,25-$(OH)_2D_3$ increases transport of Aβ across the BBB by regulating expression of amyloid transporters, such as LRP-1, via its nuclear receptor VDR only, or by binding heterodimeric complexes of VDR with RXR [222].

3.15. Sirtuin

Sirtuin 1 (SIRT 1), one of NRs that has recently emerged as a crucial protein that may play protective roles in AD and other NDs, including PD and MND (for review see [223]). SIRT1 belongs to the family of sirtuins (Sir2, silent information regulator 2 protein) that was shown to regulate lifespan in lower organisms and affect diseases of aging in mammals. SIRT1 is a nicotinamide adenine dinucleotide (NAD^+)-dependent histone deacetylase involved in calorie restriction (CR) (reviewed in [224]). Calorie restriction promotes mammalian cell survival by inducing the SIRT1 deacetylase. As it was mentioned above, it was proposed that AD may be described as new form of diabetes or "type 3 diabetes". The resistance to insulin and insulin-like growth factor are thought to be crucial for the progression of AD [225]. SIRT1 deficiency is also ascribed to be responsible for the increased risk of insulin resistance, obesity and diabetes, including type 3 diabetes, whereas low-calorie diet and nutrition reverse type 3 diabetes and accelerated aging linked to global chronic diseases [226].

SIRT1 is involved in neurodevelopment, including axon elongation, neurite outgrowth and dendritic branching [223]. Furthermore, this NR is also essential for normal cognitive function and synaptic plasticity. It was demonstrated that SIRT1 attenuates amyloidogenic processing of APP by increasing α-secretase activity via SIRT1-coupled retinoic acid receptor-β (RARβ) activation [227].

Upregulation of α-secretase shifts APP processing to non-amyloidogenic cleavage of APP and reduces the pathological accumulation of the toxic Aβ species that results from β- and γ-secretase activity. It may be confirmed by the fact that a significant decrease in SIRT1 level, both mRNA and protein, was observed in the cortex of AD patients [228]. SIRT1 reduction paralleled tau accumulation in the AD brain and may be closely associated with deposition of Aβ in the cerebral cortex of patients with AD.

Although it is difficult to determine when exactly SIRT1 loss occurs in AD, it was suggested that it may be rather a relatively late event. A significant correlation was observed between SIRT1 and the duration of AD symptoms, accumulation of tau, as well as Aβ42 deposition [228]. Furthermore, it seems that AβOs toxicity and their binding to AβO receptors are rather primary toxic effects, than secondary to NRs disturbances with consecutive AβOs receptor interactions.

4. Conclusions

Since its first description over hundred years ago, Alzheimer's disease is one of the diseases in modern biomedicine that have garnered most scientific attention. Within these 100 years there have emerged various hypotheses in order to explain underlying pathology. The dominant model of AD pathogenesis is amyloid hypothesis, although its details were changing over this time, indicating increasing role of oligomeric amyloid beta species as the main toxic factors leading to damage of neurons and loss of synapses. Recent studies have identified that soluble Aβ oligomers interact with certain receptor proteins.

In conclusion, a variety of specific receptors could be responsible for mediating the synaptotoxicity caused by AβOs in AD (Table 1). The AβO-associated receptors include ionotropic and metabotropic glutamate receptors NMDAR, AMPAR, and mGluR, their co-receptor—cellular prion protein PrPc, ephrin receptors EphB2 and EphA4, RAGE, immunoglobulin and immunoglobulin-like receptors FcγRIIB and PirB/LiL2R, neurotrophin receptor p75NTR, β-adrenergic as well as acetylcholine receptors a7nAChRs. Despite over twenty various protein receptors proposed within over twenty years of amyloid hypothesis, no single candidate receptor has been revealed to be necessary and sufficient to account for all features of AβO toxic activity. Taken together, it seems that among this abundancy glutamate and Eph receptors could explain most of the pathophysiological defects and structural changes observed in central nervous system. However, further studies are needed to determine the relevance and contribution of each of these molecules to the pathogenesis of this disease.

Table 1. Cellular receptors related to amyloid β oligomer (AβO) activity.

Name of the Receptor	Abbreviation	References
N-Methyl-d-aspartate receptor	NMDAR	[83–89,94,96,97]
α-Amino-3-hydroxy-5-methyl-4-isoxazolepropionic acid receptor	AMPAR	[100,101,103–110]
Metabotropic glutamate receptor	mGluR	[112,113,116]
Cellular prion protein	PrP^c	[113,115,120–122]
β2-Adrenergic receptor	β2AR	[128,140–148]
α7 nicotinic acetylcholine receptor	α7nAChR	[149–154]
Insulin receptor	IR	[158,159]
p75 neutrotrophin receptor	p75NTR	[96,163–166]
Human leukocyte immunoglobulin-like receptor B2	LilrB2	[167,168]
Paired immunoglobulin-like receptor B	PirB	[168,172]
Fragment crystallizable gamma receptor II b	FcγRIIb	[175,176]
Triggering receptor expressed on myeloid cells 2	TREM2	[181,182]
Tyrosine kinase ephrin type-A receptor 4	Eph4A	[183,202,203]
Tyrosine kinase ephrin type-B receptor 2	EphB2	[183,188–194]
Receptor for advanced glycation endproducts	RAGE	[204,206–209]
Megalin (glycoprotein 330, low density lipoprotein-related protein 2)	gp330, LRP2	[210–216]

Int. J. Mol. Sci. **2018**, *19*, 1884

Funding: This research was funded by grants for neurodegenerative diseases, Medical University of Białystok, Poland. This research received no external funding.

Acknowledgments: B.M. has received consultation and/or lecture honoraria from Roche, Cormay and Biameditek. P.L. received research support from the Innovative Medicines Initiative Joint Undertaking under grant agreement n° 115372, resources of which are composed of financial contribution from the European Union's Seventh Framework Programme (FP7/2007–2013) and EFPIA companies' in kind contribution, and he received consultation and lectures honoraria from Innogenetics/Fujirebio Europe, IBL International, AJ Roboscreen, and Roche.

Conflicts of Interest: The authors declare no conflict of interest.

Abbreviations

AβOs	Amyloid β oligomers
AD	Alzheimer's disease
NDs	Neurodegenerative diseases
Aβ	Amyloid beta
NFTs	Neurofibrillary tangles
PD	Parkinson's disease
MND	Motor neuron diseases
HD	Huntington's disease
SCA	Spinocerebellar ataxia
SMA	Spinal muscular atrophy
pTau	Tau protein
CSF	Cerebrospinal fluid
PS	Presenilin
APP	Amyloid precursor protein
FAD	Familial Alzheimer's disease
APOE	Apolipoprotein E
PSP	Progressive supranuclear palsy
CBD	Corticobasal degeneration
FTDP-17	Frontotemporal dementia and Parkinsonism linked to chromosome 17
LTP	Long-term potentiation
LTD	Long-term synaptic depression
LMW	Low molecular weight
HMW	High molecular weight
CNS	Central nervous system
NMDAR	N-methyl-d-aspartate receptor
ROS	reactive oxygen species
CaMKII	Ca^{2+}/calmodulin-dependent kinase II
PSD-95	Postsynaptic density-95
AMPK	AMP-activated kinase
GLUTs	Glucose transporters
AMPAR	α-amino-3-hydroxy-5-methyl-4-isoxazolepropionic acid receptor
mGluRs	Metabotropic glutamate receptors
PrPC	Cellular prion protein
AICD	Amyloid intracellular domain
BACE1	β-site APP cleaving enzyme-1
GPI	Glycosylphosphatidylinositol
β2ARs	β2-adrenergic receptors
α7nAChR	α7 nicotinic acetylcholine receptor
IR	Insulin receptor
NGF	Nerve growth factor
p75NTR	p75 neurotrophin receptor
TNF	Tumor necrosis factor
IGF-1R	Insulin-like growth factor 1 receptor

PI3K	Phosphatidylinositide 3-kinase
LilrB2	Leukocyte immunoglobulin-like receptor B2
LIR	Leukocyte immunoglobulin-like receptors
MHC	Major histocompatibility complex
PirB	Paired immunoglobulin-like receptor B
ODP	Ocular dominance plasticity
FcγR	Fc-gamma receptors
IgG	Immunoglobulin gamma
TREM2	Triggering receptor expressed on myeloid cells 2
DAP12	DNAX-activating protein of 12 kDa
Eph	Erythropoietin-producing hepatocellular carcinoma
FN	Fibronectin
RAGE	Receptor for advanced glycation endproducts
LRP	Low density lipoprotein-related protein
RARβ	Retinoic acid receptor-β
HMGB-1	High mobility group box 1 protein
AGE	Advanced glycation endproducts
NF-κB	Nuclear factor kappa-light-chain-enhancer of activated B cells
gp330	Glycoprotein 330
LDLR	Low density lipoprotein receptor
NR	Nuclear receptor
VDR	Vitamin D receptor
LBD	Ligand-binding/dimerization domain
DBD	DNA-binding/weak dimerization domain
AF-1	Activation function 1
RXR	Retinoid X receptor
SIRT 1	Sirtuin 1

References

1. World Alzheimer Report 2015. Available online: https://www.alz.co.uk/research/WorldAlzheimerReport2015.pdf (accessed on 12 May 2018).
2. JPND Research. Available online: http://www.neurodegenerationresearch.eu/about/what/ (accessed on 12 May 2018).
3. Serrano-Pozo, A.; Frosch, M.P.; Masliah, E.; Hyman, B.T. Neuropathological Alterations in Alzheimer Disease. *Cold Spring Harb. Perspect. Med.* **2011**, *1*, a006189. [CrossRef] [PubMed]
4. De Leon, M.J.; Golomb, J.; George, A.E.; Convit, A.; Tarshish, C.Y.; McRae, T.; de Santi, S.; Smith, G.; Ferris, S.H.; Noz, M.; et al. The radiologic prediction of Alzheimer disease: The atrophic hippocampal formation. *AJNR Am. J. Neuroradiol.* **1993**, *14*, 897–906. [PubMed]
5. Braak, H.; Braak, E. Evolution of the neuropathology of Alzheimer's disease. *Acta Neurol. Scand. Suppl.* **1996**, *165*, 3–12. [CrossRef] [PubMed]
6. Beason-Held, L.L.; Goh, J.O.; An, Y.; Kraut, M.A.; O'Brien, R.J.; Ferrucci, L.; Resnick, S.M. Changes in brain function occur years before the onset of cognitive impairment. *J. Neurosci.* **2013**, *33*, 18008–18014. [CrossRef] [PubMed]
7. Jack, C.R.; Knopman, D.S.; Jagust, W.J.; Petersen, R.C.; Weiner, M.W.; Aisen, P.S.; Shaw, L.M.; Vemuri, P.; Wiste, H.J.; Weigand, S.D.; et al. Tracking pathophysiological processes in Alzheimer's disease: An updated hypothetical model of dynamic biomarkers. *Lancet Neurol.* **2013**, *12*, 207–216. [CrossRef]
8. Reitz, C.; Mayeux, R. Alzheimer disease: Epidemiology, Diagnostic Criteria, Risk Factors and Biomarkers. *Biochem. Pharmacol.* **2014**, *88*, 640–651. [CrossRef] [PubMed]
9. Hardy, J.; Gwinn-Hardy, K. Genetic classification of primary neurodegenerative disease. *Science* **1998**, *282*, 1075–1079. [CrossRef] [PubMed]
10. Holtzman, D.M.; Herz, J.; Bu, G. Apolipoprotein E and apolipoprotein E receptors: Normal biology and roles in Alzheimer disease. *Cold Spring Harb. Perspect. Med.* **2012**, *2*, a006312. [CrossRef] [PubMed]
11. Spinney, L. Alzheimer's disease: The forgetting gene. *Nature* **2014**, *510*, 26–28. [CrossRef] [PubMed]

12. Alzheimer, A. Über eine eigenartige Erkrankung der Hirnrinde. *Allgemeine Zeitschrift fur Psychiatrie und Psychisch-Gerichtliche Medizin* **1907**, *64*, 146–148.

13. Selkoe, D.J.; Hardy, J. The amyloid hypothesis of Alzheimer's disease at 25 years. *EMBO Mol. Med.* **2016**, *8*, 595–608. [CrossRef] [PubMed]

14. Mahley, R.W.; Weisgraber, K.H.; Huang, Y. Apolipoprotein E4: A causative factor and therapeutic target in neuropathology, including Alzheimer's disease. *Proc. Natl. Acad. Sci. USA* **2006**, *103*, 5644–5651. [CrossRef] [PubMed]

15. Strittmatter, W.J.; Saunders, A.M.; Schmechel, D.; Pericak-Vance, M.; Enghild, J.; Salvesen, G.S.; Roses, A.D. Apolipoprotein E: High-avidity binding to beta-amyloid and increased frequency of type 4 allele in late-onset familial Alzheimer disease. *Proc. Natl. Acad. Sci. USA* **1993**, *90*, 1977–1981. [CrossRef] [PubMed]

16. Ma, J.; Yee, A.; Brewer, H.B.; Das, S.; Potter, H. The amyloid-associated proteins a1-antichymotrypsin and apolipoprotein E promote the assembly of the Alzheimer b-protein into filaments. *Nature* **1994**, *372*, 92–94. [CrossRef] [PubMed]

17. Wisniewski, T.; Castano, A.M.; Golabek, A.; Vogel, T.; Frangione, B. Acceleration of Alzheimer's fibril formation by apolipoprotein E in vitro. *Am. J. Pathol.* **1994**, *145*, 1030–1035. [PubMed]

18. Evans, K.C.; Berger, E.P.; Cho, C.G.; Weisgraber, K.H.; Lansbury, P.T. Apolipoprotein E is a kinetic but not a thermodynamic inhibitor of amyloid formation: Implications for the pathogenesis and treatment of Alzheimer disease. *Proc. Natl. Acad. Sci. USA* **1995**, *92*, 763–767. [CrossRef] [PubMed]

19. Polvikoski, T.; Sulkava, R.; Haltia, M.; Kainulainen, K.; Vuorio, A.; Verkkoniemi, A.; Niinistö, L.; Halonen, P.; Kontula, K. Apolipoprotein E, dementia, and cortical deposition of beta-amyloid protein. *N. Engl. J. Med.* **1995**, *333*, 1242–1247. [CrossRef] [PubMed]

20. Rebeck, G.W.; Reiter, J.S.; Strickland, D.K.; Hyman, B.T. Apolipoprotein E in sporadic Alzheimer's disease: Allelic variation and receptor interactions. *Neuron* **1993**, *11*, 575–580. [CrossRef]

21. Castellano, J.M.; Kim, J.; Stewart, F.R.; Jiang, H.; DeMattos, R.B.; Patterson, B.W.; Fagan, A.M.; Morris, J.C.; Mawuenyega, K.G.; Cruchaga, C.; et al. Human apoE isoforms differentially regulate brain amyloid-b peptide clearance. *Sci. Transl. Med.* **2011**, *3*, 89ra57. [CrossRef] [PubMed]

22. Waring, S.C.; Rosenberg, R.N. Genome-Wide Association Studies in Alzheimer Disease. *Arch. Neurol.* **2008**, *65*, 329–334. [CrossRef] [PubMed]

23. Priller, C.; Bauer, T.; Mitteregger, G.; Krebs, B.; Kretzschmar, H.A.; Herms, J. Synapse formation and function is modulated by the amyloid precursor protein. *J. Neurosci.* **2006**, *26*, 7212–7221. [CrossRef] [PubMed]

24. Turner, P.R.; O'Connor, K.; Tate, W.P.; Abraham, W.C. Roles of amyloid precursor protein and its fragments in regulating neural activity, plasticity and memory. *Prog. Neurobiol.* **2003**, *70*, 1–32. [CrossRef]

25. Duce, J.A.; Tsatsanis, A.; Cater, M.A.; James, S.A.; Robb, E.; Wikhe, K.; Leong, S.L.; Perez, K.; Johanssen, T.; Greenough, M.A.; et al. Iron-export ferroxidase activity of β-amyloid precursor protein is inhibited by zinc in Alzheimer's disease. *Cell* **2010**, *142*, 857–867. [CrossRef] [PubMed]

26. De Strooper, B.; Annaert, W. Proteolytic processing and cell biological functions of the amyloid precursor protein. *J. Cell Sci.* **2000**, *113*, 1857–1870. [PubMed]

27. Zheng, H.; Koo, E.H. The amyloid precursor protein: Beyond amyloid. *Mol. Neurodegener.* **2006**, *1*, 5. [CrossRef] [PubMed]

28. Chen, F.; Hasegawa, H.; Schmitt-Ulms, G.; Kawarai, T.; Bohm, C.; Katayama, T.; Gu, Y.; Sanjo, N.; Glista, M.; Rogaeva, E.; et al. TMP21 is a presenilin complex component that modulates gamma-secretase but not epsilon-secretase activity. *Nature* **2006**, *440*, 1208–1212. [CrossRef] [PubMed]

29. Ehehalt, R.; Keller, P.; Haass, C.; Thiele, C.; Simons, K. Amyloidogenic processing of the Alzheimer beta-amyloid precursor protein depends on lipid rafts. *J. Cell Biol.* **2003**, *160*, 113–123. [CrossRef] [PubMed]

30. Vetrivel, K.S.; Cheng, H.; Lin, W.; Sakurai, T.; Li, T.; Nukina, N.; Wong, P.C.; Xu, H.; Thinakaran, G. Association of gamma-secretase with lipid rafts in post-Golgi and endosome membranes. *J. Biol. Chem.* **2004**, *279*, 44945–44954. [CrossRef] [PubMed]

31. Riddell, D.R.; Christie, G.; Hussain, I.; Dingwall, C. Compartmentalization of beta-secretase (Asp2) into low-buoyant density, noncaveolar lipid rafts. *Curr. Biol.* **2001**, *11*, 1288–1293. [CrossRef]

32. Selkoe, D.J. Translating cell biology into therapeutic advances in Alzheimer's disease. *Nature* **1999**, *399*, A23–A31. [CrossRef] [PubMed]

33. Lemere, C.A.; Blustzjan, J.K.; Yamaguchi, H.; Wisniewski, T.; Saido, T.C.; Selkoe, D.J. Sequence of deposition of heterogeneous amyloid b-peptides and Apo E in Down syndrome: Implications for initial events in amyloid plaque formation. *Neurobiol. Dis.* **1996**, *3*, 16–32. [CrossRef] [PubMed]

34. Lemere, C.A.; Lopera, F.; Kosik, K.S.; Lendon, C.L.; Ossa, J.; Saido, T.C.; Yamaguchi, H.; Ruiz, A.; Martinez, A.; Madrigal, L.; et al. The E280A presenilin 1 Alzheimer mutation produces increased Ab42 deposition and severe cerebellar pathology. *Nat. Med.* **1996**, *2*, 1146–1150. [CrossRef] [PubMed]

35. Bateman, R.J.; Xiong, C.; Benzinger, T.L.; Fagan, A.M.; Goate, A.; Fox, N.C.; Marcus, D.S.; Cairns, N.J.; Xie, X.; Blazey, T.M.; et al. Clinical and biomarker changes in dominantly inherited Alzheimer's disease. *N. Engl. J. Med.* **2012**, *367*, 795–804. [CrossRef] [PubMed]

36. Kim, J.; Onstead, L.; Randle, S.; Price, R.; Smithson, L.; Zwizinski, C.; Dickson, D.W.; Golde, T.; McGowan, E. Abeta40 inhibits amyloid deposition in vivo. *J. Neurosci.* **2007**, *27*, 627–633. [CrossRef] [PubMed]

37. Borchelt, D.R.; Thinakaran, G.; Eckman, C.B.; Lee, M.K.; Davenport, F.; Ratovitsky, T.; Prada, C.M.; Kim, G.; Seekins, S.; Yager, D.; et al. Familial Alzheimer's Disease–Linked Presenilin 1 Variants Elevate $A\beta 1$–42/1–40 Ratio In Vitro and In Vivo. *Neuron* **1996**, *17*, 1005–1013. [CrossRef]

38. Shioi, J.; Georgakopoulos, A.; Mehta, P.; Kouchi, Z.; Litterst, C.M.; Baki, L.; Robakis, N.K. FAD mutants unable to increase neurotoxic Abeta 42 suggest that mutation effects on neurodegeneration may be independent of effects on Abeta. *J. Neurochem.* **2007**, *101*, 674–681. [CrossRef] [PubMed]

39. Nistor, M.; Don, M.; Parekh, M.; Sarsoza, F.; Goodus, M.; Lopez, G.E.; Kawas, C.; Leverenz, J.; Doran, E.; Lott, I.T.; Hill, M.; Head, E. Alpha- and beta-secretase activity as a function of age and beta-amyloid in Down syndrome and normal brain. *Neurobiol. Aging* **2007**, *28*, 1493–1506. [CrossRef] [PubMed]

40. Lott, I.T.; Head, E. Alzheimer disease and Down syndrome: Factors in pathogenesis. *Neurobiol. Aging* **2005**, *26*, 383–389. [CrossRef] [PubMed]

41. Jonsson, T.; Atwal, J.K.; Steinberg, S.; Snaedal, J.; Jonsson, P.V.; Bjornsson, S.; Stefansson, H.; Sulem, P.; Gudbjartsson, D.; Maloney, J.; et al. A mutation in APP protects against Alzheimer's disease and age-related cognitive decline. *Nature* **2012**, *488*, 96–99. [CrossRef] [PubMed]

42. Chételat, G. Alzheimer disease: Aβ-independent processes-rethinking preclinical AD. *Nat. Rev. Neurol.* **2013**, *9*, 123–124. [CrossRef] [PubMed]

43. Chételat, G. Reply: The amyloid cascade is not the only pathway to AD. *Nat. Rev. Neurol.* **2013**, *9*, 356. [CrossRef] [PubMed]

44. Vishnu, V.Y. Can tauopathy shake the amyloid cascade hypothesis? *Nat. Rev. Neurol.* **2013**, *9*, 356. [CrossRef] [PubMed]

45. Mudher, A.; Lovestone, S. Alzheimer's disease-do tauists and baptists finally shake hands? *Trends Neurosci.* **2002**, *25*, 22–26. [CrossRef]

46. Goedert, M.; Spillantini, M.G.; Jakes, R.; Rutherford, D.; Crowther, R.A. Multiple isoforms of human microtubule-associated protein tau: Sequences and localization in neurofibrillary tangles in Alzheimer's disease. *Neuron* **1989**, *3*, 519–526. [CrossRef]

47. Goedert, M.; Spillantini, M.G.; Crowther, R.A. Tau proteins and neurofibrillary degeneration. *Brain Pathol.* **1991**, *1*, 279–286. [CrossRef] [PubMed]

48. Iqbal, K.; del C Alonso, A.; Chen, S.; Chohan, M.O.; El-Akkad, E.; Gong, C.X.; Khatoon, S.; Li, B.; Liu, F.; Rahman, A.; et al. Tau pathology in Alzheimer disease and other tauopathies. *Biochim. Biophys. Acta* **2005**, *1739*, 198–210. [CrossRef] [PubMed]

49. Chun, W.; Johnson, G.V. The role of tau phosphorylation and cleavage in neuronal cell death. *Front. Biosci.* **2007**, *12*, 733–756. [CrossRef] [PubMed]

50. Maruyama, M.; Shimada, H.; Suhara, T.; Shinotoh, H.; Ji, B.; Maeda, J.; Zhang, M.R.; Trojanowski, J.Q.; Lee, V.M.; Ono, M.; et al. Imaging of tau pathology in a tauopathy mouse model and in Alzheimer patients compared to normal controls. *Neuron* **2013**, *79*, 1094–1108. [CrossRef] [PubMed]

51. Shankar, G.M.; Li, S.; Mehta, T.H.; Garcia-Munoz, A.; Shepardson, N.E.; Smith, I.; Brett, F.M.; Farrell, M.A.; Rowan, M.J.; Lemere, C.A.; et al. Amyloid-beta protein dimers isolated directly from Alzheimer's brains impair synaptic plasticity and memory. *Nat. Med.* **2008**, *14*, 837–842. [CrossRef] [PubMed]

52. Weksler, M.E.; Relkin, N.; Turkenich, R.; LaRusse, S.; Zhou, L.; Szabo, P. Patients with Alzheimer disease have lower levels of serum anti-amyloid peptide antibodies than healthy elderly individuals. *Exp. Gerontol.* **2002**, *37*, 943–948. [CrossRef]

53. Su, B.; Wang, X.; Nunomura, A.; Moreira, P.I.; Lee, H.G.; Perry, G.; Smith, M.A.; Zhu, X. Oxidative stress signaling in Alzheimer's disease. *Curr. Alzheimer Res.* **2008**, *5*, 525–532. [CrossRef] [PubMed]

54. Kastenholz, B.; Garfin, D.E.; Horst, J.; Nagel, K.A. Plant metal chaperones: A novel perspective in dementia therapy. *Amyloid* **2009**, *16*, 81–83. [CrossRef] [PubMed]

55. Johnson, L.V.; Leitner, W.P.; Rivest, A.J.; Staples, M.K.; Radeke, M.J.; Anderson, D.H. The Alzheimer's Aß-peptide is deposited at sites of complement activation in pathologic deposits associated with aging and age-related macular degeneration. *Proc. Natl. Acad. Sci. USA* **2002**, *99*, 11830–11835. [CrossRef] [PubMed]

56. Tuppo, E.E.; Arias, H.R. The role of inflammation in Alzheimer's disease. *Int. J. Biochem. Cell Biol.* **2005**, *37*, 289–305. [CrossRef] [PubMed]

57. Rebeck, G.W. The role of APOE on lipid homeostasis and inflammation in normal brains. *J. Lipid Res.* **2017**, *58*, 1493–1499. [CrossRef] [PubMed]

58. Vitek, M.P.; Brown, C.M.; Colton, C.A. APOE genotype-specific differences in the innate immune response. *Neurobiol. Aging* **2009**, *30*, 1350–1360. [CrossRef] [PubMed]

59. Lansbury, P.T. Evolution of amyloid: What normal protein folding may tell us about fibrillogenesis and disease. *Proc. Natl. Acad. Sci. USA* **1999**, *96*, 3342–3344. [CrossRef] [PubMed]

60. Kayed, R.; Head, E.; Thompson, J.L.; McIntire, T.M.; Milton, S.C.; Cotman, C.W.; Glabe, C.G. Common structure of soluble amyloid oligomers implies common mechanism of pathogenesis. *Science* **2003**, *300*, 486–489. [CrossRef] [PubMed]

61. Scheff, S.W.; Price, D.A.; Schmitt, F.A.; DeKosky, S.T.; Mufson, E.J. Synaptic alterations in CA1 in mild Alzheimer disease and mild cognitive impairment. *Neurology* **2007**, *68*, 1501–1508. [CrossRef] [PubMed]

62. Lanz, T.A.; Carter, D.B.; Merchant, K.M. Dendritic spine loss in the hippocampus of young PDAPP and Tg2576 mice and its prevention by the ApoE2 genotype. *Neurobiol. Dis.* **2003**, *13*, 246–253. [CrossRef]

63. Koffie, R.M.; Meyer-Luehmann, M.; Hashimoto, T.; Adams, K.W.; Mielke, M.L.; Garcia-Alloza, M.; Micheva, K.D.; Smith, S.J.; Kim, M.L.; Lee, V.M.; et al. Oligomeric amyloid beta associates with postsynaptic densities and correlates with excitatory synapse loss near senile plaques. *Proc. Natl. Acad. Sci. USA* **2009**, *106*, 4012–4017. [CrossRef] [PubMed]

64. Lambert, M.P.; Barlow, A.K.; Chromy, B.A.; Edwards, C.; Freed, R.; Liosatos, M.; Morgan, T.E.; Rozovsky, I.; Trommer, B.; Viola, K.L.; et al. Diffusible, nonfibrillar ligands derived from A 1-42 are potent central nervous system neurotoxins. *Proc. Natl. Acad. Sci. USA* **1998**, *95*, 6448–6453. [CrossRef] [PubMed]

65. Hsia, A.Y.; Masliah, E.; McConlogue, L.; Yu, G.Q.; Tatsuno, G.; Hu, K.; Kholodenko, D.; Malenka, R.C.; Nicoll, R.A.; Mucke, L. Plaque-independent disruption of neural circuits in Alzheimer's disease mouse models. *Proc. Natl. Acad. Sci. USA* **1999**, *96*, 3228–3233. [CrossRef] [PubMed]

66. Nath, S.; Agholme, L.; Kurudenkandy, F.R.; Granseth, B.; Marcusson, J.; Hallbeck, M. Spreading of neurodegenerative pathology via neuron-to-neuron transmission of beta-amyloid. *J. Neurosci.* **2012**, *32*, 8767–8777. [CrossRef] [PubMed]

67. Ye, L.; Fritschi, S.K.; Schelle, J.; Obermuller, U.; Degenhardt, K.; Kaeser, S.A.; Eisele, Y.S.; Walker, L.C.; Baumann, F.; Staufenbiel, M.; et al. Persistence of Abeta seeds in APP null mouse brain. *Nat. Neurosci.* **2015**, *18*, 1559–1561. [CrossRef] [PubMed]

68. Domert, J.; Rao, S.B.; Agholme, L.; Brorsson, A.C.; Marcusson, J.; Hallbeck, M.; Nath, S. Spreading of amyloid-beta peptides via neuritic cell-to-cell transfer is dependent on insufficient cellular clearance. *Neurobiol. Dis.* **2014**, *65*, 82–92. [CrossRef] [PubMed]

69. Condello, C.; Stöehr, J. Aβ propagation and strains: Implications for the phenotypic diversity in Alzheimer's disease. *Neurobiol. Dis.* **2018**, *109*, 191–200. [CrossRef] [PubMed]

70. Benilova, I.; Karran, E.; De Strooper, B. The toxic Abeta oligomer and Alzheimer's disease: An emperor in need of clothes. *Nat. Neurosci.* **2012**, *15*, 349–357. [CrossRef] [PubMed]

71. Kostylev, M.A.; Kaufman, A.C.; Nygaard, H.B.; Patel, P.; Haas, L.T.; Gunther, E.C.; Vortmeyer, A.; Strittmatter, S.M. Prion-protein-interacting amyloid-beta oligomers of high molecular weight are tightly correlated with memory impairment in multiple Alzheimer mouse models. *J. Biol. Chem.* **2015**, *290*, 17415–17438. [CrossRef] [PubMed]

72. Sepulveda, F.J.; Parodi, J.; Peoples, R.W.; Opazo, C.; Aguayo, L.G. Synaptotoxicity of Alzheimer Beta Amyloid Can Be Explained by Its Membrane Perforating Property. *PLoS ONE* **2010**, *5*, e11820. [CrossRef] [PubMed]

73. Wilcox, K.C.; Marunde, M.R.; Das, A.; Velasco, P.T.; Kuhns, B.D.; Marty, M.T.; Klein, W.L. Nanoscale Synaptic Membrane Mimetic Allows Unbiased High Throughput Screen That Targets Binding Sites for Alzheimer's-Associated Aβ Oligomers. *PLoS ONE* **2015**, *10*, e0125263. [CrossRef] [PubMed]

74. Kayed, R.; Lasagna-Reeves, C.A. Molecular mechanisms of amyloid oligomers toxicity. *J. Alzheimers Dis.* **2013**, *33*, S67–S78. [CrossRef] [PubMed]
75. Chromy, B.A.; Nowak, R.J.; Lambert, M.P.; Viola, K.L.; Chang, L.; Velasco, P.T.; Jones, B.W.; Fernandez, S.J.; Lacor, P.N.; Horowitz, P.; et al. Self-assembly of Abeta(1-42) into globular neurotoxins. *Biochemistry* **2003**, *42*, 12749–12760. [CrossRef] [PubMed]
76. Lambert, M.P.; Viola, K.L.; Chromy, B.A.; Chang, L.; Morgan, T.E.; Yu, J.; Venton, D.L.; Krafft, G.A.; Finch, C.E.; Klein, W.L. Vaccination with soluble Abeta oligomers generates toxicity-neutralizing antibodies. *J. Neurochem.* **2001**, *79*, 595–605. [CrossRef] [PubMed]
77. Gong, Y.; Chang, L.; Viola, K.L.; Lacor, P.N.; Lambert, M.P.; Finch, C.E.; Krafft, G.A.; Klein, W.L. Alzheimer's disease-affected brain: Presence of oligomeric A beta ligands (ADDLs) suggests a molecular basis for reversible memory loss. *Proc. Natl. Acad. Sci. USA* **2003**, *100*, 10417–10422. [CrossRef] [PubMed]
78. Walsh, D.M.; Klyubin, I.; Fadeeva, J.V.; Cullen, W.K.; Anwyl, R.; Wolfe, M.S.; Rowan, M.J.; Selkoe, D.J. Naturally secreted oligomers of amyloid beta protein potently inhibit hippocampal long-term potentiation in vivo. *Nature* **2002**, *416*, 535–539. [CrossRef] [PubMed]
79. Wang, H.W.; Pasternak, J.F.; Kuo, H.; Ristic, H.; Lambert, M.P.; Chromy, B.; Viola, K.L.; Klein, W.L.; Stine, W.B.; Krafft, G.A. Soluble oligomers of beta amyloid (1-42) inhibit long-term potentiation but not long-term depression in rat dentate gyrus. *Brain Res.* **2002**, *924*, 133–140. [CrossRef]
80. Kim, H.J.; Chae, S.C.; Lee, D.K. Selective neuronal degeneration induced by soluble oligomeric amyloid beta protein. *FASEB J.* **2003**, *17*, 118–120. [CrossRef] [PubMed]
81. Jarosz-Griffiths, H.H.; Noble, E.; Rushworth, J.V.; Hooper, N.M. Amyloid-β Receptors: The Good, the Bad, and the Prion Protein. *J. Biol. Chem.* **2016**, *291*, 3174–3183. [CrossRef] [PubMed]
82. Zhang, D.; Qi, Y.; Klyubin, I.; Ondrejcak, T.; Sarell, C.J.; Cuello, A.C.; Collinge, J.; Rowan, M.J. Targeting glutamatergic and cellular prion protein mechanisms of amyloid β-mediated persistent synaptic plasticity disruption: Longitudinal studies. *Neuropharmacology* **2017**, *15*, 231–246. [CrossRef] [PubMed]
83. Li, F.; Tsien, J.Z. Memory and the NMDA Receptors. *N. Engl. J. Med.* **2009**, *361*, 302–303. [CrossRef] [PubMed]
84. Snyder, E.M.; Nong, Y.; Almeida, C.G.; Paul, S.; Moran, T.; Choi, E.Y.; Nairn, A.C.; Salter, M.W.; Lombroso, P.J.; Gouras, G.K.; et al. Regulation of NMDA receptor trafficking by amyloid-beta. *Nat. Neurosci.* **2005**, *8*, 1051–1058. [CrossRef] [PubMed]
85. Shankar, G.M.; Bloodgood, B.L.; Townsend, M.; Walsh, D.M.; Selkoe, D.J.; Sabatini, B.L. Natural oligomers of the Alzheimer amyloid-beta protein induce reversible synapse loss by modulating an NMDA-type glutamate receptordependent signaling pathway. *J. Neurosci.* **2007**, *27*, 2866–2875. [CrossRef] [PubMed]
86. Tamburri, A.; Dudilot, A.; Licea, S.; Bourgeois, C.; Boehm, J. NMDA-receptor activation but not ion flux is required for amyloid-beta induced synaptic depression. *PLoS ONE* **2013**, *8*, e65350. [CrossRef] [PubMed]
87. Wei, W.; Nguyen, L.N.; Kessels, H.W.; Hagiwara, H.; Sisodia, S.; Malinow, R. Amyloid beta from axons and dendrites reduces local spine number and plasticity. *Nat. Neurosci.* **2010**, *13*, 190–196. [CrossRef] [PubMed]
88. Yamin, G. NMDA receptor-dependent signaling pathways that underlie amyloid beta-protein disruption of LTP in the hippocampus. *J. Neurosci. Res.* **2009**, *87*, 1729–1736. [CrossRef] [PubMed]
89. De Felice, F.G.; Velasco, P.T.; Lambert, M.P.; Viola, K.; Fernandez, S.J.; Ferreira, S.T.; Klein, W.L. Abeta oligomers induce neuronal oxidative stress through an N-methyl-D-aspartate receptor-dependent mechanism that is blocked by the Alzheimer drug memantine. *J. Biol. Chem.* **2007**, *282*, 11590–11601. [CrossRef] [PubMed]
90. Takahashi-Ito, K.; Makino, M.; Okado, K.; Tomita, T. Memantine inhibits β-amyloid aggregation and disassembles preformed β-amyloid aggregates. *Biochem. Biophys. Res. Commun.* **2017**, *493*, 158–163. [CrossRef] [PubMed]
91. Demuro, A.; Mina, E.; Kayed, R.; Milton, S.C.; Parker, I.; Glabe, C.G. Calcium dysregulation and membrane disruption as a ubiquitous neurotoxic mechanism of soluble amyloid oligomers. *J. Biol. Chem.* **2005**, *280*, 17294–17300. [CrossRef] [PubMed]
92. Alberdi, E.; Sánchez-Gómez, M.V.; Cavaliere, F.; Pérez-Samartín, A.; Zugaza, J.L.; Trullas, R.; Domercq, M.; Matute, C. Amyloid beta oligomers induce Ca²⁺ dysregulation and neuronal death through activation of ionotropic glutamate receptors. *Cell Calcium* **2010**, *47*, 264–272. [CrossRef] [PubMed]
93. Liang, J.; Kulasiri, D.; Samarasinghe, S. Computational investigation of Amyloid-β-induced location- and subunit-specific disturbances of NMDAR at hippocampal dendritic spine in Alzheimer's disease. *PLoS ONE* **2017**, *12*, e0182743. [CrossRef] [PubMed]

94. Popugaeva, E.; Pchitskaya, E.; Bezprozvanny, I. Dysregulation of neuronal calcium homeostasis in Alzheimer's disease—A therapeutic opportunity? *Biochem. Biophys. Res. Commun.* **2017**, *483*, 998–1004. [CrossRef] [PubMed]

95. Zhao, D.; Watson, J.B.; Xie, C.W. Amyloid beta prevents activation of calcium/calmodulin-dependent protein kinase II and AMPA receptor phosphorylation during hippocampal long-term potentiation. *J. Neurophysiol.* **2004**, *92*, 2853–2858. [CrossRef] [PubMed]

96. Sivanesan, S.; Tan, A.; Rajadas, J. Pathogenesis of Abeta oligomers in synaptic failure. *Curr. Alzheimer Res.* **2013**, *10*, 316–323. [CrossRef] [PubMed]

97. Seixas da Silva, G.S.; Melo, H.M.; Lourenco, M.V.; Lyra, E.; Silva, N.M.; de Carvalho, M.B.; Alves-Leon, S.V.; de Souza, J.M.; Klein, W.L.; da-Silva, W.S.; et al. Amyloid-β oligomers transiently inhibit AMP-activated kinase and cause metabolic defects in hippocampal neurons. *J. Biol. Chem.* **2017**, *292*, 7395–7406. [CrossRef] [PubMed]

98. Boehm, J.; Kang, M.G.; Johnson, R.C.; Esteban, J.; Huganir, R.L.; Malinow, R. Synaptic incorporation of AMPA receptors during LTP is controlled by a PKC phosphorylation site on GluR1. *Neuron* **2006**, *51*, 213–225. [CrossRef] [PubMed]

99. Maren, S.; Tocco, G.; Standley, S.; Baudry, M.; Thompson, R.F. Postsynaptic factors in the expression of long-term potentiation (LTP): Increased glutamate receptor binding following LTP induction in vivo. *Proc. Natl. Acad. Sci. USA* **1993**, *90*, 9654–9658. [CrossRef] [PubMed]

100. Banke, T.G.; Bowie, D.; Lee, H.; Huganir, R.L.; Schousboe, A.; Traynelis, S.F. Control of GluR1 AMPA receptor function by cAMP-dependent protein kinase. *J. Neurosci.* **2000**, *20*, 89–102. [CrossRef] [PubMed]

101. Hsieh, H.; Boehm, J.; Sato, C.; Iwatsubo, T.; Tomita, T.; Sisodia, S.; Malinow, R. AMPAR removal underlies Abeta-induced synaptic depression and dendritic spine loss. *Neuron* **2006**, *52*, 831–843. [CrossRef] [PubMed]

102. Zhang, Y.; Kurup, P.; Xu, J.; Anderson, G.M.; Greengard, P.; Nairn, A.C.; Lombroso, P.J. Reduced levels of the tyrosine phosphatase STEP block β amyloid-mediated GluA1/GluA2 receptor internalization. *J. Neurochem.* **2011**, *119*, 664–672. [CrossRef] [PubMed]

103. Rui, Y.; Gu, J.; Yu, K.; Hartzell, H.C.; Zheng, J.Q. Inhibition of AMPA receptor trafficking at hippocampal synapses by beta-amyloid oligomers: The mitochondrial contribution. *Mol. Brain* **2010**, *3*, 10. [CrossRef] [PubMed]

104. Guntupalli, S.; Jang, S.E.; Zhu, T.; Huganir, R.L.; Widagdo, J.; Anggono, V. GluA1 subunit ubiquitination mediates amyloid-β-induced loss of surface α-amino-3-hydroxy-5-methyl-4-isoxazolepropionic acid (AMPA) receptors. *J. Biol. Chem.* **2017**, *292*, 8186–8194. [CrossRef] [PubMed]

105. Miñano-Molina, A.J.; España, J.; Martín, E. Soluble oligomers of amyloid-β peptide disrupt membrane trafficking of α-amino-3-hydroxy-5-methylisoxazole-4-propionic acid receptor contributing to early synapse dysfunction. *J. Biol. Chem.* **2011**, *286*, 27311–27321. [CrossRef] [PubMed]

106. Zhao, W.Q.; Santini, F.; Breese, R.; Ross, D.; Zhang, X.D.; Stone, D.J.; Ferrer, M.; Townsend, M.; Wolfe, A.L.; Seager, M.A.; et al. Inhibition of calcineurin-mediated endocytosis and alpha-amino-3-hydroxy-5-methyl-4-isoxazolepropionic acid (AMPA) receptors prevents amyloid beta oligomer-induced synaptic disruption. *J. Biol. Chem.* **2010**, *285*, 7619–7632. [CrossRef] [PubMed]

107. Renner, M.C.; Albers, E.H.; Gutierrez-Castellanos, N.; Reinders, N.R.; van Huijstee, A.N.; Xiong, H.; Lodder, T.R.; Kessels, H.W. Synaptic plasticity through activation of GluA3-containing AMPA-receptors. *eLife* **2017**, *6*, e25462. [CrossRef] [PubMed]

108. Gutierrez-Castellanos, N.; Da Silva-Matos, C.M.; Zhou, K.; Canto, C.B.; Renner, M.C.; Koene, L.M.C.; Ozyildirim, O.; Sprengel, R.; Kessels, H.W.; De Zeeuw, C.I. Motor Learning Requires Purkinje Cell Synaptic Potentiation through Activation of AMPA-Receptor Subunit GluA3. *Neuron* **2017**, *93*, 409–424. [CrossRef] [PubMed]

109. Reinders, N.R.; Pao, Y.; Renner, M.C.; da Silva-Matos, C.M.; Lodder, T.R.; Malinow, R.; Kessels, H.W. Amyloid-β effects on synapses and memory require AMPA receptor subunit GluA3. *Proc. Natl. Acad. Sci. USA* **2016**, *113*, E6526–E6534. [CrossRef] [PubMed]

110. Miller, E.C.; Teravskis, P.J.; Dummer, B.W.; Zhao, X.; Huganir, R.L.; Liao, D. Tau phosphorylation and tau mislocalization mediate soluble Aβ oligomer-induced AMPA glutamate receptor signaling deficits. *Eur. J. Neurosci.* **2014**, *39*, 1214–1224. [CrossRef] [PubMed]

111. Minakami, R.; Katsuki, F.; Yamamoto, T.; Nakamura, K.; Sugiyama, H. Molecular cloning and the functional expression of two isoforms of human metabotropic glutamate receptor subtype 5. *Biochem. Biophys. Res. Commun.* **1994**, *199*, 1136–1143. [CrossRef] [PubMed]

112. Gilman, A.G. G proteins: Transducers of receptor-generated signals. *Annu. Rev. Biochem.* **1987**, *56*, 615–649. [CrossRef] [PubMed]

113. Um, J.W.; Kaufman, A.C.; Kostylev, M.; Heiss, J.K.; Stagi, M.; Takahashi, H.; Kerrisk, M.E.; Vortmeyer, A.; Wisniewski, T.; Koleske, A.J.; et al. Metabotropic glutamate receptor 5 is a coreceptor for Alzheimer aβ oligomer bound to cellular prion protein. *Neuron* **2013**, *79*, 887–902. [CrossRef] [PubMed]

114. Chen, S.; Yadav, S.P.; Surewicz, W.K. Interaction between human prion protein and amyloid-beta (Abeta) oligomers: Role OF N-terminal residues. *J. Biol. Chem.* **2010**, *285*, 26377–26383. [CrossRef] [PubMed]

115. Lauren, J.; Gimbel, D.A.; Nygaard, H.B.; Gilbert, J.W.; Strittmatter, S.M. Cellular prion protein mediates impairment of synaptic plasticity by amyloid-beta oligomers. *Nature* **2009**, *457*, 1128–1132. [CrossRef] [PubMed]

116. Lazzari, C.; Kipanyula, M.J.; Agostini, M.; Pozzan, T.; Fasolato, C. Aβ42 oligomers selectively disrupt neuronal calcium release. *Neurobiol. Aging* **2015**, *36*, 877–885. [CrossRef] [PubMed]

117. Parkin, E.T.; Watt, N.T.; Hussain, I.; Eckman, E.A.; Eckman, C.B.; Manson, J.C.; Baybutt, H.N.; Turner, A.J.; Hooper, N.M. Cellular prion protein regulates beta-secretase cleavage of the Alzheimer's amyloid precursor protein. *Proc. Natl. Acad. Sci. USA* **2007**, *104*, 11062–11067. [CrossRef] [PubMed]

118. Vincent, B.; Sunyach, C.; Orzechowski, H.D.; St George-Hyslop, P.; Checler, F. p53-Dependent transcriptional control of cellular prion by presenilins. *J. Neurosci.* **2009**, *29*, 6752–6760. [CrossRef] [PubMed]

119. Younan, N.D.; Sarell, C.J.; Davies, P.; Brown, D.R.; Viles, J.H. The cellular prion protein traps Alzheimer's Abeta in an oligomeric form and disassembles amyloid fibers. *FASEB J.* **2013**, *27*, 1847–1858. [CrossRef] [PubMed]

120. Bove-Fenderson, E.; Urano, R.; Straub, J.E.; Harris, D.A. Cellular prion protein targets amyloid-β fibril ends via its C-terminal domain to prevent elongation. *J. Biol. Chem.* **2017**, *292*, 16858–16871. [CrossRef] [PubMed]

121. Beraldo, F.H.; Ostapchenko, V.G.; Caetano, F.A.; Guimaraes, A.L.; Ferretti, G.D.; Daude, N.; Bertram, L.; Nogueira, K.O.; Silva, J.L.; Westaway, D.; et al. Regulation of Amyloid β Oligomer Binding to Neurons and Neurotoxicity by the Prion Protein-mGluR5 Complex. *J. Biol. Chem.* **2016**, *291*, 21945–21955. [CrossRef] [PubMed]

122. Del Río, J.A.; Ferrer, I.; Gavín, R. Role of cellular prion protein in interneuronal amyloidtransmission. *Prog. Neurobiol.* **2018**. [CrossRef] [PubMed]

123. Beraldo, F.H.; Arantes, C.P.; Santos, T.G.; Queiroz, N.G.; Young, K.; Rylett, R.J.; Markus, R.P.; Prado, M.A.; Martins, V.R. Role of alpha7 nicotinic acetylcholine receptor in calcium signaling induced by prion protein interaction with stress-inducible protein 1. *J. Biol. Chem.* **2010**, *285*, 36542–36550. [CrossRef] [PubMed]

124. Beraldo, F.H.; Arantes, C.P.; Santos, T.G.; Machado, C.F.; Roffe, M.; Hajj, G.N.; Lee, K.S.; Magalhães, A.C.; Caetano, F.A.; Mancini, G.L.; et al. Metabotropic glutamate receptors transduce signals for neurite outgrowth after binding of the prion protein to laminin gamma1 chain. *FASEB J.* **2011**, *25*, 265–279. [CrossRef] [PubMed]

125. Haas, L.T.; Salazar, S.V.; Kostylev, M.A.; Um, J.W.; Kaufman, A.C.; Strittmatter, S.M. Metabotropic glutamate receptor 5 couples cellular prion protein to intracellular signalling in Alzheimer's disease. *Brain* **2016**, *139*, 526–546. [CrossRef] [PubMed]

126. You, H.; Tsutsui, S.; Hameed, S.; Kannanayakal, T.J.; Chen, L.; Xia, P.; Engbers, J.D.; Lipton, S.A.; Stys, P.K.; Zamponi, G.W. Abeta neurotoxicity depends on interactions between copper ions, prion protein, and N-methyl-D-aspartate receptors. *Proc. Natl. Acad. Sci. USA* **2012**, *109*, 1737–1742. [CrossRef] [PubMed]

127. Chin, J.; Palop, J.J.; Yu, G.Q.; Kojima, N.; Masliah, E.; Mucke, L. Fyn kinase modulates synaptotoxicity, but not aberrant sprouting, in human amyloid precursor protein transgenic mice. *J. Neurosci.* **2004**, *24*, 4692–4697. [CrossRef] [PubMed]

128. Daly, C.J.; McGrath, J.C. Previously unsuspected widespread cellular and tissue distribution of beta-adrenoceptors and its relevance to drug action. *Trends Pharmacol. Sci.* **2011**, *32*, 219–226. [CrossRef] [PubMed]

129. Gibbs, M.E.; Summers, R.J. Role of adrenoceptor subtypes in memory consolidation. *Prog. Neurobiol.* **2002**, *67*, 345–391. [CrossRef]

130. McIntyre, C.K.; McGaugh, J.L.; Williams, C.L. Interacting Brain Systems Modulate Memory Consolidation. *Neurosci. Biobehav. Rev.* **2012**, *36*, 1750–1762. [CrossRef] [PubMed]

131. Connor, S.A.; Wang, Y.T.; Nguyen, P.V. Activation of β-adrenergic receptors facilitates heterosynaptic translation-dependent long-term potentiation. *J. Physiol.* **2011**, *589*, 4321–4340. [CrossRef] [PubMed]

132. Gelinas, J.N.; Nguyen, P.V. Beta-adrenergic receptor activation facilitates induction of a protein synthesis-dependent late phase of long-term potentiation. *J. Neurosci.* **2005**, *25*, 3294–3303. [CrossRef] [PubMed]

133. Lin, Y.W.; Min, M.Y.; Chiu, T.H.; Yang, H.W. Enhancement of associative long-term potentiation by activation of beta-adrenergic receptors at CA1 synapses in rat hippocampal slices. *J. Neurosci.* **2003**, *23*, 74173–74181. [CrossRef]

134. Qian, H.; Matt, L.; Zhang, M.; Nguyen, M.; Patriarchi, T.; Koval, O.M.; Anderson, M.E.; He, K.; Lee, H.K.; Hell, J.W. β2-Adrenergic receptor supports prolonged theta tetanus-induced LTP. *J. Neurophysiol.* **2012**, *107*, 2703–2712. [CrossRef] [PubMed]

135. Thomas, M.J.; Moody, T.D.; Makhinson, M.; O'Dell, T.J. Activity-dependent beta-adrenergic modulation of low frequency stimulation induced LTP in the hippocampal CA1 region. *Neuron* **1996**, *17*, 475–482. [CrossRef]

136. Walling, S.G.; Harley, C.W. Locus ceruleus activation initiates delayed synaptic potentiation of perforant path input to the dentate gyrus in awake rats: A novel beta-adrenergic- and protein synthesis-dependent mammalian plasticity mechanism. *J. Neurosci.* **2004**, *24*, 598–604. [CrossRef] [PubMed]

137. Joiner, M.L.; Lise, M.F.; Yuen, E.Y.; Kam, A.Y.; Zhang, M.; Hall, D.D.; Malik, Z.A.; Qian, H.; Chen, Y.; Ulrich, J.D.; et al. Assembly of a beta2-adrenergic receptor-GluR1 signalling complex for localized cAMP signalling. *EMBO J.* **2010**, *29*, 482–495. [CrossRef] [PubMed]

138. Wang, D.; Govindaiah, G.; Liu, R.; De Arcangelis, V.; Cox, C.L.; Xiang, Y.K. Binding of amyloid beta peptide to beta2 adrenergic receptor induces PKA-dependent AMPA receptor hyperactivity. *FASEB J.* **2010**, *24*, 3511–3521. [CrossRef] [PubMed]

139. Chai, G.; Wang, Y.; Yasheng, A.; Zhao, P. Beta 2-adrenergic receptor activation enhances neurogenesis in Alzheimer's disease mice. *Neural Regen. Res.* **2016**, *11*, 1617–1624. [CrossRef] [PubMed]

140. Marien, M.R.; Colpaert, F.C.; Rosenquist, A.C. Noradrenergic mechanisms in neurodegenerative diseases: A theory. *Brain Res. Brain Res. Rev.* **2004**, *45*, 38–78. [CrossRef] [PubMed]

141. Szot, P.; White, S.S.; Greenup, J.L.; Leverenz, J.B.; Peskind, E.R.; Raskind, M.A. Compensatory changes in the noradrenergic nervous system in the locus ceruleus and hippocampus of postmortem subjects with Alzheimer's disease and dementia with Lewy bodies. *J. Neurosci.* **2006**, *26*, 467–478. [CrossRef] [PubMed]

142. Manaye, K.F.; Mouton, P.R.; Xu, G.; Drew, A.; Lei, D.L.; Sharma, Y.; Rebeck, G.W.; Turner, S. Age-related loss of noradrenergic neurons in the brains of triple transgenic mice. *Age* **2013**, *35*, 139–147. [CrossRef] [PubMed]

143. Ni, Y.; Zhao, X.; Bao, G.; Zou, L.; Teng, L.; Wang, Z.; Song, M.; Xiong, J.; Bai, Y.; Pei, G. Activation of beta2-adrenergic receptor stimulates gamma-secretase activity and accelerates amyloid plaque formation. *Nat. Med.* **2006**, *12*, 1390–1396. [CrossRef] [PubMed]

144. Yu, N.N.; Wang, X.X.; Yu, J.T.; Wang, ND.; Lu, RC.; Miao, D.; Tian, Y.; Tan, L. Blocking beta2-adrenergic receptor attenuates acute stress-induced amyloid beta peptides production. *Brain Res.* **2010**, *1317*, 305–310. [CrossRef] [PubMed]

145. Branca, C.; Wisely, E.V.; Hartman, L.K.; Caccamo, A.; Oddo, S. Administration of a selective β2 adrenergic receptor antagonist exacerbates neuropathology and cognitive deficits in a mouse model of Alzheimer's disease. *Neurobiol. Aging* **2014**, *35*, 2726–2735. [CrossRef] [PubMed]

146. Wang, D.; Yuen, E.Y.; Zhou, Y.; Yan, Z.; Xiang, Y.K. Amyloid beta peptide-(1–42) induces internalization and degradation of beta2 adrenergic receptors in prefrontal cortical neurons. *J. Biol. Chem.* **2011**, *286*, 31852–31863. [CrossRef] [PubMed]

147. Li, S.; Jin, M.; Zhang, D.; Yang, T.; Koeglsperger, T.; Fu, H.; Selkoe, D.J. Environmental novelty activates beta2-adrenergic signaling to prevent the impairment of hippocampal LTP by Abeta oligomers. *Neuron* **2013**, *77*, 929–941. [CrossRef] [PubMed]

148. Yang, T.; Li, S.; Xu, H.; Walsh, D.M.; Selkoe, D.J. Large Soluble Oligomers of Amyloid β-Protein from Alzheimer Brain Are Far Less Neuroactive Than the Smaller Oligomers to Which They Dissociate. *J. Neurosci.* **2017**, *37*, 152–163. [CrossRef] [PubMed]

149. Wang, H.Y.; Lee, D.H.; D'Andrea, M.R.; Peterson, P.A.; Shank, R.P.; Reitz, A.B. beta-Amyloid(1-42) binds to alpha7 nicotinic acetylcholine receptor with high affinity. Implications for Alzheimer's disease pathology. *J. Biol. Chem.* **2000**, *275*, 5626–5632. [CrossRef] [PubMed]

150. Nagele, R.G.; D'Andrea, M.R.; Anderson, W.J.; Wang, H.Y. Intracellular accumulation of beta-amyloid(1-42) in neurons is facilitated by the alpha 7 nicotinic acetylcholine receptor in Alzheimer's disease. *Neuroscience* **2002**, *110*, 199–211. [CrossRef]

151. Baker-Nigh, A.; Vahedi, S.; Davis, E.G.; Weintraub, S.; Bigio, E.H.; Klein, W.L.; Geula, C. Neuronal amyloid-β accumulation within cholinergic basal forebrain in ageing and Alzheimer's disease. *Brain* **2015**, *138*, 1722–1737. [CrossRef] [PubMed]

152. Oddo, S.; Caccamo, A.; Green, K.N.; Liang, K.; Tran, L.; Chen, Y.; Leslie, F.M.; LaFerla, F.M. Chronic nicotine administration exacerbates tau pathology in a transgenic model of Alzheimer's disease. *Proc. Natl. Acad. Sci. USA* **2005**, *102*, 3046–3051. [CrossRef] [PubMed]

153. Hernandez, C.M.; Kayed, R.; Zheng, H.; Sweatt, J.D.; Dineley, K.T. Loss of alpha7 nicotinic receptors enhances beta-amyloid oligomer accumulation, exacerbating earlystage cognitive decline and septohippocampal pathology in a mouse model of Alzheimer's disease. *J. Neurosci.* **2010**, *30*, 2442–2453. [CrossRef] [PubMed]

154. Kroker, K.S.; Moreth, J.; Kussmaul, L.; Rast, G.; Rosenbrock, H. Restoring long-term potentiation impaired by amyloid-beta oligomers: Comparison of an acetylcholinesterase inhibitior and selective neuronal nicotinic receptor agonists. *Brain Res. Bull.* **2013**, *96*, 28–38. [CrossRef] [PubMed]

155. De Felice, F.G.; Lourenco, M.V.; Ferreira, S.T. How does brain insulin resistance develop in Alzheimer's disease? *Alzheimers Dement. J. Alzheimers Assoc.* **2014**, *10*, S26–S32. [CrossRef] [PubMed]

156. De la Monte, S.M.; Wands, J.R. Alzheimer's disease is type 3 diabetes-evidence reviewed. *J. Diabetes Sci. Technol.* **2008**, *2*, 1101–1113. [CrossRef] [PubMed]

157. De la Monte, S.M. Type 3 diabetes is sporadic Alzheimers disease: Mini-review. *Eur. Neuropsychopharmacol.* **2014**, *24*, 1954–1960. [CrossRef] [PubMed]

158. De Felice, F.G.; Vieira, M.N.; Bomfim, T.R.; Decker, H.; Velasco, P.T.; Lambert, M.P.; Viola, K.L.; Zhao, W.Q.; Ferreira, S.T.; Klein, W.L. Protection of synapses against Alzheimer's-linked toxins: Insulin signaling prevents the pathogenic binding of Abeta oligomers. *Proc. Natl. Acad. Sci. USA* **2009**, *106*, 1971–1976. [CrossRef] [PubMed]

159. Townsend, M.; Mehta, T.; Selkoe, D.J. Soluble Abeta inhibits specific signal transduction cascades common to the insulin receptor pathway. *J. Biol. Chem.* **2007**, *282*, 33305–33312. [CrossRef] [PubMed]

160. Yamamoto, N.; Matsubara, E.; Maeda, S.; Minagawa, H.; Takashima, A.; Maruyama, W.; Michikawa, M.; Yanagisawa, K. A ganglioside-induced toxic soluble Abeta assembly. Its enhanced formation from Abeta bearing the Arctic mutation. *J. Biol. Chem.* **2007**, *282*, 2646–2655. [CrossRef] [PubMed]

161. Zhang, Y.; Hong, Y.; Bounhar, Y.; Blacker, M.; Roucou, X.; Tounekti, O.; Vereker, E.; Bowers, W.J.; Federoff, H.J.; Goodyer, C.G.; et al. p75 neurotrophin receptor protects primary cultures of human neurons against extracellular amyloid beta peptide cytotoxicity. *J. Neurosci.* **2003**, *23*, 7385–7394. [CrossRef] [PubMed]

162. Costantini, C.; Rossi, F.; Formaggio, E.; Bernardoni, R.; Cecconi, D.; Della-Bianca, V. Characterization of the signalling pathway downstream p75 neurotrophin receptor involved in beta-amyloid peptide-dependent cell death. *J. Mol. Neurosci.* **2005**, *25*, 141–156. [CrossRef]

163. Ito, S.; Ménard, M.; Atkinson, T.; Gaudet, C.; Brown, L.; Whitfield, J.; Chakravarthy, B. Involvement of insulin-like growth factor 1 receptor signaling in the amyloid-β peptide oligomers-induced p75 neurotrophin receptor protein expression in mouse hippocampus. *J. Alzheimers Dis.* **2012**, *31*, 493–506. [CrossRef] [PubMed]

164. Dechant, G.; Barde, Y.A. The neurotrophin receptor p75(NTR): Novel functions and implications for diseases of the nervous system. *Nat. Neurosci.* **2002**, *5*, 1131–1136. [CrossRef] [PubMed]

165. Perini, G.; Della-Bianca, V.; Politi, V.; Della Valle, G.; Dal-Pra, I.; Rossi, F.; Armato, U. Role of p75 neurotrophin receptor in the neurotoxicity by beta-amyloid peptides and synergistic effect of inflammatory cytokines. *J. Exp. Med.* **2002**, *195*, 907–918. [CrossRef] [PubMed]

166. Costantini, C.; Della-Bianca, V.; Formaggio, E.; Chiamulera, C.; Montresor, A.; Rossi, F. The expression of p75 neurotrophin receptor protects against the neurotoxicity of soluble oligomers of beta-amyloid. *Exp. Cell Res.* **2005**, *311*, 126–134. [CrossRef] [PubMed]

167. Boulanger, L.M.; Shatz, C.J. Immune signalling in neural development, synaptic plasticity and disease. *Nat. Rev. Neurosci.* **2004**, *5*, 521–531. [CrossRef] [PubMed]

168. Kim, T.; Vidal, G.S.; Djurisic, M.; William, C.M.; Birnbaum, M.E.; Garcia, K.C.; Hyman, B.T.; Shatz, C.J. Human LilrB2 Is a β-Amyloid Receptor and Its Murine Homolog PirB Regulates Synaptic Plasticity in an Alzheimer's Model. *Science* **2013**, *341*, 1399–1404. [CrossRef] [PubMed]

169. Kubagawa, H.; Burrows, P.D.; Cooper, M.D. A novel pair of immunoglobulin-like receptors expressed by B cells and myeloid cells. *Proc. Natl. Acad. Sci. USA* **1997**, *94*, 5261–5266. [CrossRef] [PubMed]

170. Liu, J.; Wang, Y.; Fu, W. Axon regeneration impediment: The role of paired immunoglobulin-like receptor B. *Neural Regen. Res.* **2015**, *10*, 1338–1342. [CrossRef] [PubMed]

171. Atwal, J.K.; Pinkston-Gosse, J.; Syken, J.; Stawicki, S.; Wu, Y.; Shatz, C.; Tessier-Lavigne, M. PirB is a functional receptor for myelin inhibitors of axonal regeneration. *Science* **2008**, *322*, 967–970. [CrossRef] [PubMed]

172. VanGuilder Starkey, H.D.; Van Kirk, C.A.; Bixler, G.V.; Imperio, C.G.; Kale, V.P.; Serfass, J.M.; Farley, J.A.; Yan, H.; Warrington, J.P.; Han, S.; et al. Neuroglial expression of the MHCI pathway and PirB receptor is upregulated in the hippocampus with advanced aging. *J. Mol. Neurosci.* **2012**, *48*, 111–126. [CrossRef] [PubMed]

173. William, C.M.; Andermann, M.L.; Goldey, G.J.; Roumis, DK.; Reid, RC.; Shatz, CJ.; Albers, MW.; Frosch, MP.; Hyman, BT. Synaptic plasticity defect following visual deprivation in Alzheimer's disease model transgenic mice. *J. Neurosci.* **2012**, *32*, 8004–8011. [CrossRef] [PubMed]

174. Syken, J.; Grandpre, T.; Kanold, P.O.; Shatz, C.J. PirB restricts ocular dominance plasticity in visual cortex. *Science* **2006**, *313*, 1795–1800. [CrossRef] [PubMed]

175. Kam, T.I.; Song, S.; Gwon, Y.; Park, H.; Yan, J.J.; Im, I.; Choi, J.W.; Choi, T.Y.; Kim, J.; Song, D.K.; et al. FcγRIIb mediates amyloid-β neurotoxicity and memory impairment in Alzheimer's disease. *J. Clin. Investig.* **2013**, *123*, 2791–2802. [CrossRef] [PubMed]

176. Maverakis, E.; Kim, K.; Shimoda, M.; Gershwin, M.E.; Patel, F.; Wilken, R.; Lebrilla, C.B. Glycans in the Immune system and The Altered Glycan Theory of Autoimmunity: A Critical Review. *J. Autoimmun.* **2015**, *57*, 1–13. [CrossRef] [PubMed]

177. Efthymiou, A.G.; Goate, A.M. Late onset Alzheimer's disease genetics implicates microglial pathways in disease risk. *Mol. Neurodegener.* **2017**, *12*, 43. [CrossRef] [PubMed]

178. Song, W.; Hooli, B.; Mullin, K.; Jin, S.C.; Cella, M.; Ulland, T.K.; Wang, Y.; Tanzi, R.E.; Colonna, M. Alzheimer's disease-associated TREM2 variants exhibit either decreased or increased ligand-dependent activation. *Alzheimers Dement.* **2017**, *4*, 381–387. [CrossRef] [PubMed]

179. Lue, L.F.; Schmitz, C.; Walker, D.G. What happens to microglial TREM2 in Alzheimer's disease: Immunoregulatory turned into immunopathogenic? *Neuroscience* **2015**, *302*, 138–150. [CrossRef] [PubMed]

180. Jin, S.C.; Benitez, B.A.; Karch, C.M.; Cooper, B.; Skorupa, T.; Carrell, D.; Cruchaga, C. Coding variants in TREM2 increase risk for Alzheimer's disease. *Hum. Mol. Gen.* **2014**, *23*, 5838–5846. [CrossRef] [PubMed]

181. Zhao, Y.; Wu, X.; Li, X.; Jiang, L.L.; Gui, X.; Liu, Y.; Sun, Y.; Zhu, B.; Piña-Crespo, J.C.; Zhang, M.; et al. TREM2 Is a Receptor for β-Amyloid that Mediates Microglial Function. *Neuron* **2018**, *97*, 1023–1031.e7. [CrossRef] [PubMed]

182. Yeh, F.L.; Hansen, D.V.; Sheng, M. TREM2, Microglia, and Neurodegenerative Diseases. *Trends Mol. Med.* **2017**, *23*, 512–533. [CrossRef] [PubMed]

183. Vargas, L.M.; Cerpa, W.; Muñoz, F.J.; Zanlungo, S.; Alvarez, A.R. Amyloid-β oligomers synaptotoxicity: The emerging role of EphA4/c-Abl signaling in Alzheimer's disease. *Biochim. Biophys. Acta* **2018**, *1864*, 1148–1159. [CrossRef] [PubMed]

184. Klein, R. Bidirectional modulation of synaptic functions by Eph/ephrin signaling. *Nat. Neurosci.* **2009**, *12*, 15–20. [CrossRef] [PubMed]

185. Hruska, M.; Dalva, M.B. Ephrin regulation of synapse formation, function and plasticity. *Mol. Cell. Neurosci.* **2012**, *50*, 35–44. [CrossRef] [PubMed]

186. Yamaguchi, Y.; Pasquale, E.B. Eph receptors in the adult brain. *Curr. Opin. Neurobiol.* **2004**, *14*, 288–296. [CrossRef] [PubMed]

187. Henkemeyer, M.; Itkis, O.S.; Ngo, M.; Hickmott, P.W.; Ethell, I.M. Multiple EphB receptor tyrosine kinases shape dendritic spines in the hippocampus. *J. Cell Biol.* **2003**, *163*, 1313–1326. [CrossRef] [PubMed]

188. Attwood, B.K.; Patel, S.; Pawlak, R. Ephs and ephrins: Emerging therapeutic targets in neuropathology. *Int. J. Biochem. Cell Biol.* **2012**, *44*, 578–581. [CrossRef] [PubMed]

189. Lacor, P.N.; Buniel, M.C.; Furlow, P.W.; Clemente, AS.; Velasco, PT.; Wood, M.; Viola, KL.; Klein, WL. Aβ Oligomer-Induced Aberrations in Synapse Composition, Shape, and Density Provide a Molecular Basis for Loss of Connectivity in Alzheimer's Disease. *J. Neurosci.* **2007**, *27*, 796–807. [CrossRef] [PubMed]

190. Takasu, M.A.; Dalva, M.B.; Zigmond, R.E.; Greenberg, M.E. Modulation of NMDA receptor-dependent calcium influx and gene expression through EphB receptors. *Science* **2002**, *295*, 491–495. [CrossRef] [PubMed]

191. Cisse, M.; Halabisky, B.; Harris, J.; Devidze, N.; Dubal, D.B.; Sun, B.; Orr, A.; Lotz, G.; Kim, D.H.; Hamto, P.; et al. Reversing EphB2 depletion rescues cognitive functions in Alzheimer model. *Nature* **2011**, *469*, 47–52. [CrossRef] [PubMed]

192. Shi, X.D.; Sun, K.; Hu, R.; Liu, X.Y.; Hu, Q.M.; Sun, X.Y.; Yao, B.; Sun, N.; Hao, J.R.; Wei, P.; et al. Blocking the Interaction between EphB2 and ADDLs by a Small Peptide Rescues Impaired Synaptic Plasticity and Memory Deficits in a Mouse Model of Alzheimer's Disease. *J. Neurosci.* **2016**, *36*, 11959–11973. [CrossRef] [PubMed]

193. Miyamoto, T.; Kim, D.; Knox, J.A.; Johnson, E.; Mucke, L. Increasing the receptor tyrosine kinase EphB2 prevents amyloid-beta-induced depletion of cell surface glutamate receptors by a mechanism that requires the PDZ-binding motif of EphB2 and neuronal activity. *J. Biol. Chem.* **2016**, *291*, 1719–1734. [CrossRef] [PubMed]

194. Simon, A.M.; de Maturana, R.L.; Ricobaraza, A.; Escribano, L.; Schiapparelli, L.; Cuadrado-Tejedor, M.; Perez-Mediavilla, A.; Avila, J.; Del Rio, J.; Frechilla, D. Early changes in hippocampal Eph receptors precede the onset of memory decline in mouse models of Alzheimer's disease. *J. Alzheimers Dis.* **2009**, *17*, 773–786. [CrossRef] [PubMed]

195. Murai, K.K.; Nguyen, L.N.; Irie, F.; Yamaguchi, Y.; Pasquale, E.B. Control of hippocampal dendritic spine morphology through ephrin-A3/EphA4 signaling. *Nat. Neurosci.* **2003**, *6*, 153–160. [CrossRef] [PubMed]

196. Fu, A.K.; Hung, K.W.; Fu, W.Y.; Shen, C.; Chen, Y.; Xia, J.; Lai, K.O.; Ip, N.Y. APC(Cdh1) mediates EphA4-dependent downregulation of AMPA receptors in homeostatic plasticity. *Nat. Neurosci.* **2011**, *14*, 181–189. [CrossRef] [PubMed]

197. Peng, Y.R.; Hou, Z.H.; Yu, X. The kinase activity of EphA4 mediates homeostatic scaling-down of synaptic strength via activation of Cdk5. *Neuropharmacology* **2013**, *65*, 232–243. [CrossRef] [PubMed]

198. Van Hoecke, A.; Schoonaert, L.; Lemmens, R.; Timmers, M.; Staats, K.A.; Laird, A.S.; Peeters, E.; Philips, T.; Goris, A.; Dubois, B.; et al. EPHA4 is a disease modifier of amyotrophic lateral sclerosis in animal models and in humans. *Nat. Med.* **2012**, *18*, 1418–1422. [CrossRef] [PubMed]

199. Zhao, J.; Boyd, A.W.; Bartlett, P.F. The identification of a novel isoform of EphA4 and its expression in SOD1(G93A) mice. *Neuroscience* **2017**, *347*, 11–21. [CrossRef] [PubMed]

200. Rosenberger, A.F.; Rozemuller, A.J.; van der Flier, W.M.; Scheltens, P.; van der Vies, S.M.; Hoozemans, J.J. Altered distribution of the EphA4 kinase in hippocampal brain tissue of patients with Alzheimer's disease correlates with pathology. *Acta Neuropathol. Commun.* **2014**, *2*, 79. [CrossRef] [PubMed]

201. Williams, C.; Mehrian Shai, R.; Wu, Y.; Hsu, Y.-H.; Sitzer, T.; Spann, B.; Miller, C.A. Transcriptome Analysis of Synaptoneurosomes Identifies Neuroplasticity Genes Overexpressed in Incipient Alzheimer's Disease. *PLoS ONE* **2009**, *4*, e4936. [CrossRef] [PubMed]

202. Fu, A.K.; Hung, K.W.; Huang, H.; Gu, S.; Shen, Y.; Cheng, E.Y.; Ip, F.C.; Huang, X.; Fu, W.Y.; Ip, N.Y. Blockade of EphA4 signaling ameliorates hippocampal synaptic dysfunctions in mouse models of Alzheimer's disease. *Proc. Natl. Acad. Sci. USA* **2014**, *111*, 9959–9964. [CrossRef] [PubMed]

203. Vargas, L.M.; Leal, N.; Estrada, L.D.; Gonzalez, A.; Serrano, F.; Araya, K.; Gysling, K.; Inestrosa, N.C.; Pasquale, E.B.; Alvarez, A.R. EphA4 activation of c-Abl mediates synaptic loss and LTP blockade caused by amyloid-beta oligomers. *PLoS ONE* **2014**, *9*, e92309. [CrossRef] [PubMed]

204. Sturchler, E.; Galichet, A.; Weibel, M.; Leclerc, E.; Heizmann, C.W. Site-Specific Blockade of RAGE-Vd Prevents Amyloid-β Oligomer Neurotoxicity. *J. Neurosci.* **2008**, *28*, 5149–5158. [CrossRef] [PubMed]

205. Neeper, M.; Schmidt, A.M.; Brett, J.; Yan, SD.; Wang, F.; Pan, YC.; Elliston, K.; Stern, D.; Shaw, A. Cloning and expression of a cell surface receptor for advanced glycosylation end products of proteins. *J. Biol. Chem.* **1992**, *267*, 14998–15004. [PubMed]

206. Yan, S.D.; Chen, X.; Fu, J.; Chen, M.; Zhu, H.; Roher, A.; Slattery, T.; Zhao, L.; Nagashima, M.; Morser, J.; et al. RAGE and amyloid-beta peptide neurotoxicity in Alzheimer's disease. *Nature* **1996**, *382*, 685–691. [CrossRef] [PubMed]

207. Bierhaus, A.; Schiekofer, S.; Schwaninger, M.; Andrassy, M.; Humpert, PM.; Chen, J.; Hong, M.; Luther, T.; Henle, T.; Klöting, I.; et al. Diabetes-associated sustained activation of the transcription factor nuclear factor-kappaB. *Diabetes* **2001**, *50*, 2792–2808. [CrossRef] [PubMed]

208. Deane, R.; Wu, Z.; Zlokovic, B.V. RAGE (yin) versus LRP (yang) balance regulates alzheimer amyloid beta-peptide clearance through transport across the blood-brain barrier. *Stroke* **2004**, *35*, 2628–2631. [CrossRef] [PubMed]

209. Deane, R.; Bell, R.; Sagare, A.; Zlokovic, B. Clearance of amyloid-β peptide across the blood-brain barrier: Implication for therapies in Alzheimer's disease. *CNS Neurol. Disord. Drug Targets* **2009**, *8*, 16–30. [CrossRef] [PubMed]

210. Saito, A.; Pietromonaco, S.; Loo, A.K.-C.; Farquhar, M.G. Complete cloning and sequencing of rat gp330/'megalin', a distinctive member of the low density lipoprotein receptor gene family. *Proc. Natl. Acad. Sci. USA* **1994**, *91*, 9725–9729. [CrossRef] [PubMed]

211. Marzolo, M.P.; Farfán, P. New insights into the roles of megalin/LRP2 and the regulation of its functional expression. *Biol. Res.* **2011**, *44*, 89–105. [CrossRef] [PubMed]

212. Christensen, E.I.; Birn, H. Megalin and cubilin: Multifunctional endocytic receptors. *Nat. Rev. Mol. Cell Biol.* **2002**, *3*, 256–266. [CrossRef] [PubMed]

213. Larsson, M.; Hjälm, G.; Sakwe, A.M.; Engström, A.; Höglund, AS.; Larsson, E.; Robinson, RC.; Sundberg, C.; Rask, L. Selective interaction of megalin with postsynaptic density-95 (PSD-95)-like membrane-associated guanylate kinase (MAGUK) proteins. *Biochem. J.* **2003**, *373*, 381–391. [CrossRef] [PubMed]

214. Kounnas, M.Z.; Loukinova, E.B.; Stefansson, S.; Harmony, J.A.; Brewer, B.; Strickland, D.K.; Argraves, W.S. Identification of glycoprotein 330 as an endocytic receptor for apolipoprotein J/clusterin. *J. Biol. Chem.* **1995**, *270*, 13070–13075. [CrossRef] [PubMed]

215. Zlokovic, B.V.; Martel, C.L.; Matsubara, E.; McComb, JG.; Zheng, G.; McCluskey, RT.; Frangione, B.; Ghiso, J. Glycoprotein 330/megalin: Probable role in receptor-mediated transport of apolipoprotein J alone and in a complex with Alzheimer disease amyloid beta at the blood-brain and blood-cerebrospinal fluid barriers. *Proc. Natl. Acad. Sci. USA* **1996**, *93*, 4229–4234. [CrossRef] [PubMed]

216. Hammad, S.M.; Ranganathan, S.; Loukinova, E.; Twal, W.O.; Argraves, W.S. Interaction of apolipoprotein J-amyloid beta-peptide complex with low density lipoprotein receptor-related protein-2/megalin. A mechanism to prevent pathological accumulation of amyloid beta-peptide. *J. Biol. Chem.* **1997**, *272*, 18644–18649. [CrossRef] [PubMed]

217. Mangelsdorf, D.J.; Thummel, C.; Beato, M.; Herrlich, P.; Schütz, G.; Umesono, K.; Blumberg, B.; Kastner, P.; Mark, M.; Chambon, P.; et al. The nuclear receptor superfamily: The second decade. *Cell* **1995**, *83*, 835–839. [CrossRef]

218. Woods, C.G.; Heuvel, J.P.; Rusyn, I. Genomic profiling in nuclear receptor-mediated toxicity. *Toxicol. Pathol.* **2007**, *35*, 474–494. [CrossRef] [PubMed]

219. Burris, T.P.; Busby, S.A.; Griffin, P.R. Targeting orphan nuclear receptors for treatment of metabolic diseases and autoimmunity. *Chem. Biol.* **2012**, *19*, 51–59. [CrossRef] [PubMed]

220. Patel, P.; Shah, J. Role of Vitamin D in Amyloid clearance via LRP-1 upregulation in Alzheimer's disease: A potential therapeutic target? *J. Chem. Neuroanat.* **2017**, *85*, 36–42. [CrossRef] [PubMed]

221. Gezen-Ak, D.; Yılmazer, S.; Dursun, E. Why Vitamin D in Alzheimer's disease? The hypothesis. *J. Alzheimers Dis.* **2014**, *40*, 257–269. [CrossRef] [PubMed]

222. Ito, S.; Ohtsuki, S.; Nezu, Y.; Koitabashi, Y.; Murata, S.; Terasaki, T. 1α,25-Dihydroxyvitamin D3 enhances cerebral clearance of human amyloid-β peptide(1-40) from mouse brain across the blood-brain barrier. *Fluids Barriers CNS* **2011**, *8*, 20. [CrossRef] [PubMed]

223. Herskovits, A.Z.; Guarente, L. SIRT1 in neurodevelopment and brain senescence. *Neuron* **2014**, *81*, 471–483. [CrossRef] [PubMed]

224. Martins, I.J.; Calderón, A.M. Diet and Nutrition reverse Type 3 Diabetes and Accelerated Aging linked to Global chronic diseases. *J. Diabetes Res. Ther.* **2016**, *2*. [CrossRef]

225. Talbot, K. Brain insulin resistance in Alzheimer's disease and its potential treatment with GLP-1 analogs. *Neurodegener. Dis. Manag.* **2014**, *4*, 31–40. [CrossRef] [PubMed]

226. Liang, F.; Kume, S.; Koya, D. SIRT1 and insulin resistance. *Nat. Rev. Endocrinol.* **2009**, *5*, 367–373. [CrossRef] [PubMed]

227. Lee, H.R.; Shin, H.K.; Park, S.Y.; Kim, H.Y.; Lee, W.S.; Rhim, B.Y.; Hong, K.W.; Kim, C.D. Cilostazol suppresses β-amyloid production by activating a disintegrin and metalloproteinase 10 via the upregulation of SIRT1-coupled retinoic acid receptor-β. *J. Neurosci. Res.* **2014**, *92*, 1581–1590. [CrossRef] [PubMed]

228. Julien, C.; Tremblay, C.; Emond, V.; Lebbadi, M.; Salem, N., Jr.; Bennett, D.A.; Calon, F. Sirtuin 1 reduction parallels the accumulation of tau in Alzheimer disease. *J. Neuropathol. Exp. Neurol.* **2009**, *68*, 48–58. [CrossRef] [PubMed]

International Journal of
Molecular Sciences

MDPI

Review

Preventive Effects of Dairy Products on Dementia and the Underlying Mechanisms

Yasuhisa Ano [1,*] and Hiroyuki Nakayama [2]

[1]   Research Laboratories for Health Science & Food Technologies, Kirin Company Ltd.,
      1-13-5 Fukuura Kanazawa-ku, Yokohama-shi, Kanagawa 236-0004, Japan
[2]   Laboratory of Veterinary Pathology, Graduate School of Agricultural and Life Sciences,
      the University of Tokyo, 1-1-1 Yayoi, Bunkyo-ku, Tokyo 113-8657, Japan; anakaya@mail.ecc.u-tokyo.ac.jp
*   Correspondence: Yasuhisa_Ano@kirin.co.jp; Tel.: +81-45-330-9007

Received: 24 May 2018; Accepted: 29 June 2018; Published: 30 June 2018

Abstract: Alongside the rapid population aging occurring worldwide, the prevention of age-related memory decline and dementia has become a high priority. Dairy products have many physiological effects owing to their contents of lactic acid bacteria and the fatty acids and peptides generated during their fermentation. In particular, several recent studies have elucidated the effects of fermented dairy products on cognitive function. Epidemiological and clinical evidence has indicated that fermented dairy products have preventive effects against dementia, including Alzheimer's disease. Recent preclinical studies have identified individual molecules generated during fermentation that are responsible for those preventive effects. Oleamide and dehydroergosterol have been identified as the agents responsible for reducing microglial inflammatory responses and neurotoxicity. In this review, the protective effects of fermented dairy products and their components on cognitive function, the mechanisms underlying those effects, and the prospects for their future clinical development will be discussed.

Keywords: amyloid β; Alzheimer's disease; cognitive function; dairy products; dementia; inflammation; microglia

1. Introduction

With the rapid aging of the population worldwide, cognitive decline and dementia are becoming increasing burdens not only on patients and their families, but also on national healthcare systems. Dementia is a general term for memory loss and other mental abilities severe enough to interfere with daily life. It is caused by physical changes in the brain. The most common type of dementia is Alzheimer's disease, which comprises 50–70% of dementia cases. Alzheimer's disease is histopathologically characterized by the presence of amyloid beta (Aβ) plaques and intracellular neurofibrillary tangles (NFTs) consisting of hyperphosphorylated tau. Both Aβ senile plaques and NFTs are formed of insoluble, densely-packed protein filaments. The accumulation of Aβ plaques and NFTs correlates with the symptoms of Alzheimer's disease and results in neuronal damage and death [1–4]. Other common types of dementia include vascular dementia [5], Lewy body dementia [6], and front temporal dementia [7]. Numerous reports have demonstrated that the formation of Aβ plaques is followed by inflammation in the brain, which is closely associated with NFTs and accelerates the development and progression of Alzheimer's disease [3,4,8,9]. The mechanisms that regulate Aβ plaques, NFTs, and inflammation are important targets for the therapy and prevention of Alzheimer's disease. However, owing to the lack of a disease-modifying therapy for dementia, preventive approaches such as diet, exercise, and active learning are being explored. In addition, high blood pressure, smoking, diabetes, and obesity are risk factors for dementia [10]. A recent study suggested that stress is also a risk factor for Alzheimer's disease [11]. Therefore, it is important to reduce these risk factors through adopting appropriate lifestyle habits.

There is now substantial evidence that dietary factors can modify the risk of dementia. Based on the results of several epidemiological investigations, the Mediterranean diet has been receiving increasing attention as a nutritional approach for lowering the risks of dementia [12–14]. In addition, specific dietary components including docosahexaenoic acid (DHA) from fish oil [15–18], resveratrol from red grapes [19–21], and curcumin from turmeric [22–24] have been evaluated for their potential protective effects against dementia or cognitive decline in several clinical trials. Recent epidemiological and clinical studies have indicated that fermented dairy products and their components, including lactic acid bacteria as well as peptides and fatty acids generated during fermentation, may also protect against dementia or cognitive decline. In this review, the recent studies investigating the effects of the intake of fermented dairy products on the risks of dementia and the underlying mechanisms demonstrated by recent studies will be discussed.

2. Epidemiological Studies on the Relationship between Fermented Dairy Product Consumption and Cognitive Function

Recent epidemiological studies have suggested that the consumption of dairy products, including yogurt and low-fat cheese, may reduce the risk of cognitive decline in the elderly and contribute to the prevention of Alzheimer's disease. Camfield et al. [25] suggested in their review that specific components of dairy products including bioactive peptides, colostrinin, proline-rich polypeptides, α-lactalbumin, vitamin B12, calcium, and probiotics might promote healthy brain function during aging. They also suggested that low-fat dairy products, when consumed regularly as part of a balanced diet, may have several beneficial outcomes for neurocognitive health during aging. However, the underlying mechanisms that benefit cognitive function have not been elucidated.

Crichton et al. [26,27] revealed that individuals who consumed low-fat dairy products, including yogurt and cheese, once a week had a higher cognitive function than those who did not. Previous reports had suggested that low-fat dairy products might lower the risks of obesity [28–30], type 2 diabetes [31,32], and cardiovascular diseases, which are all linked with the risks of cognitive decline and dementia [33–35]. They evaluated pre-existing data from 1183 participants and examined associations between dairy intake and self-reported cognitive and memory functions, self-esteem, stress, anxiety, mood, and psychological well-being. Dairy intake was calculated using a self-completed, quantified food frequency questionnaire (FFQ) and detailed information regarding the intakes of milk, cheese, ice cream, cream, yogurt, and dairy desserts as well as the fat content of each item was analyzed from the FFQ. The results revealed that the intakes of dairy and macronutrients were significantly associated with psychological health measures. In men, a higher protein intake was associated with lower perceived stress scores. Analyses of individual dairy foods found that the consumption of low-fat yogurt was positively associated with memory performance (quality of recall) in men. In women, the consumption of low-fat cheese was positively associated with social functioning and negatively associated with perceived stress levels. The intake of cheese was reported to be associated with decreased cognitive impairment in a population of late middle-aged to elderly people [36]. These results suggested that low-fat yogurt and cheese are beneficial for cognitive functions.

Ozawa et al. [37] surveyed more than 1000 Japanese subjects who were living in a local community, aged 60–79 years, and free from dementia to investigate any potential association between their diet and their risks of dementia. Their dietary patterns were surveyed using a 70-item semi-quantitative FFQ, and their average daily nutritional intakes were calculated from the weekly frequency and portion size of various foods. Their health status was monitored by several methods including a neuropsychological test. Dietary patterns associated with the risk of dementia were assessed using a reduced rank regression analysis [38]. This analysis extracted seven dietary patterns that explained 87.1% of the total variation in the intakes of the following seven nutrients, which were selected as responsible variables: saturated fatty acid, monounsaturated fatty acid, polyunsaturated fatty acid, vitamin C, potassium, calcium, and magnesium. Seven dietary factors were associated with a preventive effect against the risks of dementia that was greater than 20% in magnitude. These protective factors were soybeans and soybean products, green vegetables, other vegetables, algae, and milk and dairy products. In contrast, a high intake of rice was associated with an increase in the risks of dementia greater than 20% in magnitude. Thus, a dietary pattern characterized by high intakes of soybeans and soybean products, vegetables, algae, and milk or dairy products together with a low intake of rice was associated with a reduced risk of dementia. These results support the contention that a high intake of milk and dairy products helps to prevent cognitive decline.

The subsequent research of Ozawa et al. [39] focused on the effects of milk and dairy intake on the development of all-cause dementia and its subtypes in an elderly Japanese population. During 17 years of follow-up on individuals aged 60 years or older who did not initially have dementia (N = 1081), 303 subjects developed all-cause dementia. The age- and sex-adjusted incidences of all-cause dementia, Alzheimer's disease, and vascular dementia all showed a significant inverse correlation with the intake of milk and dairy products. After adjusting for potential confounders, the linear inverse relationship between the intake of milk and dairy products and the development of Alzheimer's disease remained significant, whereas the relationships with all-cause dementia and vascular dementia were not significant. The investigation showed that a greater intake of milk and dairy products was associated with a reduced risk of dementia, especially Alzheimer's disease, in the general Japanese population.

3. Clinical Trials for the Improvement of Cognitive Function by Dairy Products

In a clinical trial, Ogata et al. [40] found that the intake of dairy products was strongly associated with better short-term memory. The association was significant both with and without adjustment for genetic and family environment factors using a sample of twin pairs. Short-term memory was evaluated using the "logical memory I" (LM-I) scores from the Japanese version of the revised Wechsler memory scale. Participants were asked to listen to two short stories and immediately recall the details. The final analysis was performed using data from 78 men and 278 women. All individual-level analyses using generalized estimating equations showed that dairy product intake was significantly associated with the LM-I scores in men. In addition, a within-pair analysis using within-monozygotic and within–dizygotic pair-difference scores showed a significant association between the intake of dairy products and LM-I scores in men. Furthermore, a within-pair analysis using within-monozygotic pair-difference scores indicated a significant association between the intake of dairy products and LM-I scores in men. Among men, a high intake of dairy products was significantly associated with better short-term memory after adjustment for possible covariates. The authors concluded that the intake of dairy products may prevent cognitive decline regardless of genetic and family environment factors in men.

Markus et al. [41,42] demonstrated that the intakes of α-lactalbumin-rich whey protein isolate improved cognitive performance in stress-vulnerable subjects. They evaluated the effects of the intake of tryptophan-rich whey protein on the ratio of plasma tryptophan to the sum of the other large neutral amino acids (Trp-LNAA ratio) and cognitive performance in high stress-vulnerable subjects. Their double-blind, placebo-controlled, crossover study included 29 high stress-vulnerable subjects and 29 low stress-vulnerable subjects. A significantly greater increase in the plasma Trp-LNAA ratio was observed after the consumption of the α-lactalbumin-rich diet than after the consumption of the control diet. Cognitive performance was evaluated using a computerized Sternberg memory scanning task, and the subjects' reaction time and amount of errors across the different subtasks were measured. The mean reaction time showed a significant difference between the high and low stress-vulnerable subjects. Furthermore, the reaction time of the high-stress-vulnerable subjects was significantly lower after consuming the α-lactalbumin diet (758 ± 137 ms) than it was after consuming the control diet (800 ± 173 ms). An increase in the plasma Trp-LNAA ratio is considered to be an indirect indicator of increased brain serotonin function, which results in the improvement of cognitive performance.

The authors suggested that the intake of an α-lactalbumin-rich diet increases the level of tryptophan and serotonin in the brain and improves cognitive performance in stress-vulnerable subjects.

4. Preventive Effects of Dairy Products Fermented with *Penicillium candidum* against the Pathology of Alzheimer's Disease

Recent epidemiological and clinical studies have suggested that a high intake of dairy products may have preventive effects against cognitive decline and Alzheimer's disease. Following this finding, it is important to elucidate the mechanism and responsible molecular components. Using transgenic model mice, Ano et al. [43] demonstrated that the intake of Camembert cheese, which is a fermented dairy product, displayed preventive effects against Alzheimer's disease. In the experiment, 5xFAD mice were used as an Alzheimer's disease model. These mice overexpress mutant human amyloid precursor protein with the Swedish (K670N, M671L), Florida (I716V), and London (V717I) familial Alzheimer's disease mutations along with human presenilin 1 harboring two familial Alzheimer's disease mutations, namely M146L and L286V. The 5xFAD mice display Aβ depositions, plaques, and severe inflammation in the brain in addition to impaired cognitive function. 5xFAD mice aged 3–6 months were given food with or without an extract from Camembert cheese. The brain tissues of mice fed with the Camembert cheese extract showed reduced levels of inflammatory cytokines including tumor necrosis factor α (TNF-α) and macrophage inflammatory protein 1α. The amounts of Aβ$_{1-42}$ quantified by enzyme-linked immunosorbent assay (ELISA) and detected immunohistochemically were reduced in the group fed with the Camembert cheese extract. The abundances of the brain-derived and glial-derived neurotropic factors increased. These results showed that certain components of Camembert cheese contribute to the suppression of the inflammation and reduction of Aβ in the brain. In subsequent studies of the same research group, the responsible agents in Camembert cheese that contribute to the prevention of Alzheimer's disease pathology, especially inflammation in the brain, were explored.

5. Neuronal Inflammation Accelerates the Pathology of Alzheimer's Disease

Several reports have demonstrated that inflammation in the brain following the formation of Aβ plaques and NFTs is closely associated with the development and progression of Alzheimer's disease [3,8,9]. Inflammation in the brain is mainly regulated by microglia, which remove apoptotic cells and waste products such as Aβ through phagocytosis and also contribute to the host defense against virus infection in the central nervous system [44,45]. Microglia play an important role in clearing Aβ to regulate the pathology of Alzheimer's disease [46]. Recent studies revealed the crucial roles of microglia in maintaining the brain environment and cognitive function. However, in the brain tissues of patients with Alzheimer's disease, microglia infiltrate around the Aβ plaques, become excessively activated, and chronically produce inflammatory mediators such as TNF-α, macrophage inflammatory protein 1α, reactive oxygen species, and nitric oxide. These inflammatory mediators are toxic to neurons and cause neuronal cell death subsequent to Aβ deposition [47]. Under normal physiological conditions, microglia are important for the maintenance of brain environment; however, under pathological conditions, microglia can become excessively activated and negatively contribute to the brain environment.

Numerous reports have suggested that controlling the activities of microglia may contribute to the prevention and cure of Alzheimer's disease and cognitive decline. Epidemiological studies have suggested that the prolonged use of nonsteroidal anti-inflammatory drugs, including a common medication, ibuprofen, significantly reduces the risk of developing Alzheimer's disease [48–51]. Consistent with those results, a long-term ibuprofen treatment was found to significantly suppress microglial inflammation and the development of Aβ plaques in a transgenic mouse model of Alzheimer's disease [52,53].

Before Ano et al. evaluated the effects of a Camembert cheese extract using 5xFAD transgenic Alzheimer's model mice, they assessed the effects of various fermented dairy products on microglia [43]. Their results indicated that dairy products fermented with the fungi *Penicillium candidum* and *Penicillium roqueforti* suppressed the inflammatory responses of primary microglia. Their findings suggested that *Penicillium* fermentation is essential for the anti-inflammatory activity, although other dairy products that were unfermented or fermented did not suppress the microglial inflammatory responses. The intake of Camembert cheese extract reduced inflammatory responses and Aβ production in the hippocampus of 5xFAD transgenic mice, while it increased the production of the brain-derived and glial-derived neurotropic factors. Various fatty acids and peptides were also generated during the fermentation of dairy products with the *Penicillium* fungi. Ano et al., in addition, investigated the molecular components of Camembert cheese that are responsible for preventing the development of Alzheimer's disease pathology.

6. Effects of Oleamide and Dehydroergosterol Generated during the Fermentation of Dairy Products with *Penicillium* Fungi on Brain Inflammation

Ano et al. [43,54] identified oleamide and dehydroergosterol as two components of Camembert cheese that were responsible for suppressing microglial inflammation. Oleamide is an amide of oleic acid [55,56] (Figure 1) that forms naturally in the body of animals and might be synthesized during the fermentation of dairy products by *Penicillium* fungi [57]. Oleamide accumulates in the cerebrospinal fluid during sleep deprivation and induces sleep in animals; thus, it has potential applications in the treatment of mood and sleep disorders [58,59]. Oleamide not only suppresses the production of inflammatory cytokines and chemokines but also increases the microglial phagocytosis of Aβ. In addition, it induces the differentiation of microglia into the anti-inflammatory M2 type. A recent study using murine macrophages showed that oleamide suppressed the induction of iNOS and COX-2 by lipopolysaccharides (LPS) via preventing the nuclear translocation of NF-κβ by suppressing the phosphorylation of Iκβ-α [60]. Oleamide is also known as an endogenous substance that binds to cannabinoid receptor 2 (CB2), which is mainly expressed on the surface of immune cells (monocytes, macrophages, and B cells). CB2 is also expressed on microglia and contributes to the inhibition of microglia-mediated neurotoxicity by reducing the production of pro-inflammatory molecules [61]. In addition, CB2 activity facilitates the transformation of microglial cells from the M1 to M2 phenotype, which is suggested to favor phagocytosis and reparative mechanisms [62]. Several studies have proposed a direct role for CB2 in the modulation of Aβ peptide levels in the brain. Most of those studies have suggested that CB2 participates in Aβ clearance rather than in Aβ production and aggregation. In the case of amyloid precursor protein/presenilin 1 mice lacking CB2, the increased Aβ deposition observed may be related to a reduced phagocytotic activity of microglia in their brains [63], considering the role of CB2 activity in promoting microglial-induced Aβ phagocytosis [64,65]. These reports support the contention that oleamide, which is an agonist of CB2, suppresses the microglial inflammation and enhances the phagocytosis of Aβ, resulting in a preventive action against Alzheimer's disease. CB2 has been receiving increasing attention as a therapeutic target for Alzheimer's disease [66–68]. Ano et al. discovered that the concentration of oleamide in dairy products depends on the type of dairy product and the fermentation process used. Therefore, the dietary intake of oleamide can be increased via supplementation or the consumption of specific dairy products.

Dehydroergosterol was also discovered as a component of Camembert cheese that suppressed microglial inflammation. Dehydroergosterol is an analogue of ergosterol, which is a sterol found in the cell membrane of fungi [69,70] (Figure 1). Dehydroergosterol is generated by fungi during fermentation. Dehydroergosterol suppresses the LPS-induced inflammatory response (including TNF-α production) of primary microglia in a concentration-dependent manner, whereas ergosterol does not display this activity. Microglia treated with dehydroergosterol were observed to differentiate into the anti-inflammatory M2 phenotype. The number of apoptotic neurons detected by staining for caspase 3/7 decreased after co-culturing the cells with a microglial culture supernatant treated with dehydroergosterol and LPS as compared with an LPS-only control. These results suggested that dehydroergosterol suppresses the inflammatory responses of microglia and exerts a neuroprotective effect via promoting synaptic extension and neuronal survival [71]. The amount of dehydroergosterol produced during fermentation varies among different *Penicillium* strains, so it might be possible to increase the amount of dehydroergosterol in dairy products by optimizing the fermentation processes.

The mechanisms by which oleamide and dehydroergosterol regulate inflammation in the brain have been receiving increasing attention, because neural inflammation is involved not only in dementia but also in other neuronal disorders [72,73] including depression [74,75], anxiety disorder [76], and chronic fatigue [77]. A recent study demonstrated that the intakes of oleamide shows antidepressant-like effects in mice subjected to the forced swimming test via the activation of cannabinoid receptors [78]. In another demonstration, oleamide improved schizophrenia-like symptoms in mice treated with an NMDA receptor antagonist, MK-801 [79]. The balance of microglial differentiation between M1 pro-inflammatory type and M2 anti-inflammatory type is important to maintain homeostasis in the central nervous system. Oleamide has the potential to have preventive effects against the development of several brain disorders. Daily habits including nutrition, sleep, and exercise are closely related to the maintenance of the microglial balance and the prevention of neuronal disorders. Future studies will further elucidate the effects of active ingredients of fermented dairy products against depression and other brain diseases.

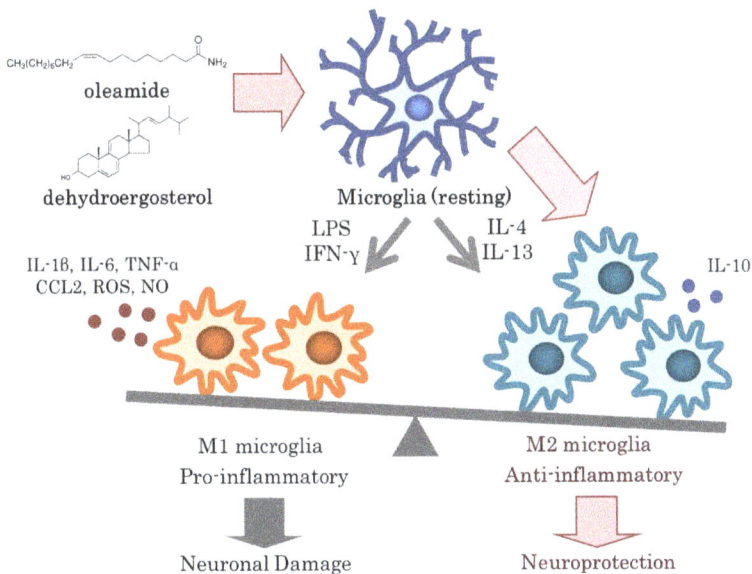

Figure 1. Modulation of microglial activation by dairy products.

7. Conclusions

This review introduced recent advances regarding the protective effects of dairy product intake against dementia and cognitive decline. The reports regarding these issues will help with the development of new approaches for the prevention of dementia. Oleamide and dehydroergosterol were discussed in this review as responsible agents for these protective effects of dairy products (Figure 1). However, the functions of the other various fatty acids and peptides generated during fermentation have not yet been elucidated. Future studies are expected to elucidate the mechanisms underlying the physiological benefits of dairy products in more detail. Based on the current evidence, the regular intake of dairy products and their molecular or microbial components seems to have the potential to contribute to the prevention of dementia and cognitive decline.

Int. J. Mol. Sci. **2018**, *19*, 1927

Oleamide and dehydroergosterol identified from Camembert cheese induce microglia into the M2 anti-inflammatory phenotype, leading to neuroprotection. The mechanisms that regulate microglial activation and inflammation in Alzheimer's disease are important targets for disease prevention. The regulation of microglia via daily lifestyle habits has been receiving increasing attention. The intake of neuroprotective and anti-inflammatory compounds including oleamide and dehydroergosterol in meals is safe and easy, so nutritional approaches are promising for the prevention of neurodegenerative disorders.

Funding: This research received no external funding.

Conflicts of Interest: The authors declare no conflict of interest.

Abbreviations

Aβ	Amyloid β
CB2	Cannabinoid receptor 2
FAD	Familial Alzheimer's disease
FFQ	Food frequency questionnaire
LM-I	Logical memory I
LPS	Lipopolysaccharides
NFT	Neurofibrillary tangle
TNF-α	Tumor necrosis factor α

References

1. Bloom, G.S. Amyloid-β and tau: The trigger and bullet in Alzheimer disease pathogenesis. *JAMA Neurol.* **2014**, *71*, 505–508. [CrossRef] [PubMed]
2. Polanco, J.C.; Li, C.; Bodea, L.G.; Martinez-Marmol, R.; Meunier, F.A.; Gotz, J. Amyloid-β and tau complexity—Towards improved biomarkers and targeted therapies. *Nat. Rev. Neurol.* **2018**, *14*, 22–39. [CrossRef] [PubMed]
3. Takashima, A. Amyloid-β, tau, and dementia. *J. Alzheimer Dis. JAD* **2009**, *17*, 729–736. [CrossRef] [PubMed]
4. Congdon, E.E.; Sigurdsson, E.M. Tau-targeting therapies for Alzheimer disease. *Nat. Rev. Neurol.* **2018**. [CrossRef] [PubMed]
5. Kalaria, R.N. The pathology and pathophysiology of vascular dementia. *Neuropharmacology* **2018**, *134 Pt B*, 226–239. [CrossRef] [PubMed]
6. Mueller, C.; Ballard, C.; Corbett, A.; Aarsland, D. The prognosis of dementia with Lewy bodies. *Lancet Neurol.* **2017**, *16*, 390–398. [CrossRef]
7. Silveri, M.C. Frontotemporal dementia to Alzheimer's disease. *Dialogues Clin. Neurosci.* **2007**, *9*, 153–160. [PubMed]
8. Heppner, F.L.; Ransohoff, R.M.; Becher, B. Immune attack: The role of inflammation in Alzheimer disease. *Nat. Rev. Neurosci.* **2015**, *16*, 358–372. [CrossRef] [PubMed]
9. Ransohoff, R.M. Specks of insight into Alzheimer's disease. *Nature* **2017**, *552*, 342–343. [CrossRef] [PubMed]
10. Livingston, G.; Sommerlad, A.; Orgeta, V.; Costafreda, S.G.; Huntley, J.; Ames, D.; Ballard, C.; Banerjee, S.; Burns, A.; Cohen-Mansfield, J.; et al. Dementia prevention, intervention, and care. *Lancet* **2017**, *390*, 2673–2734. [CrossRef]
11. Caruso, A.; Nicoletti, F.; Mango, D.; Saidi, A.; Orlando, R.; Scaccianoce, S. Stress as risk factor for Alzheimer's disease. *Pharmacol. Res.* **2018**, *132*, 130–134. [CrossRef] [PubMed]
12. Petersson, S.D.; Philippou, E. Mediterranean Diet, Cognitive Function, and Dementia: A Systematic Review of the Evidence. *Adv. Nutr.* **2016**, *7*, 889–904. [CrossRef] [PubMed]
13. Singh, B.; Parsaik, A.K.; Mielke, M.M.; Erwin, P.J.; Knopman, D.S.; Petersen, R.C.; Roberts, R.O. Association of mediterranean diet with mild cognitive impairment and Alzheimer's disease: A systematic review and meta-analysis. *J. Alzheimer Dis. JAD* **2014**, *39*, 271–282. [CrossRef] [PubMed]
14. Lourida, I.; Soni, M.; Thompson-Coon, J.; Purandare, N.; Lang, I.A.; Ukoumunne, O.C.; Llewellyn, D.J. Mediterranean diet, cognitive function, and dementia: A systematic review. *Epidemiology* **2013**, *24*, 479–489. [CrossRef] [PubMed]

15. Freitas, H.R.; Ferreira, G.D.C.; Trevenzoli, I.H.; Oliveira, K.J.; de Melo Reis, R.A. Fatty Acids, Antioxidants and Physical Activity in Brain Aging. *Nutrients* **2017**, *9*. [CrossRef] [PubMed]

16. Heras-Sandoval, D.; Pedraza-Chaverri, J.; Perez-Rojas, J.M. Role of docosahexaenoic acid in the modulation of glial cells in Alzheimer's disease. *J. Neuroinflamm.* **2016**, *13*, 61. [CrossRef] [PubMed]

17. Dyall, S.C. Long-chain omega-3 fatty acids and the brain: A review of the independent and shared effects of EPA, DPA and DHA. *Front. Aging Neurosci.* **2015**, *7*, 52. [CrossRef] [PubMed]

18. Crupi, R.; Marino, A.; Cuzzocrea, S. n-3 fatty acids: Role in neurogenesis and neuroplasticity. *Curr. Med. Chem.* **2013**, *20*, 2953–2963. [CrossRef] [PubMed]

19. Sawda, C.; Moussa, C.; Turner, R.S. Resveratrol for Alzheimer's disease. *Ann. N. Y. Acad. Sci.* **2017**, *1403*, 142–149. [CrossRef] [PubMed]

20. Bastianetto, S.; Menard, C.; Quirion, R. Neuroprotective action of resveratrol. *Biochim. Biophys. Acta* **2015**, *1852*, 1195–1201. [CrossRef] [PubMed]

21. Turner, R.S.; Thomas, R.G.; Craft, S.; van Dyck, C.H.; Mintzer, J.; Reynolds, B.A.; Brewer, J.B.; Rissman, R.A.; Raman, R.; Aisen, P.S. A randomized, double-blind, placebo-controlled trial of resveratrol for Alzheimer disease. *Neurology* **2015**, *85*, 1383–1391. [CrossRef] [PubMed]

22. Goozee, K.G.; Shah, T.M.; Sohrabi, H.R.; Rainey-Smith, S.R.; Brown, B.; Verdile, G.; Martins, R.N. Examining the potential clinical value of curcumin in the prevention and diagnosis of Alzheimer's disease. *Br. J. Nutr.* **2016**, *115*, 449–465. [CrossRef] [PubMed]

23. Hemachandra Reddy, P.; Manczak, M.; Yin, X.; Grady, M.C.; Mitchell, A.; Tonk, S.; Kuruva, C.S.; Bhatti, J.S.; Kandimalla, R.; Vijayan, M.; et al. Protective Effects of Indian Spice Curcumin Against Amyloid-β in Alzheimer's Disease. *J. Alzheimer Dis. JAD* **2018**, *61*, 843–866. [CrossRef] [PubMed]

24. Botchway, B.O.A.; Moore, M.K.; Akinleye, F.O.; Iyer, I.C.; Fang, M. Nutrition: Review on the Possible Treatment for Alzheimer's Disease. *J. Alzheimer Dis. JAD* **2018**, *61*, 867–883. [CrossRef] [PubMed]

25. Camfield, D.A.; Owen, L.; Scholey, A.B.; Pipingas, A.; Stough, C. Dairy constituents and neurocognitive health in ageing. *Br. J. Nutr.* **2011**, *106*, 159–174. [CrossRef] [PubMed]

26. Crichton, G.E.; Murphy, K.J.; Bryan, J. Dairy intake and cognitive health in middle-aged South Australians. *Asia Pac. J. Clin. Nutr.* **2010**, *19*, 161–171. [PubMed]

27. Crichton, G.E.; Bryan, J.; Murphy, K.J.; Buckley, J. Review of dairy consumption and cognitive performance in adults: Findings and methodological issues. *Dement. Geriatr. Cogn. Disord.* **2010**, *30*, 352–361. [CrossRef] [PubMed]

28. Major, G.C.; Chaput, J.P.; Ledoux, M.; St-Pierre, S.; Anderson, G.H.; Zemel, M.B.; Tremblay, A. Recent developments in calcium-related obesity research. *Obes. Rev.* **2008**, *9*, 428–445. [CrossRef] [PubMed]

29. Zemel, M.B. Role of calcium and dairy products in energy partitioning and weight management. *Am. J. Clin. Nutr.* **2004**, *79*, 907S–912S. [CrossRef] [PubMed]

30. Azadbakht, L.; Mirmiran, P.; Esmaillzadeh, A.; Azizi, T.; Azizi, F. Beneficial effects of a Dietary Approaches to Stop Hypertension eating plan on features of the metabolic syndrome. *Diabetes Care* **2005**, *28*, 2823–2831. [CrossRef] [PubMed]

31. Choi, H.K.; Willett, W.C.; Stampfer, M.J.; Rimm, E.; Hu, F.B. Dairy consumption and risk of type 2 diabetes mellitus in men: A prospective study. *Arch. Intern. Med.* **2005**, *165*, 997–1003. [CrossRef] [PubMed]

32. Liu, S.; Choi, H.K.; Ford, E.; Song, Y.; Klevak, A.; Buring, J.E.; Manson, J.E. A prospective study of dairy intake and the risk of type 2 diabetes in women. *Diabetes Care* **2006**, *29*, 1579–1584. [CrossRef] [PubMed]

33. Dik, M.G.; Jonker, C.; Comijs, H.C.; Deeg, D.J.; Kok, A.; Yaffe, K.; Penninx, B.W. Contribution of metabolic syndrome components to cognition in older individuals. *Diabetes Care* **2007**, *30*, 2655–2660. [CrossRef] [PubMed]

34. Crichton, G.E.; Elias, M.F.; Buckley, J.D.; Murphy, K.J.; Bryan, J.; Frisardi, V. Metabolic syndrome, cognitive performance, and dementia. *J. Alzheimer Dis. JAD* **2012**, *30* (Suppl. 2), S77–S87. [CrossRef] [PubMed]

35. Raffaitin, C.; Gin, H.; Empana, J.P.; Helmer, C.; Berr, C.; Tzourio, C.; Portet, F.; Dartigues, J.F.; Alperovitch, A.; Barberger-Gateau, P. Metabolic syndrome and risk for incident Alzheimer's disease or vascular dementia: The Three-City Study. *Diabetes Care* **2009**, *32*, 169–174. [CrossRef] [PubMed]

36. Rahman, A.; Sawyer Baker, P.; Allman, R.M.; Zamrini, E. Dietary factors and cognitive impairment in community-dwelling elderly. *J. Nutr. Health Aging* **2007**, *11*, 49–54. [PubMed]

37. Ozawa, M.; Ninomiya, T.; Ohara, T.; Doi, Y.; Uchida, K.; Shirota, T.; Yonemoto, K.; Kitazono, T.; Kiyohara, Y. Dietary patterns and risk of dementia in an elderly Japanese population: The Hisayama Study. *Am. J. Clin. Nutr.* **2013**, *97*, 1076–1082. [CrossRef] [PubMed]

38. Hoffmann, K.; Schulze, M.B.; Schienkiewitz, A.; Nothlings, U.; Boeing, H. Application of a new statistical method to derive dietary patterns in nutritional epidemiology. *Am. J. Epidemiol.* **2004**, *159*, 935–944. [CrossRef] [PubMed]

39. Ozawa, M.; Ohara, T.; Ninomiya, T.; Hata, J.; Yoshida, D.; Mukai, N.; Nagata, M.; Uchida, K.; Shirota, T.; Kitazono, T.; et al. Milk and dairy consumption and risk of dementia in an elderly Japanese population: The Hisayama Study. *J. Am. Geriatr. Soc.* **2014**, *62*, 1224–1230. [CrossRef] [PubMed]

40. Ogata, S.; Tanaka, H.; Omura, K.; Honda, C.; Osaka Twin Research, G.; Hayakawa, K. Association between intake of dairy products and short-term memory with and without adjustment for genetic and family environmental factors: A twin study. *Clin. Nutr.* **2016**, *35*, 507–513. [CrossRef] [PubMed]

41. Markus, C.R.; Olivier, B.; Panhuysen, G.E.; Van Der Gugten, J.; Alles, M.S.; Tuiten, A.; Westenberg, H.G.; Fekkes, D.; Koppeschaar, H.F.; de Haan, E.E. The bovine protein alpha-lactalbumin increases the plasma ratio of tryptophan to the other large neutral amino acids, and in vulnerable subjects raises brain serotonin activity, reduces cortisol concentration, and improves mood under stress. *Am. J. Clin. Nutr.* **2000**, *71*, 1536–1544. [CrossRef] [PubMed]

42. Markus, C.R.; Olivier, B.; de Haan, E.H. Whey protein rich in alpha-lactalbumin increases the ratio of plasma tryptophan to the sum of the other large neutral amino acids and improves cognitive performance in stress-vulnerable subjects. *Am. J. Clin. Nutr.* **2002**, *75*, 1051–1056. [CrossRef] [PubMed]

43. Ano, Y.; Ozawa, M.; Kutsukake, T.; Sugiyama, S.; Uchida, K.; Yoshida, A.; Nakayama, H. Preventive effects of a fermented dairy product against Alzheimer's disease and identification of a novel oleamide with enhanced microglial phagocytosis and anti-inflammatory activity. *PLoS ONE* **2015**, *10*, e0118512. [CrossRef] [PubMed]

44. Sarlus, H.; Heneka, M.T. Microglia in Alzheimer's disease. *J. Clin. Investig.* **2017**, *127*, 3240–3249. [CrossRef] [PubMed]

45. Visan, I. Alzheimer's disease microglia. *Nat. Immunol.* **2017**, *18*, 876. [CrossRef] [PubMed]

46. Lee, C.Y.; Landreth, G.E. The role of microglia in amyloid clearance from the AD brain. *J. Neural Transm.* **2010**, *117*, 949–960. [CrossRef] [PubMed]

47. Hansen, D.V.; Hanson, J.E.; Sheng, M. Microglia in Alzheimer's disease. *J. Cell Biol.* **2018**, *217*. [CrossRef] [PubMed]

48. Stewart, W.F.; Kawas, C.; Corrada, M.; Metter, E.J. Risk of Alzheimer's disease and duration of NSAID use. *Neurology* **1997**, *48*, 626–632. [CrossRef] [PubMed]

49. Wang, J.; Tan, L.; Wang, H.F.; Tan, C.C.; Meng, X.F.; Wang, C.; Tang, S.W.; Yu, J.T. Anti-inflammatory drugs and risk of Alzheimer's disease: An updated systematic review and meta-analysis. *J. Alzheimer Dis. JAD* **2015**, *44*, 385–396. [CrossRef] [PubMed]

50. Imbimbo, B.P.; Solfrizzi, V.; Panza, F. Are NSAIDs useful to treat Alzheimer's disease or mild cognitive impairment? *Front. Aging Neurosci.* **2010**, *2*. [CrossRef] [PubMed]

51. Miguel-Alvarez, M.; Santos-Lozano, A.; Sanchis-Gomar, F.; Fiuza-Luces, C.; Pareja-Galeano, H.; Garatachea, N.; Lucia, A. Non-steroidal anti-inflammatory drugs as a treatment for Alzheimer's disease: A systematic review and meta-analysis of treatment effect. *Drugs Aging* **2015**, *32*, 139–147. [CrossRef] [PubMed]

52. Varvel, N.H.; Bhaskar, K.; Kounnas, M.Z.; Wagner, S.L.; Yang, Y.; Lamb, B.T.; Herrup, K. NSAIDs prevent, but do not reverse, neuronal cell cycle reentry in a mouse model of Alzheimer disease. *J. Clin. Investig.* **2009**, *119*, 3692–3702. [CrossRef] [PubMed]

53. Wyss-Coray, T.; Mucke, L. Ibuprofen, inflammation and Alzheimer disease. *Nat. Med.* **2000**, *6*, 973–974. [CrossRef] [PubMed]

54. Ano, Y.; Kutsukake, T.; Hoshi, A.; Yoshida, A.; Nakayama, H. Identification of a novel dehydroergosterol enhancing microglial anti-inflammatory activity in a dairy product fermented with *Penicillium candidum*. *PLoS ONE* **2015**, *10*, e0116598. [CrossRef] [PubMed]

55. Awasthi, N.P.; Singh, R.P. Lipase-catalyzed synthesis of fatty acid amide (erucamide) using fatty acid and urea. *J. Oleo Sci.* **2007**, *56*, 507–509. [CrossRef] [PubMed]

56. Slotema, W.F.; Sandoval, G.; Guieysse, D.; Straathof, A.J.; Marty, A. Economically pertinent continuous amide formation by direct lipase-catalyzed amidation with ammonia. *Biotechnol. Bioeng.* **2003**, *82*, 664–669. [CrossRef] [PubMed]

57. Cravatt, B.F.; Giang, D.K.; Mayfield, S.P.; Boger, D.L.; Lerner, R.A.; Gilula, N.B. Molecular characterization of an enzyme that degrades neuromodulatory fatty-acid amides. *Nature* **1996**, *384*, 83–87. [CrossRef] [PubMed]

58. Boger, D.L.; Henriksen, S.J.; Cravatt, B.F. Oleamide: An endogenous sleep-inducing lipid and prototypical member of a new class of biological signaling molecules. *Curr. Pharm. Des.* **1998**, *4*, 303–314. [PubMed]

59. Prospero-Garcia, O.; Amancio-Belmont, O.; Becerril Melendez, A.L.; Ruiz-Contreras, A.E.; Mendez-Diaz, M. Endocannabinoids and sleep. *Neurosci. Biobehav. Rev.* **2016**, *71*, 671–679. [CrossRef] [PubMed]

60. Moon, S.M.; Lee, S.A.; Hong, J.H.; Kim, J.S.; Kim, D.K.; Kim, C.S. Oleamide suppresses inflammatory responses in LPS-induced RAW264.7 murine macrophages and alleviates paw edema in a carrageenan-induced inflammatory rat model. *Int. Immunopharmacol.* **2018**, *56*, 179–185. [CrossRef] [PubMed]

61. Cabral, G.A.; Griffin-Thomas, L. Emerging role of the cannabinoid receptor CB2 in immune regulation: Therapeutic prospects for neuroinflammation. *Expert Rev. Mol. Med.* **2009**, *11*, e3. [CrossRef] [PubMed]

62. Mecha, M.; Feliu, A.; Carrillo-Salinas, F.J.; Rueda-Zubiaurre, A.; Ortega-Gutierrez, S.; de Sola, R.G.; Guaza, C. Endocannabinoids drive the acquisition of an alternative phenotype in microglia. *Brain Behave. Immun.* **2015**, *49*, 233–245. [CrossRef] [PubMed]

63. Aso, E.; Andres-Benito, P.; Carmona, M.; Maldonado, R.; Ferrer, I. Cannabinoid Receptor 2 Participates in Amyloid-β Processing in a Mouse Model of Alzheimer's Disease but Plays a Minor Role in the Therapeutic Properties of a Cannabis-Based Medicine. *J. Alzheimer Dis. JAD* **2016**, *51*, 489–500. [CrossRef] [PubMed]

64. Ehrhart, J.; Obregon, D.; Mori, T.; Hou, H.; Sun, N.; Bai, Y.; Klein, T.; Fernandez, F.; Tan, J.; Shytle, R.D. Stimulation of cannabinoid receptor 2 (CB2) suppresses microglial activation. *J. Neuroinflamm.* **2005**, *2*, 29. [CrossRef] [PubMed]

65. Tolon, R.M.; Nunez, E.; Pazos, M.R.; Benito, C.; Castillo, A.I.; Martinez-Orgado, J.A.; Romero, J. The activation of cannabinoid CB2 receptors stimulates in situ and in vitro β-amyloid removal by human macrophages. *Brain Res.* **2009**, *1283*, 148–154. [CrossRef] [PubMed]

66. Navarro, G.; Morales, P.; Rodriguez-Cueto, C.; Fernandez-Ruiz, J.; Jagerovic, N.; Franco, R. Targeting Cannabinoid CB2 Receptors in the Central Nervous System. Medicinal Chemistry Approaches with Focus on Neurodegenerative Disorders. *Front. Neurosci.* **2016**, *10*, 406. [CrossRef] [PubMed]

67. Mecha, M.; Carrillo-Salinas, F.J.; Feliu, A.; Mestre, L.; Guaza, C. Microglia activation states and cannabinoid system: Therapeutic implications. *Pharmacol. Ther.* **2016**, *166*, 40–55. [CrossRef] [PubMed]

68. Cassano, T.; Calcagnini, S.; Pace, L.; De Marco, F.; Romano, A.; Gaetani, S. Cannabinoid Receptor 2 Signaling in Neurodegenerative Disorders: From Pathogenesis to a Promising Therapeutic Target. *Front. Neurosci.* **2017**, *11*, 30. [CrossRef] [PubMed]

69. Wustner, D. Fluorescent sterols as tools in membrane biophysics and cell biology. *Chem. Phys. Lipids* **2007**, *146*, 1–25. [CrossRef] [PubMed]

70. Tachibana, Y. Synthesis and structure-activity relationships of bioactive compounds using sterols. *Yakugaku Zasshi* **2006**, *126*, 1139–1154. [CrossRef] [PubMed]

71. Wu, Y.; Dissing-Olesen, L.; MacVicar, B.A.; Stevens, B. Microglia: Dynamic Mediators of Synapse Development and Plasticity. *Trends Immunol.* **2015**, *36*, 605–613. [CrossRef] [PubMed]

72. Amor, S.; Peferoen, L.A.; Vogel, D.Y.; Breur, M.; van der Valk, P.; Baker, D.; van Noort, J.M. Inflammation in neurodegenerative diseases—An update. *Immunology* **2014**, *142*, 151–166. [CrossRef] [PubMed]

73. Griffin, W.S. Inflammation and neurodegenerative diseases. *Am. J. Clin. Nutr.* **2006**, *83*, 470S–474S. [CrossRef] [PubMed]

74. Miller, A.H.; Raison, C.L. The role of inflammation in depression: From evolutionary imperative to modern treatment target. *Nat. Rev. Immunol.* **2016**, *16*, 22–34. [CrossRef] [PubMed]

75. Kohler, O.; Krogh, J.; Mors, O.; Benros, M.E. Inflammation in Depression and the Potential for Anti-Inflammatory Treatment. *Curr. Neuropharmacol.* **2016**, *14*, 732–742. [CrossRef] [PubMed]

76. Salim, S.; Chugh, G.; Asghar, M. Inflammation in anxiety. *Adv. Protein Chem. Struct. Boil.* **2012**, *88*, 1–25.

77. Komaroff, A.L. Inflammation correlates with symptoms in chronic fatigue syndrome. *Proc. Natl. Acad. Sci. USA* **2017**, *114*, 8914–8916. [CrossRef] [PubMed]

78. Kruk-Slomka, M.; Michalak, A.; Biala, G. Antidepressant-like effects of the cannabinoid receptor ligands in the forced swimming test in mice: Mechanism of action and possible interactions with cholinergic system. *Behav. Brain Res.* **2015**, *284*, 24–36. [CrossRef] [PubMed]
79. Kruk-Slomka, M.; Budzynska, B.; Slomka, T.; Banaszkiewicz, I.; Biala, G. The Influence of the CB1 Receptor Ligands on the Schizophrenia-Like Effects in Mice Induced by MK-801. *Neurotox. Res.* **2016**, *30*, 658–676. [CrossRef] [PubMed]

International Journal of
Molecular Sciences

MDPI

Review

Epigenetic Factors in Late-Onset Alzheimer's Disease: *MTHFR* and *CTH* Gene Polymorphisms, Metabolic Transsulfuration and Methylation Pathways, and B Vitamins

Gustavo C. Román [1,2,*,†], Oscar Mancera-Páez [3,4,†] and Camilo Bernal [3]

1. Department of Neurology, Methodist Neurological Institute, Institute for Academic Medicine Houston Methodist Research Institute, Houston Methodist Hospital, Houston, TX 77030, USA
2. Weill Cornell Medical College, Department of Neurology, Cornell University, New York, NY 10065, USA
3. Universidad Nacional de Colombia, Hospital Universitario Nacional, Faculty of Medicine, Department of Neurology, Bogotá ZC 57, Colombia; ogmancerap@unal.edu.co (O.M.-P.); camilobernalmd@hotmail.com (C.B.)
4. David Cabello International Alzheimer Disease Scholarship Fund, Houston Methodist Hospital, Houston, TX77030, USA
* Correspondence: GCRoman@HoustonMethodist.org; Tel.: +1-713-441-1150; Fax: +1-713-790-4990
† These authors contributed equally to this work.

Received: 20 December 2018; Accepted: 11 January 2019; Published: 14 January 2019

Abstract: DNA methylation and other epigenetic factors are important in the pathogenesis of late-onset Alzheimer's disease (LOAD). Methylenetetrahydrofolate reductase (*MTHFR*) gene mutations occur in most elderly patients with memory loss. MTHFR is critical for production of S-adenosyl-L-methionine (SAM), the principal methyl donor. A common mutation (1364T/T) of the cystathionine-γ-lyase (*CTH*) gene affects the enzyme that converts cystathionine to cysteine in the transsulfuration pathway causing plasma elevation of total homocysteine (tHcy) or hyperhomocysteinemia—a strong and independent risk factor for cognitive loss and AD. Other causes of hyperhomocysteinemia include aging, nutritional factors, and deficiencies of B vitamins. We emphasize the importance of supplementing vitamin B_{12} (methylcobalamin), vitamin B_9 (folic acid), vitamin B_6 (pyridoxine), and SAM to patients in early stages of LOAD.

Keywords: Alzheimer's disease; cystathionine-γ-lyase *CTH* gene; DNA methylation; epigenetics; epigenome-wide association study; methylome; methylenetetrahydrofolate reductase *MTHFR* gene; nutrition; S-adenosylmethionine; vitamin B complex

---

1. Introduction

Most genetic research on late-onset Alzheimer's disease (LOAD) has focused on genome-wide association studies (GWAS) that have provided low effect size results in general, with the exception of apolipoprotein E (ApoE) [1,2]. Studies of monozygotic twins with Alzheimer's disease (AD) showed discordance in onset and progression indicating a role for nongenetic factors in disease pathogenesis [3]. For these reasons, genetic research turned to epigenetic modifications using epigenome-wide association studies (EWAS) in the last few years [4,5]. Bonasio et al. [6] defined epigenetics as "the study of molecular signatures that provide a memory of previously experienced stimuli, without irreversible changes in the genetic information". Therefore, epigenetic refers to potentially heritable and nonheritable modifications in gene expression induced by environmental factors without changes in DNA base sequences [1]. These epigenetic processes include DNA methylation, histone modification and expression of long noncoding RNAs and noncoding microRNAs (miRNAs) that primarily repress target messenger RNAs (mRNAs) [7–10].

This review focuses on DNA methylation dynamics and other epigenetic changes, including the role of methylenetetrahydrofolate reductase (*MTHFR*) gene polymorphisms and its metabolic pathways particularly in aging and LOAD pathology [11]. We also review polymorphisms of the cystathionine-gamma(γ)-lyase (*CTH*) gene [12], the enzyme that converts cystathionine to cysteine in the transsulfuration pathway and is responsible for plasma elevation of total homocysteine (tHcy). The role of relevant nutritional factors including the B-vitamins folate, vitamin B_{12}, and vitamin B_6 status is summarized. Elevation of Hcy is important in oxidative stress contributing to the decrease of S-adenosyl-L-methionine (SAM) levels, which induce demethylation of DNA resulting in overexpression of genes involved in AD pathology such as presenilin (*PSEN1*) and beta-secretase (*BACE1*), the β-site amyloid precursor protein (APP)-cleaving enzyme that increases hypomethylation and $A\beta_{1-42}$ deposition [9]. Moreover, epigenetic markers have also been demonstrated to be critical regulatory factors of brain function [9], not only in AD but also in other neurodegenerative diseases [1,2] as well as in aging [9]. Experimental antiaging epigenetic interventions attempt to reverse age-related changes in DNA methylation [10].

2. DNA Methylation Studies

2.1. 5-Cytosine Methylation and DNA Methyltransferases

Methylation at the 5-position of the cytosine base (5mC) is considered a critical phase of epigenetic regulation in pathways related with neuronal development. Methylation and demethylation of cytosine-phosphate-guanine (CpG) islands is associated with alterations in local chromatin producing a long-term regulation of transcription tagging genome into active and inactive territories introducing a "masking" function [13]. Decreased levels of 5mC [1] and targeted mutations of DNA methyltransferases introduced into the germline produce severe developmental restriction [13] and finally a lethal phenotype in mice [14]. Cytosine base methylation occurs mainly at CpG dinucleotides [1]. Gene regulation is achieved by 5mC silencing gene expression via high-density CpG areas, known as CpG islands, which remain largely unmethylated [13]. In humans, genomic DNA methylation of cytosine results from the addition of a methyl group from SAM to the cytosine, catalyzed by DNA methyltransferases (DNMT1, DNMT3A, and DNMT3B) [9]. In addition to 5mC, hydroxymethylation at the 5-position of the cytosine base (5hmC) derived from the oxidation of methylated cytosines by ten-eleven translocation (TET) enzymes is another epigenetic regulatory mechanism, which is particularly abundant in the brain. The TET family of enzymes catalyzes Fe (II)- and alpha-ketoglutarate (α-KG)–dependent oxidation reactions, [9] and produces the initial step of oxidation of 5mC to 5hmC. TET enzymes also participate in the conversion of 5-formylcytosine (5fC) to 5-carboxylcytosine (5caC); this cycle ends when 5caC is excised by a thymine-DNA glycosylase (TDG) [9].

In humans, DNA methyltransferases are involved in tumor transformation and progression resulting in genome-wide hypomethylation of tumor cells and silencing of tumor-suppressor genes [15]; also, *DNMT3A* mutations have been associated with poor prognosis in acute myeloid leukemia [15]. *DNMT1* mutations occur in hereditary sensory and autonomic neuropathy type 1 (HSAN1) [15]. In mice, *DNMT1* mutations induce global hypomethylation along with cortical and hippocampal neuronal dysfunction causing neurodegeneration with severe deficits in learning, memory and behavior [16]. Hypomethylated excitatory neurons have postnatal maturation defects including abnormal dendritic arborization and impaired neuronal excitability [16]. Grossi et al. [17] used artificial neural network analysis to illustrate how low cobalamin; low folate and high Hcy are linked to AD. Low *PSEN1* methylation was linked to low folate levels and low promoter methylation of *BACE1* and *DNMT* genes. High levels of folate-vitamin B_{12} and low Hcy promoted methylation of genes required for DNA methylation reactions (*DNMT1*, *DNMT3A*, *DNMT3B*, and *MTHFR*) [18].

2.2. DNA Methylation in Alzheimer's Disease

Early studies of DNA methylation in LOAD from peripheral blood lymphocytes [19,20], brain biopsies and autopsy material [21–29] demonstrated variable results of cytosine methylation at CpG dinucleotides. Wang and coworkers [30] studied postmortem prefrontal cortex tissue and peripheral lymphocytes of AD patients and showed that specific loci in *MTHFR* gene promoter regions were hypermethylated compared to healthy controls. Ellison and collaborators [31] using gas chromatography/mass spectrometry found abnormal levels of 5mC and 5hmC in the superior and middle temporal gyri, hippocampus and parahippocampal gyrus in early stages of AD, as well as in frontotemporal lobe degeneration and Lewy body dementia; these global values returned to control levels as the disease progressed suggesting that methylation changes occur in early stages of neurodegenerative dementias. Chouliaras et al. [32] confirmed the presence of significant decreases in levels of 5mC and 5hmC in the hippocampus of AD patients compared with negative controls. Levels of 5mC were inversely proportional to the deposition of neurofibrillary tangles in the same hippocampal cells. Hernández et al. [33] studied DNA methylation patterns of cortical pyramidal layers in 32 brains of patients with LOAD demonstrating hypermethylation of synaptic genes and genes related to oxidative stress including *HOXA3, GSTP1, CXXC1-3* and *BIN1*.

One of the major problems of initial methylation studies was the small sample size. This was solved by De Jager and collaborators [4] utilizing one of the largest clinicopathological studies to date, the Religious Orders Study, with 708 brains to assess the methylation state of the brain's DNA correlated with AD pathology. Almost a half-million CpGs were interrogated including CpGs in the *ABCA7* and *BIN1* regions. The authors also identified genes whose RNA expression was altered in AD including *ANK1, CDH23, DIP2A, RHBDF2, RPL13, SERPINF1* and *SERPINF2*. *ANKYRIN 1* (*ANK1*) and *RHOMBOID5* (*RHBDF2*) genes are involved in the protein tyrosine kinase 2-beta (*PTK2B*) gene network, a LOAD gene that is a key element of the calcium-induced signaling cascade involved in modulating the activation of microglia and macrophages, as well as in the transport of TNFα converting enzyme (ADAM17) from the cell surface.

Absence of *RHBDF2* in mice impacts the normal release of TNFα [4] activated astrocytes in the vicinity of neuritic plaques that overexpress *CADHERIN23* (*CDH23*) gene. DIP2A functions as a cell surface protein and connects directly to the known *SORTILIN RELATED RECEPTOR 1* (*SORL1*) susceptibility gene that is involved in the APP susceptibility network and amyloid processing [4]. Both *SERPIN PEPTIDASE INHIBITORS* (*SERPINF1* and *SERPINF2*) interact with elements of amyloid processing. *SERPINF1* mRNA expression is reduced in LOAD and when knocked-out in vitro leads to reduced neurite outgrowth [4].

A Religious Orders companion study by Lunnon and coworkers [5] found robust association between differences in methylation, mRNA levels, and Braak & Braak staging. The severity of Alzheimer's disease is defined in neuropathology by the presence of tau-based neurofibrillary tangles ranging from early stages (I and II) to extensive neocortical involvement in Braak & Braak stages V and VI in advanced disease. Dysregulation of DNA methylation occurred earlier in brain areas affected at onset by AD and appeared to have stronger effects (28.7%) than the combination of ApoE and other risk genes (13.9%) identified by GWAS [1,2], indicating the importance of epigenetic changes in AD. Additional studies by Yu et al. [34] confirmed the association of DNA methylation in *SORL1, ABCA7, HLA-DRB5, SLC24A4,* and *BIN1* genes with pathological diagnosis of AD including both Aβ load and tau tangle density. RNA expression of transcripts of *SORL1* and *ABCA7* was associated with tau tangle density, and the expression of *BIN1* was associated with Aβ load [34]. Moreover, Lunnon et al. [5] found hypermethylation of the *ANK1* gene in the entorhinal cortex, superior temporal gyrus and prefrontal cortex in LOAD. These findings confirm that AD involves significant disruption of DNA methylation. Epigenetic age-associated alterations of DNA methylation have also been reported in animal models of AD, in particular global DNA hypomethylation in the J20 model and DNA hypermethylation in the triple transgenic 3xTg-AD model [35].

3. miRNAs Epigenetic Effects

Long noncoding RNAs and noncoding microRNAs (miRNAs) that primarily repress target messenger RNAs (mRNAs) play a pivotal role in oncology, cardiovascular diseases and dementia [7,8]. In AD, the miRNA-125b is overexpressed enhancing neuronal apoptosis and tau phosphorylation by activation of cyclin-dependent kinase 5 (CDK5) and p35/25. *FORKHEAD BOX Q1 (FOXQ1)* is the direct target gene of miR-125b [7]. Patrick and coworkers [8] studied the role of miRNA-132, miRNA-129 and miRNA-99 in the dorsolateral prefrontal cortex of more than 500 brain samples demonstrating a small number of specific alterations on target genes such as *EP300* that encodes p300, a histone acetyltransferase that regulates transcription in the cortex of subjects with AD.

4. Transsulfuration Metabolic Pathways and Remethylation Defects

The metabolism of sulfur-containing amino acids in the transsulfuration pathway involves the transfer of the sulfur atom of methionine to serine to produce cysteine (Figure 1). Methionine first reacts with ATP to form S-adenosyl-L-methionine (SAM), then S-adenosyl-homocysteine (SAH) and finally, homocysteine. Plasma elevation of total homocysteine (tHcy) or hyperhomocysteinemia may result from congenital deficiency of cystathionine β-synthase (CBS) leading to homocystinuria, or more frequently from polymorphisms of the cystathionine γ lyase (*CTH*) gene (OMIM *607657; EC 4.4.1.1.) in chromosome 1 (1p31.1) [36]. CTH is the enzyme that converts cystathionine to cysteine, the last step in the transsulfuration pathway. Wang et al [12] demonstrated that a single nucleotide polymorphism (SNP), namely c.1364G > T in exon 12 of the *CTH* gene causes elevation of tHcy and cystathioninuria. Caucasian subjects homozygous for the *CTH* 1364T/T SNP showed elevation of tHcy that reached effects sizes similar to those caused by the 677C > T *MTHFR* gene polymorphism [12].

Closely related to the transsulfuration pathway are the remethylation defects resulting from the failure to convert homocysteine to the amino acid methionine (Figure 1). This pathway requires the integrity of the gene encoding methylenetetrahydrofolate reductase (*MTHFR*) required for the interaction of folate and cobalamin (vitamin B_{12}). Folate provides the methyl group required for the remethylation pathway (Figure 1) to finally produce SAM, the main methyl donor for epigenetic processes. The human *MTHFR* gene (OMIM *607093; EC 1.5.1.20) is localized in chromosome 1 (1p36.3) and it encodes for 5,10-methylenetetrahydrofolate reductase (MTHFR) [37]. This enzyme catalyzes the conversion of 5,10-methylenetetrahydrofolate to 5-methyltetrahydrofolate, a co-substrate with vitamin B_{12} for the remethylation of homocysteine to methionine [11]. Mutations of this gene occur in 10–15% of the population and the resulting MTHFR deficiency affects the production of methionine and SAM. Patients with AD have low levels of SAM in the CSF [38].

MTHFR gene polymorphisms cause enzyme thermolability and involve C-to-T substitution at nucleotide 667 and A-to-C at nucleotide 1298; these *MTHFR* mutations have been associated with homocystinuria, neural tube defects, preeclampsia, cleft lip and cleft palate, cerebrovascular disease, and psychiatric disorders including susceptibility to depression and schizophrenia [39,40]. Population-based international studies showed no increased risk of dementia in subjects with *MTHFR* polymorphisms [41,42]. In Japan, Nishiyama et al. [43] found a slight association of the *MTHFR*-C667T polymorphism with senile cognitive decline in men but not with AD. In Australia, a causal link between high tHcy and incident dementia was demonstrated [44] but the study lacked power to determine an effect of the *MTHFR*-C667T genotype. In contrast, de Lau and collaborators [45] in the normal elderly population of the Rotterdam Study observed that the *MTHFR*-C665T genotype was associated with elevated tHcy but not with cognitive loss or white matter lesions. In a small patient population in Tunisia [46], the *MTHFR*-A1298C mutation was associated with susceptibility to AD. As mentioned earlier, Román [11] found a very high frequency (above 90%) of *MTHFR* gene mutations in an elderly population attending a memory clinic in the USA, with diagnoses ranging from mild cognitive impairment (MCI) to LOAD; about 65% had single mutations; the *MTHFR*-C667T mutation was found in 58.5% of the patients and 41.5% had the *MTHFR*-A1298C mutation whereas 20% were compound heterozygous for both mutations [11].

MTHFR and Epigenetic Drift

In 2005, a multinational study of identical twins by Fraga and collaborators [47] first demonstrated that whereas DNA methylation and histone acetylation in young identical twins are indistinguishable, older identical twins showed substantial differences; epigenetic changes were up to four times greater than those of young twin pairs. The authors concluded that this "epigenetic drift" was associated with aging [47]. Epigenetic drift of identical twins with aging also occurs among a large number of animal species [48] following a non-Mendelian pattern. In identical twins with AD, the prognosis and onset of AD can differ by more than ten years [3,49,50]; young identical twin pairs are essentially indistinguishable in their epigenetic markings while older identical twin pairs show substantial variations. Breitner et al. [50] suggested that twins with a history of systemic infection developed AD at an earlier onset than their identical twin. Epigenetic drift can be caused by lifestyle, diet, infections, folate status, homocysteine status or toxic exposure [51]. Wang et al. [30] demonstrated that the *MTHFR* gene promoter in the brain displayed high interindividual variance in DNA methylation among twins. The methylation level of *MTHFR* and *APOE* in individuals 30 years of age apart decreased by 10.6%, whereas in patients with AD the methylation level increased by 6.8%. The epigenetic drift increases with age particularly in genes that play pivotal roles in removing β-amyloid such as *APOE* and among methylation genes such as *MTHFR* and *DNMT1* [9,52].

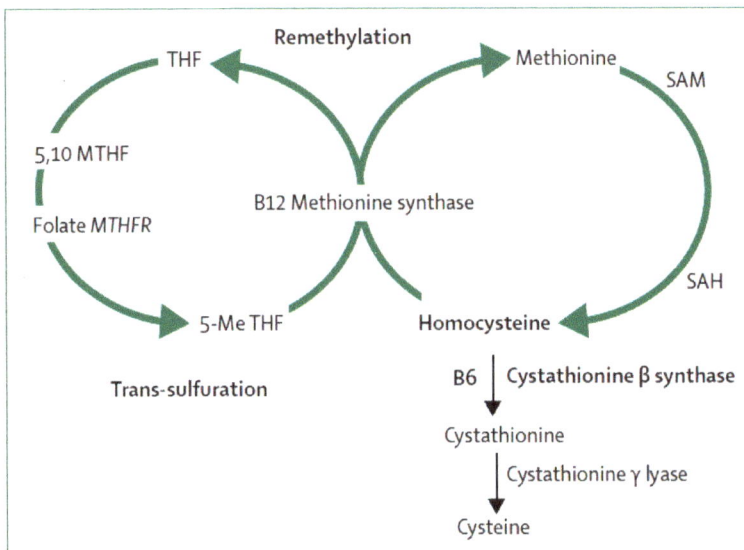

Figure 1. Homocysteine metabolism: B12 = cobalamin (vitamin B12). B6 = pyridoxine (vitamin B6). MTH = methylenetetrahydrofolate. MTHFR = methylenetetrahydrofolate reductase. SAM = S-adenosyl- methionine. SAH = S-adenosylhomocysteine. 5-Me THF = 5-methyl tetrahydrofolate. (From [53]).

5. Homocysteine (Hcy): A Risk Factor for Cognitive Loss and Dementia

Hcy is a sulfur-containing amino acid produced in the transsulfuration pathway (Figure 1) from the reaction of methionine with ATP to form SAM, then SAH and finally homocysteine.   Homocystinuria due to congenital deficiency of the *CBS* gene causes hyperhomocysteinemia. Polymorphisms of the *CTH* and *MTHFR* genes are common genetic causes of hyperhomocysteinemia [38,39]. The remethylation pathway (Figure 1) involves reactions enzymatically mediated by MTHFR requiring as co-substrates the B-group vitamins folic acid (vitamin B$_9$) and cobalamin (vitamin B_{12}) for the remethylation of homocysteine to methionine. Pyridoxine (vitamin B_6) is required by CBS for the conversion of homocysteine to cysteine (Figure 1).

5.1. Hyperhomocysteinemia is an Independent Vascular Risk Factor

Elevation of plasma or serum tHcy (hyperhomocysteinemia) is an independent vascular risk factor linked to coronary disease, peripheral vascular disease, stroke and small-vessel cerebrovascular disease [53]. More importantly, elevated tHcy is considered a risk factor for dementia and cognitive decline in the elderly, particularly in association with low levels of folate and cobalamin [54,55]. A number of studies in cognitively normal elderly subjects, demonstrated that baseline tHcy is a strong and independent predictor of cognitive decline after observation periods ranging from 3 years (USA, $n = 321$ men [55] and Sydney, Australia, $n = 889$ [56]), 4 years (France, $n = 1241$) [57], 5 years (Wales, UK, $n = 32$) [58], 6 years (Norway, $n = 2189$) [59], 7 years (Finland $n = 274$) [60] and up to 10 years (UK, $n = 691$) [61]. In the Finland cohort [60], the magnetic resonance imaging (MRI) study demonstrated the association of higher baseline vitamin B_{12} and holotranscobalamin levels with a decreased rate of total brain volume loss during 8 years of the study period [62]. Increased tHcy levels were associated with faster rates of total brain volume loss and with progression of white matter hyperintensities among participants with hypertension (systolic blood pressure > 140 mm Hg) [62].

Regarding the risk of AD associated to elevated tHcy, in the Framingham Study, Seshadri and colleagues [63] demonstrated in elderly subjects (mean age, 76 years) that raised tHcy above 14 µmol/L nearly doubled the risk of LOAD over a period of 8 years. Similar findings were corroborated in two large Finnish [60,64] and Australian [65] cohorts. In 2008, Smith [66] performed a comprehensive review of cross-sectional and prospective studies involving >46,000 subjects and confirmed the association between elevated tHcy and cognitive deficit or dementia.

According to a recent international consensus statement [67], moderately raised homocysteine (>11 µmol/L) increases the relative risk of dementia in the elderly 1.15 to 2.5 fold, and the Population Attributable risk from 4.3 to 31% [67]. From the Public Health viewpoint, homocysteine-lowering treatment with B vitamins that markedly slows down the rate of brain atrophy and cognitive decline in the elderly offers the possibility that, in addition to folic acid fortification, mandatory methylcobalamin supplementation should also be considered for the prevention of LOAD [67].

5.2. Genetic and Nongenetic Causes of Hyperhomocysteinemia

Elevation of tHcy is caused by numerous factors including advancing age, diet, supplementation of B-vitamins, obstructive sleep apnea, smoking, *Helicobacter pylori* infection, and renal failure, among others [53,54]. As indicated earlier, both *CBS* gene polymorphisms and the C667T and the A1298C SNPs in the *MTHFR* gene decrease the activity of the MTHFR enzyme leading to hyperhomocysteinemia. Minagawa et al. [68] found that elevated Hcy inhibits the dimerization of ApoE3 and reduces ApoE3-mediated high-density lipoprotein (HDL) concentrations involved in degradation of soluble Aβ within microglia. ApoE4 was not affected; in patients with hyperhomocysteinemia the CSF levels of ApoE3 dimers were significantly lower than in controls. Minagawa and colleagues [68] suggested that the effects of elevated Hcy on ApoE3 contribute to the pathogenesis of AD. Smith and Refsum [54] reviewed the proposed mechanisms responsible for the harmful cognitive effects of hyperhomocysteinemia (Table 1). These include impaired endothelial function with reduced inducible nitric oxide synthase; augmented oxidative stress and decreased activity of key antioxidant enzymes; raised generation of the superoxide anion; alterations of lipid metabolism with increased cholesterol synthesis and reduced synthesis of apolipoprotein 1; and, carotid stenosis and induction of thrombosis [69,70].

Table 1. Harmful effects of homocysteine on vascular function and cognition (modified from Smith & Refsum [54]).

Proposed Mechanisms	
Vascular Mechanisms	
1	Impairs endothelial function reducing inducible NO synthase
2	NO-mediated endothelial dysfunction in brain vasculature
3	Causes a leaky blood-brain barrier
4	Induces thrombosis
5	Cerebrovascular ischemia leading to neuronal death and tau tangle deposition
6	Affects lipid metabolism increasing cholesterol synthesis
7	Reduces synthesis of apolipoprotein 1
8	Causes cerebral amyloid angiopathy
Neuronal Mechanisms	
1	Direct activation of NMDA receptor causes excitotoxic neuronal death
2	Homocysteic acid and cysteine sulfinic acid activate NMDA receptor causing neuronal death by excitotoxicity
3	Oxidative stress induced by generating superoxide and reactive oxygen species
4	Decreased activity of antioxidant enzymes
5	Formation and deposition of β-amyloid
6	Potentiates neurotoxic effects of β-amyloid by itself or via homocysteic acid
7	Activates tau kinases, such as Cdk5, causing tau tangle deposition
8	Triggers the cell cycle in neurons, leading to tangle formation and cell death
9	Causes DNA damage, limits DNA repair, leading to apoptosis
10	Increases SAH inhibiting methylation reactions, such as DNA cytosine methylation in promoters for amyloid genes, causing epigenetic effects
11	Inhibits PP2A activity leading to tau tangle deposition
12	Inhibits methylation of phosphatidyletanolamine
13	Stimulates endoplasmic reticulum stress response leading to amyloid formation
14	Activates the immune system
15	Decreases SAM-dependent synthesis of catecholamines and other neurotransmitters

Hyperhomocysteinemia induces a decrease in the SAM-dependent synthesis of catecholamines including dopamine, norepinephrine, and epinephrine, as well as non-catecholamine neurotransmitters such as melatonin and serotonin (5-HT) that contribute to development of depression [69]. Moreover, elevated tHcy produces two neurotoxic products, homocysteic acid (HCA) and cysteine sulfinic acid (CSA), which are agonists of the N-methyl-D-aspartate (NMDA) glutamate receptor, with neurotoxic effects on dopaminergic neurons derived from excessive Ca^{++} influx and reactive oxygen generation [70]. The beneficial effects of B-group vitamins on elevated tHcy will be reviewed next.

6. Folate Metabolism

Vitamin B_9 or folic acid (from the Latin *folium*, leaf) is abundantly found in green leafy vegetables. Folate is vital for cell development and growth given its role in numerous biochemical one-carbon (methyl-group, $-CH_3$) reactions, many of them critical for cognition. The Nun Study [71] first provided epidemiological and neuropathological data demonstrating that limited lifetime consumption of salads with low blood folate levels increased the risk of cognitive decline and dementia. Also, the severity of the atrophy in the neocortex and of the Alzheimer disease lesions were strongly correlated with low serum folate levels; none of 18 other nutrients, lipoproteins, or nutritional markers measured in the study correlated with the atrophy [71]. Further studies confirmed that normal cognitive scores were highly associated with elevated blood folate despite the neuropathological evidence of LOAD brain lesions [72].

The primary methyl-group donor for DNA methylation reactions is 5-methyl-tetrahydrofolate (CH_3-THF) required for the transformation of homocysteine into methionine mediated by methionine synthase with cobalamin (vitamin B_{12}) as a cosubstrate (Figure 1), leading to the synthesis of SAM.

Also, CH_3-THF is critical in the de novo purine synthesis to convert dUMP (deoxyuridylate) into dTMP (thymidylate) for DNA and RNA synthesis, DNA repair or replication. Several forms of cancer are associated with epigenetic differential methylation causing disturbances in nucleotide synthesis; for instance, hypermethylation may inhibit tumor suppressors. Folate, therefore, is a B-vitamin that plays an important role as a precursor in the epigenetic regulation of gene expression, DNA stability, DNA integrity and mutagenesis. Abnormal folate status has been associated with neural tube defects, cardiovascular and cerebrovascular diseases, cleft lip and palate, neurodegenerative diseases, schizophrenia and depression [40,73,74].

Telomeres and Folate Levels

Telomeres protect chromosomes from abnormal combination and degradation. The shortening of telomeres' cap serve as a signature of cell division history, acting as biomarker of aging. In peripheral leukocytes, short telomere length is associated with increased risk of cognitive decline and LOAD [75,76]. Low folate levels are associated with short telomeres due to DNA damage in the telomeric region. Telomere length is epigenetically regulated by DNA methylation and directly influenced by folate status, a process independent of DNA damage due to uracil incorporation. Shorter telomeres occur with age, infection, stress, and chronic diseases including LOAD [75].

Paul and collaborators [76] observed that decreased plasma folate concentration to <11.6 µmol/L was correlated with a decrease in mean telomere length. In this population, homozygous carriers of the MTHFR-C677T gene mutation showed decreased levels of plasma folate [77]. Decreased serum folate induces anomalous integration of uracil in place of thymidine in DNA [78], a mechanism corrected by folic acid supplementation. Troesch, Weber and Mohajeri [79] summarized the importance for the development of LOAD of reduced SAM-dependent methylation reactions due to genetic factors along with reduction of folate, vitamin B_6 and vitamin B_{12} levels. The resulting elevation of Hcy levels and the reduced capacity to synthetize, methylate and repair DNA, along with the impaired modulation of neurotransmission, appears to favor the development of AD particularly when combined with increased oxidative stress, particularly in ApoE ε4 carriers [80].

7. Vitamin B_{12} Deficiency and β-amyloid Deposition

Smith, Warren and Refsum [81] have recently provided a comprehensive review of vitamin B_{12}. Only bacteria can biosynthesize vitamin B_{12}; in humans B_{12} from the diet is a cofactor for the enzymes methionine synthase and L-methyl-malonyl-CoA mutase. B_{12} deficiency results in build-up of homocysteine and lack of interaction with folate that is trapped as CH_3-THF leading to depletion of tetrahydrofolates used in thymidylate and purine synthesis blocking DNA for the production of red cells in the bone marrow. B_{12} deficiency impedes cellular proliferation and protein synthesis and thereby causes development of megaloblastic anemia [81].

7.1. Clinical Manifestations of Vitamin B_{12} Deficiency

In 1920, pernicious anemia—a fatal form of a megaloblastic anemia—was successfully treated by adding liver to the diet. In 1955, Dorothy Hodgkin used crystallography to first identify the molecular structure of cyanocobalamin or vitamin B_{12} from the deep-red cyanide-containing pigment isolated from liver tissue. Pernicious anemia was the first disease to be identified as caused by vitamin B_{12} deficiency [81].

Stabler [82] reviewed the clinical manifestations of vitamin B_{12} deficiency. In addition to megaloblastic anemia, acidemia from elevation of serum methylmalonic acid (MMA), and methylmalonic aciduria, the neurological manifestations of pernicious anemia include memory loss and cognitive decline, visual disturbances from optic nerve neuropathy, burning and painful sensations in hands and feet from peripheral neuropathy, and spinal cord involvement with subacute combined degeneration resulting in loss of proprioception from dorsal column involvement and pyramidal tract symptoms such as paralysis and incontinence.

7.2. Measuring Total Serum B_{12} Levels

Dietary sources of B_{12} include liver, meat, fish, shellfish and dairy products; vegans are prone to B_{12} deficiency [81,82]. Vitamin B_{12} deficiency occurs from inborn metabolic errors, alterations of B_{12}-binding proteins including *haptocorrin* (HC) found in saliva, *intrinsic factor* (IF) produced by parietal cells in the stomach (pernicious anemia is associated with anti-parietal-cell and anti-IF auto-antibodies), and *transcobalamin* (TC), which binds B_{12} to facilitate uptake by the cells [81]. According to Stabler [82], measurement of total serum B_{12} levels is unsatisfactory because it reflects B_{12} that is bound to either HC or TC, and up to 60% of bound materials are cobalamin analogues (corrinoids). Therefore, "normal" total serum B_{12} levels can mask deficiency if serum contains relatively large amounts of cobalamin analogues [83]. Levels below 200 pg/mL usually indicate biochemical B_{12} insufficiency. Serum B_{12} < 350 pg/mL along with tHcy > 14 μmol/L indicate metabolic B_{12} deficiency [81,82]. For this reason, holotranscobalamin, MMA and tHcy levels should be included in the evaluation of a patient suspected of having B_{12} deficiency [83].

7.3. Causes of Vitamin B_{12} Deficiency

Other than pernicious anemia resulting from presence of anti-parietal-cell and anti-IF autoantibodies, other causes of B_{12} deficiency include atrophic body gastritis, *Helicobacter pylori* infection, malabsorption of vitamin B_{12}, gastrectomy, gastric bypass or other bariatric surgery, inflammatory bowel disease, tropical sprue, use of metformin, anticonvulsants, proton-pump inhibitors and other drugs to block stomach acid, and vegetarian diets low in meat and dairy products. Hemodialysis patients, nitrous oxide inhalation, and cholinesterase inhibitors in LOAD patients [84] also increase the risk of vitamin B_{12} deficiency.

Epidemiological studies have shown that prevalence of vitamin B_{12} deficiency increases with age [85,86], due to decreased saliva (e.g., dry eyes-dry mouth of Sjögren syndrome) [87] and gastric atrophy with deficits respectively of haptocorrin and intrinsic factor. Andrès and colleagues [88] have emphasized that as many as 20% of elderly people may have unrecognized B_{12} deficiency due to food-cobalamin malabsorption plus insufficient dietary intake. According to Spence [89], metabolic B_{12} deficiency occurs in 30% of vascular patients older than 71 years, increasing to as many as 40% in patients above age 80 years; these patients usually have plasma levels of tHcy >14 μmol/L resulting from B_{12} deficiency. Inadequate supply of B_{12} and folic acid is not only a strong and independent vascular risk factor particularly for subcortical ischemic small-vessel disease [90], a common and important contributor to cognitive impairment and memory complaints in the elderly, but also enhancing the development of LOAD [91]. Animal experimental data confirms the importance of B-vitamin deprivation in the expression of AD [92].

7.4. Effects of B-Group Vitamins on Cognition: Negative Clinical Trials

An international consensus [67] provided a comprehensive explanation of the negative results of meta-analyses [93] based on reviews of the results from a number of inadequately controlled clinical trials; most participants in those trials were enrolled in post-hoc studies which were not designed primarily to assess cognition. Usually, these were short-duration trials without baseline cognitive assessment and results were based on post-hoc brief cognitive assessments; only a few of these studies assessed the incidence of dementia or mild cognitive impairment.

In contrast, solid positive results were obtained in the Oxford Project to Investigate Memory and Ageing (OPTIMA) trial [94–96] that used comprehensive neuropsychological evaluations plus brain imaging end-points. The results of this trial indicate that supplementation of B_{12}, pyridoxine, and folic acid in subjects with MCI and hyperhomocysteinemia decreases tHcy resulting in improved episodic memory and global cognition [95], and most importantly, brain imaging demonstration of slowing of the progression of the brain atrophy in areas affected by AD [96]. Current recommendation is to provide oral supplementation of methylcobalamin 1000 μg/d, folic acid 800 μg/d and pyridoxine 100 mg/d.

8. SAM in Depression and Cognitive Loss

As described above (Figure 1) SAM is the main methyl-group donor for the methylation reactions reviewed here; as well as for synthesis of neurotransmitters, proteins, nucleic acids, phospholipids, and myelin. SAM has been used as an adjuvant for the treatment of depression [97]. Linnebank et al. [38] demonstrated a decrease of SAM in the cerebrospinal fluid (CSF) of patients with LOAD, affecting mainly ApoE ε4 carriers. According to Dayon et al. [98], plasma levels of one-carbon metabolites predicted cognitive decline. Despite the enhancing effects of SAM on antidepressants, no conclusive clinical trials of SAM have been reported [99].

9. Conclusions

It is established that the damaging effects of deficiencies of folate and cobalamin and the resulting elevation of tHcy contribute to the development of LOAD [67]. The numerous detrimental effects of elevated tHcy include, among others, endothelial and cerebrovascular damage of large-vessels as well as small-vessel disease [90]; activation of tau kinases; inhibition of methylation reactions; epigenetic effects on the β-amyloid pathway; reduced protein phosphatase-2A; and, impaired formation of phosphatidylcholine. Adequate supply of B-vitamins in the elderly, particularly in subjects with *MTHFR* and *CTH* gene mutations, appears to be critical to prevent the development of cognitive decline and to halt the progression of LOAD.

Author Contributions: Conceptualization, G.C.R. and O.M.-P.; methodology, G.C.R.; bibliographic investigation, G.C.R., O.M.-P. and C.B.; writing—original draft preparation, G.C.R., O.M.-P. and C.B.; writing—review and editing, G.C.R. and O.M.-P.; funding acquisition, G.C.R.

Funding: This research was funded by the Jack Blanton Presidential Distinguished Chair, the Fondren Fund and the Wareing Family Fund at Houston Methodist Hospital to G.C.R.

Acknowledgments: The authors would like to thank Houston Methodist Hospital for constant support to clinical research. Mancera-Páez was supported by the David Cabello International Alzheimer Disease Scholarship Fund. J David Spence, London, ON Canada provided valuable comments and bibliography.

Conflicts of Interest: The authors declare no conflict of interest.

References

1. Roubroeks, J.A.Y.; Smith, R.G.; van den Hove, D.L.A.; Lunnon, K. Epigenetics and DNA methylomic profiling in Alzheimer's disease and other neurodegenerative diseases. *J. Neurochem.* **2017**, *143*, 158–170. [CrossRef]
2. Millan, M.J. An epigenetic framework for neurodevelopmental disorders: From pathogenesis to potential therapy. *Neuropharmacology* **2013**, *68*, 2–82. [CrossRef] [PubMed]
3. Gatz, M.; Pedersen, N.L.; Berg, S.; Johansson, B.; Johansson, K.; Mortimer, J.A.; Posner, S.F.; Viitanen, M.; Winblad, B.; Ahlbom, A. Heritability for Alzheimer's disease: The study of dementia in Swedish twins. *J. Gerontol. A Biol. Sci. Med. Sci.* **1997**, *52*, M117–M125. [CrossRef] [PubMed]
4. De Jager, P.L.; Srivastava, G.; Lunnon, K.; Burgess, J.; Schalkwyk, L.C.; Yu, L.; Eaton, M.L.; Keenan, B.T.; Ernst, J.; McCabe, C.; et al. Alzheimer's disease: Early alterations in brain DNA methylation at *ANK1*, *BIN1*, *RHBDF2* and other loci. *Nat. Neurosci.* **2014**, *17*, 1156–1163. [CrossRef] [PubMed]
5. Lunnon, K.; Smith, R.; Hannon, E.; De Jager, P.L.; Srivastava, G.; Volta, M.; Troakes, C.; Al-Sarraj, S.; Burrage, J.; Macdonald, R.; et al. Methylomic profiling implicates cortical deregulation of *ANK1* in Alzheimer's disease. *Nat. Neurosci.* **2014**, *17*, 1164–1170. [CrossRef]
6. Bonasio, R.; Tu, S.; Reinberg, D. Molecular signals of epigenetic states. *Science* **2010**, *330*, 612–616. [CrossRef]
7. Ma, X.; Liu, L.; Meng, J. MicroRNA-125b promotes neurons cell apoptosis and tau phosphorylation in Alzheimer's disease. *Neurosci. Lett.* **2017**. [CrossRef] [PubMed]

8. Patrick, E.; Rajagopal, S.; Wong, H.A.; McCabe, C.; Xu, J.; Tang, A.; Imboywa, S.H.; Schneider, J.A.; Pochet, N.; Krichevsky, A.M.; et al. Dissecting the role of non-coding RNAs in the accumulation of amyloid and tau neuropathologies in Alzheimer's disease. *Mol. Neurodegener.* **2017**, *12*, 51. [CrossRef] [PubMed]

9. Irier, H.A.; Jin, P. Dynamics of DNA methylation in aging and Alzheimer's disease. *DNA Cell Biol.* **2012**, *31* (Suppl. 1), S42–S48. [CrossRef] [PubMed]

10. Unnikrishnan, A.; Freeman, W.M.; Jackson, J.; Wren, J.D.; Porter, H.; Richardson, A. The role of DNA methylation in epigenetics of aging. *Pharmacol. Ther.* **2018**. [CrossRef] [PubMed]

11. Román, G.C. *MTHFR* gene mutations: A potential marker of late-onset Alzheimer's disease? *J. Alzheimer's Dis.* **2015**, *47*, 323–327. [CrossRef] [PubMed]

12. Wang, J.; Huff, A.M.; Spence, J.D.; Hegele, R.A. Single nucleotide polymorphism in *CTH* associated with variation in plasma homocysteine concentration. *Clin. Genet.* **2004**, *65*, 483–486. [CrossRef] [PubMed]

13. Sharma, R.P.; Gavin, D.P.; Grayson, D.R. CpG methylation in neurons: Message, memory, or mask? *Neuropsychopharmacology* **2010**, *35*, 2009–2020. [CrossRef] [PubMed]

14. Li, E.; Bestor, T.H.; Jaenisch, R. Targeted mutation of the DNA methyltransferase gene results in embryonic lethality. *Cell* **1992**, *69*, 915–926. [CrossRef]

15. Zhang, W.; Xu, J. DNA methyltransferases and their roles in tumorigenesis. *Biomark. Res.* **2017**, *5*, 1. [CrossRef] [PubMed]

16. Hutnick, L.K.; Golshani, P.; Namihira, M.; Xue, Z.; Matynia, A.; Yang, X.W.; Silva, A.J.; Schweizer, F.E.; Fan, G. DNA hypomethylation restricted to the murine forebrain induces cortical degeneration and impairs postnatal neuronal maturation. *Hum. Mol. Genet.* **2009**, *18*, 2875–2888. [CrossRef]

17. Grossi, E.; Stoccoro, A.; Tannorella, P.; Migliore, L.; Coppedè, F. Artificial neural networks link one-carbon metabolism to gene-promoter methylation in Alzheimer's disease. *J. Alzheimers Dis.* **2016**, *53*, 1517–1522. [CrossRef] [PubMed]

18. Jones, P.A.; Liang, G. Rethinking how DNA methylation patterns are maintained. *Nat. Rev. Genet.* **2009**, *10*, 805–811. [CrossRef] [PubMed]

19. Guan, J.Z.; Guan, W.P.; Maeda, T.; Makino, N. Analysis of telomere length and subtelomeric methylation of circulating leukocytes in women with Alzheimer's disease. *Aging Clin. Exp. Res.* **2013**, *25*, 17–23. [CrossRef]

20. Piaceri, I.; Raspanti, B.; Tedde, A.; Bagnoli, S.; Sorbi, S.; Nacmias, B. Epigenetic modifications in Alzheimer's disease: Cause or effect? *J. Alzheimers Dis.* **2015**, *43*, 1169–1173. [CrossRef] [PubMed]

21. West, R.L.; Lee, J.M.; Maroun, L.E. Hypomethylation of the amyloid precursor protein gene in the brain of an Alzheimer's disease patient. *J. Mol. Neurosci.* **1995**, *6*, 141–146. [CrossRef] [PubMed]

22. Tohgi, H.; Utsugisawa, K.; Nagane, Y.; Yoshimura, M.; Genda, Y.; Ukitsu, M. Reduction with age in methylcytosine in the promoter region -224 approximately -101 of the amyloid precursor protein gene in autopsy human cortex. *Brain Res. Mol. Brain Res.* **1999**, *70*, 288–292. [CrossRef]

23. Barrachina, M.; Ferrer, I. DNA methylation of Alzheimer disease and tauopathy-related genes in postmortem brain. *J. Neuropathol. Exp. Neurol.* **2009**, *68*, 880–891. [CrossRef] [PubMed]

24. Chouliaras, L.; Rutten, B.P.; Kenis, G.; Peerbooms, O.; Visser, P.J.; Verhey, F.; van Os, J.; Steinbusch, H.W.; van den Hove, D.L. Epigenetic regulation in the pathophysiology of Alzheimer's disease. *Prog. Neurobiol.* **2010**, *90*, 498–510. [CrossRef] [PubMed]

25. Bakulski, K.M.; Dolinoy, D.C.; Sartor, M.A.; Paulson, H.L.; Konen, J.R.; Lieberman, A.P.; Albin, R.L.; Hu, H.; Rozek, L.S. Genome-wide DNA methylation differences between late-onset Alzheimer's disease and cognitively normal controls in human frontal cortex. *J. Alzheimers Dis.* **2012**, *29*, 571–588. [CrossRef] [PubMed]

26. Bradley-Whitman, M.; Lovell, M.A. Epigenetic changes in the progression of Alzheimer's disease. *Mech. Ageing Dev.* **2013**, *134*, 486–495. [CrossRef] [PubMed]

27. Coppieters, N.; Dieriks, B.V.; Lill, C.; Faull, R.L.; Curtis, M.A.; Dragunow, M. Global changes in DNA methylation and hydroxymethylation in Alzheimer's disease human brain. *Neurobiol. Aging* **2014**, *35*, 1334–1344. [CrossRef] [PubMed]

28. Iwata, A.; Nagata, K.; Hatsuta, H.; Takuma, H.; Bundo, M.; Iwamoto, K.; Tamaoka, A.; Murayama, S.; Saido, T.; Tsuji, S. Altered CpG methylation in sporadic Alzheimer's disease is associated with APP and MAPT dysregulation. *Hum. Mol. Genet.* **2014**, *23*, 648–656. [CrossRef]

29. Humphries, C.E.; Kohli, M.A.; Nathanson, L.; Whitehead, P.; Beecham, G.; Martin, E.; Mash, D.C.; Pericak-Vance, M.A.; Gilbert, J. Integrated whole transcriptome and DNA methylation analysis identifies gene networks specific to late-onset Alzheimer's disease. *J. Alzheimers Dis.* **2015**, *44*, 977–987. [CrossRef]
30. Wang, S.C.; Oelze, B.; Schumacher, A. Age-specific epigenetic drift in late-onset Alzheimer's disease. *PLoS ONE* **2008**, *3*, e2698. [CrossRef]
31. Ellison, E.M.; Abner, E.L.; Lovell, M.A. Multiregional analysis of global 5-methylcytosine and 5-hydroxymethylcytosine throughout the progression of Alzheimer's disease. *J. Neurochem.* **2017**, *140*, 383–394. [CrossRef] [PubMed]
32. Chouliaras, L.; Mastroeni, D.; Delvaux, E.; Grover, A.; Kenis, G.; Hof, P.R.; Steinbusch, H.W.; Coleman, P.D.; Rutten, B.P.; van den Hove, D.L. Consistent decrease in global DNA methylation and hydroxymethylation in the hippocampus of Alzheimer's disease patients. *Neurobiol. Aging.* **2013**, *34*, 2091–2099. [CrossRef] [PubMed]
33. Hernández, H.G.; Sandoval-Hernández, A.G.; Garrido-Gil, P.; Labandeira-Garcia, J.L.; Zelaya, M.V.; Bayon, G.F.; Fernández, A.F.; Fraga, M.F.; Arboleda, G.; Arboleda, H. Alzheimer's disease DNA methylome of pyramidal layers in frontal cortex: Laser-assisted microdissection study. *Epigenomics* **2018**. [CrossRef] [PubMed]
34. Yu, L.; Chibnik, L.B.; Srivastava, G.P.; Pochet, N.; Yang, J.; Xu, J.; Kozubek, J.; Obholzer, N.; Leurgans, S.E.; Schneider, J.A.; et al. Association of brain DNA methylation in *SORL1, ABCA7, HLA-DRB5, SLC24A4,* and *BIN1* with pathological diagnosis of Alzheimer disease. *JAMA Neurol.* **2015**, *72*, 15–24. [CrossRef] [PubMed]
35. Lardenoije, R.; van den Hove, D.L.A.; Havermans, M.; van Casteren, A.; Le, K.X.; Palmour, R.; Lemere, C.A.; Rutten, B.P.F. Age-related epigenetic changes in hippocampal subregions of four animal models of Alzheimer's disease. *Mol. Cell. Neurosci.* **2018**, *86*, 1–15. [CrossRef] [PubMed]
36. OMIM®. Online Mendelian Inheritance in Man® Cystathionine Gamma-Lyase; CTH. Available online: http://www.omim.org/entry/607657 (accessed on 12 December 2018).
37. OMIM®. Online Mendelian Inheritance in Man® 5,10-Methylenetetrahydrofolate Reductase; MTHFR. Available online: http://www.omim.org/entry/607093 (accessed on 12 December 2018).
38. Linnebank, M.; Popp, J.; Smulders, Y.; Smith, D.; Semmler, A.; Farkas, M.; Kulic, L.; Cvetanovska, G.; Blom, H.; Stoffel-Wagner, B.; et al. S-adenosylmethionine is decreased in the cerebrospinal fluid of patients with Alzheimer's disease. *Neurodegener. Dis.* **2010**, *7*, 373–378. [CrossRef] [PubMed]
39. Kirsch, S.H.; Herrmann, W.; Obeid, R. Genetic defects in folate and cobalamin pathways affecting the brain. *Clin. Chem. Lab. Med.* **2013**, *51*, 139–155. [CrossRef] [PubMed]
40. Mitchell, E.S.; Conus, N.; Kaput, J. B vitamin polymorphisms and behavior: Evidence of associations with neurodevelopment, depression, schizophrenia, bipolar disorder and cognitive decline. *Neurosci. Biobehav. Rev.* **2014**, *47*, 307–320. [CrossRef]
41. Seripa, D.; Forno, G.D.; Matera, M.G.; Gravina, C.; Margaglione, M.; Palermo, M.T.; Wekstein, D.R.; Antuono, P.; Davis, D.G.; Daniele, A.; et al. Methylenetetrahydrofolate reductase, angiotensin converting enzyme gene polymorphisms in two genetically, and diagnostically distinct cohort of Alzheimer patients. *Neurobiol. Aging* **2003**, *24*, 933–939. [CrossRef]
42. Da Silva, V.C.; da Costa Ramos, F.J.; Malaquias Freitas, E.; de Brito-Marques, P.R.; de Holanda Cavalcanti, M.N.; D'Almeida, V.; Cabral-Filho, J.E.; Cartaxo Muniz, M.T. Alzheimer's disease in Brazilian elderly has a relation with homocysteine but not with MTHFR polymorphisms. *Arq. Neuro-Psiquiatr.* **2006**, *64*, 941–945. [CrossRef]
43. Nishiyama, M.; Kato, Y.; Hashimoto, M.; Yukawa, S.; Omori, K. Apolipoprotein E, methylenetetrahydrofolate reductase (MTHFR) mutation and the risk of senile dementia—An epidemiological study using the polymerase chain reaction (PCR) method. *Epidemiology* **2000**, *10*, 163–172. [CrossRef]
44. Ford, A.H.; Flicker, L.; Alfonso, H.; Hankey, G.J.; Norman, P.E.; van Bockxmeer, F.M.; Almeida, O.P. Plasma homocysteine and MTHFRC667T polymorphism as risk factors for incident dementia. *J. Neurol. Neurosurg. Psychiatry* **2012**, *83*, 70–75. [CrossRef] [PubMed]
45. De Lau, L.M.L.; van Meurs, J.B.; Uitterlinden, A.G.; Smith, A.D.; Refsum, H.; Johnston, C.; Breteler, M.M. Genetic variation in homocysteine metabolism, cognition, and white matter lesions. *Neurobiol. Aging* **2010**, *31*, 2020–2022. [CrossRef] [PubMed]

46. Mansouri, L.; Fekih-Mrissa, N.; Klai, S.; Mansour, M.; Gritli, N.; Mrissa, R. Association of methylenetetrahydrofolate reductase polymorphisms with susceptibility to Alzheimer's disease. *Clin. Neurol. Neurosurg.* **2013**, *115*, 1693–1696. [CrossRef] [PubMed]

47. Fraga, M.F.; Ballestar, E.; Paz, M.F.; Ropero, S.; Setien, F.; Ballestar, M.L.; Heine-Suñer, D.; Cigudosa, J.C.; Urioste, M.; Benitez, J.; et al. Epigenetic differences arise during the lifetime of monozygotic twins. *Proc. Natl. Acad. Sci. USA* **2005**, *102*, 10604–10609. [CrossRef]

48. Martin, G.M. Epigenetic drift in aging identical twins. *Proc. Natl. Acad. Sci. USA* **2005**, *102*, 10413–10414. [CrossRef]

49. Cook, R.H.; Schneck, S.A.; Clark, D.B. Twins with Alzheimer's disease. *Arch. Neurol.* **1981**, *38*, 300–301. [CrossRef]

50. Breitner, J.C.; Gatz, M.; Bergem, A.L.; Christian, J.C.; Mortimer, J.A.; McClearn, G.E.; Heston, L.L.; Welsh, K.A.; Anthony, J.C.; Folstein, M.F. Use of twin cohorts for research in Alzheimer's disease. *Neurology* **1993**, *43*, 261–267. [CrossRef]

51. Nee, L.E.; Lippa, C.F. Alzheimer's disease in 22 twin pairs—13-year follow-up: Hormonal, infectious and traumatic factors. *Dement. Geriatr. Cogn. Disord.* **1999**, *10*, 148–151. [CrossRef]

52. Coppedè, F. One-carbon metabolism and Alzheimer's disease: Focus on epigenetics. *Curr. Genom.* **2010**, *11*, 246–260. [CrossRef]

53. Spence, J.D.; Yi, Q.; Hankey, G.J. B vitamins in stroke prevention: Time to reconsider. *Lancet Neurol.* **2017**, *16*, 750–760. [CrossRef]

54. Smith, A.D.; Refsum, H. Homocysteine, B vitamins, and cognitive impairment. *Annu. Rev. Nutr.* **2016**, *36*, 211–239. [CrossRef] [PubMed]

55. Tucker, K.L.; Qiao, N.; Scott, T.; Rosenberg, I.; Spiro, A. High homocysteine and low B vitamins predict cognitive decline in aging men: The Veterans Affairs Normative Aging Study. *Am. J. Clin. Nutr.* **2005**, *82*, 627–635. [CrossRef] [PubMed]

56. Lipnicki, D.M.; Sachdev, P.S.; Crawford, J.; Reppermund, S.; Kochan, N.A.; Trollor, J.N.; Draper, B.; Slavin, M.J.; Kang, K.; Lux, O.; et al. Risk factors for late-life cognitive decline and variation with age and sex in the Sydney Memory and Ageing Study. *PLoS ONE* **2013**, *8*, e65841. [CrossRef] [PubMed]

57. Dufouil, C.; Alperovitch, A.; Ducros, V.; Tzourio, C. Homocysteine, white matter hyperintensities, and cognition in healthy elderly people. *Ann. Neurol.* **2003**, *53*, 214–221. [CrossRef] [PubMed]

58. McCaddon, A.; Hudson, P.; Davies, G.; Hughes, A.; Williams, J.H.; Wilkinson, C. Homocysteine and cognitive decline in healthy elderly. *Dement. Geriatr. Cogn. Disord.* **2001**, *12*, 309–313. [CrossRef] [PubMed]

59. Nurk, E.; Refsum, H.; Tell, G.S.; Engedal, K.; Vollset, S.E.; Ueland, P.M.; Nygaard, H.A.; Smith, A.D. Plasma total homocysteine and memory in the elderly: The Hordaland Homocysteine study. *Ann. Neurol.* **2005**, *58*, 847–857. [CrossRef] [PubMed]

60. Hooshmand, B.; Solomon, A.; Kåreholt, I.; Rusanen, M.; Hänninen, T.; Leiviskä, J.; Winblad, B.; Laatikainen, T.; Soininen, H.; Kivipelto, M. Associations between serum homocysteine, holotranscobalamin, folate and cognition in the elderly: A longitudinal study. *J. Intern. Med.* **2012**, *271*, 204–221. [CrossRef] [PubMed]

61. Clarke, R.; Birks, J.; Nexo, E.; Ueland, P.M.; Schneede, J.; Scott, J.; Molloy, A.; Evans, J.G. Low vitamin B-12 status and risk of cognitive decline in older adults. *Am. J. Clin. Nutr.* **2007**, *86*, 1384–1391. [CrossRef] [PubMed]

62. Hooshmand, B.; Mangialasche, F.; Kalpouzos, G.; Solomon, A.; Kåreholt, I.; Smith, A.D.; Refsum, H.; Wang, R.; Mühlmann, M.; Ertl-Wagner, B.; et al. Association of vitamin B12, folate, and sulfur amino acids with brain magnetic resonance imaging measures in older adults: A longitudinal population-based study. *JAMA Psychiatry* **2016**, *73*, 606–613. [CrossRef] [PubMed]

63. Seshadri, S.; Beiser, A.; Selhub, J.; Jacques, P.F.; Rosenberg, I.H.; D'Agostino, R.B.; Wilson, P.W.; Wolf, P.A. Plasma homocysteine as a risk factor for dementia and Alzheimer's disease. *N. Engl. J. Med.* **2002**, *346*, 476–483. [CrossRef] [PubMed]

64. Hooshmand, B.; Solomon, A.; Kåreholt, I.; Leiviskä, J.; Rusanen, M.; Ahtiluoto, S.; Winblad, B.; Laatikainen, T.; Soininen, H.; Kivipelto, M. Homocysteine and holotranscobalamin and the risk of Alzheimer disease: A longitudinal study. *Neurology* **2010**, *75*, 1408–1414. [CrossRef]

65. Faux, N.G.; Ellis, K.A.; Porter, L.; Fowler, C.J.; Laws, S.M.; Martins, R.N.; Pertile, K.K.; Rembach, A.; Rowe, C.C.; Rumble, R.L.; et al. Homocysteine, vitamin B12, and folic acid levels in Alzheimer's disease, mild cognitive impairment, and healthy elderly: Baseline characteristics in subjects of the Australian Imaging Biomarker Lifestyle study. *J. Alzheimers Dis.* **2011**, *27*, 909–922. [CrossRef] [PubMed]

66. Smith, A.D. The worldwide challenge of the dementias: A role for B vitamins and homocysteine? *Food Nutr. Bull.* **2008**, *29*, S143–S172. [CrossRef] [PubMed]

67. Smith, A.D.; Refsum, H.; Bottiglieri, T.; Fenech, M.; Hooshmand, B.; McCaddon, A.; Miller, J.W.; Rosenberg, I.H.; Obeid, R. Homocysteine and dementia: An international consensus statement. *J. Alzheimers Dis.* **2018**, *62*, 561–570. [CrossRef] [PubMed]

68. Minagawa, H.; Watanabe, A.; Akatsu, H.; Adachi, K.; Ohtsuka, C.; Terayama, Y.; Hosono, T.; Takahashi, S.; Wakita, H.; Jung, C.G.; et al. Homocysteine, another risk factor for Alzheimer disease, impairs apolipoprotein E3 function. *J. Biol. Chem.* **2010**, *285*, 38382–38388. [CrossRef] [PubMed]

69. Bhatia, P.; Singh, N. Homocysteine excess: Delineating the possible mechanism of neurotoxicity and depression. *Fundam. Clin. Pharmacol.* **2015**, *29*, 522–528. [CrossRef]

70. Lipton, S.A.; Kim, W.K.; Choi, Y.B.; Kumar, S.; D'Emilia, D.M.; Rayudu, P.V.; Arnelle, D.R.; Stamler, J.S. Neurotoxicity associated with dual actions of homocysteine at the N-methyl-D-aspartate receptor. *Proc. Natl. Acad. Sci. USA* **1997**, *94*, 5923–5928. [CrossRef]

71. Snowdon, D.A.; Tully, C.L.; Smith, C.D.; Riley, K.P.; Markesbery, W.R. Serum folate and the severity of atrophy of the neocortex in Alzheimer disease: Findings from the Nun study. *Am. J. Clin. Nutr.* **2000**, *71*, 993–998. [CrossRef] [PubMed]

72. Wang, H.; Odegaard, A.; Thyagarajan, B.; Hayes, J.; Cruz, K.S.; Derosiers, M.F.; Tyas, S.L.; Gross, M.D. Blood folate is associated with asymptomatic or partially symptomatic Alzheimer's disease in the Nun study. *J. Alzheimers Dis.* **2012**, *28*, 637–645. [CrossRef]

73. Blom, H.J.; Smulders, Y. Overview of homocysteine and folate metabolism. With special references to cardiovascular disease and neural tube defects. *J. Inherit. Metab. Dis.* **2011**, *34*, 75–81. [CrossRef] [PubMed]

74. Nazki, F.H.; Sameer, A.S.; Ganaie, B.A. Folate: Metabolism, genes, polymorphisms and the associated diseases. *Gene* **2014**, *533*, 11–20. [CrossRef] [PubMed]

75. Panossian, L.A.; Porter, V.R.; Valenzuela, H.F.; Zhu, X.; Reback, E.; Masterman, D.; Cummings, J.L.; Effros, R.B. Telomere shortening in T cells correlates with Alzheimer's disease status. *Neurobiol. Aging* **2003**, *24*, 77–84. [CrossRef]

76. Paul, L.; Cattaneo, M.; D'Angelo, A.; Sampietro, F.; Fermo, I.; Razzari, C.; Fontana, G.; Eugene, N.; Jacques, P.F.; Selhub, J. Telomere length in peripheral blood mononuclear cells is associated with folate status in men. *J. Nutr.* **2009**, *139*, 1273–1278. [CrossRef] [PubMed]

77. Friso, S.; Choi, S.-W.; Girelli, D.; Mason, J.B.; Dolnikowski, G.G.; Bagley, P.J.; Olivieri, O.; Jacques, P.F.; Rosenberg, I.H.; Corrocher, R.; et al. A common mutation in the 5,10-methylenetetrahydrofolate reductase gene affects genomic DNA methylation through an interaction with folate status. *Proc. Natl. Acad. Sci. USA* **2002**, *99*, 5606–5811. [CrossRef] [PubMed]

78. Blount, B.C.; Mack, M.M.; Wehr, C.M.; MacGregor, J.T.; Hiatt, R.A.; Wang, G.; Wickramasinghe, S.N.; Everson, R.B.; Ames, B.N. Folate deficiency causes uracil misincorporation into human DNA and chromosome breakage: Implications for cancer and neuronal damage. *Proc. Natl. Acad. Sci. USA* **1997**, *94*, 3290–3295. [CrossRef] [PubMed]

79. Troesch, B.; Weber, P.; Mohajeri, M. Potential links between impaired one-carbon metabolism due to polymorphisms, inadequate B-vitamin status, and the development of Alzheimer's disease. *Nutrients* **2016**, *8*, 803. [CrossRef]

80. Religa, D.; Styczynska, M.; Peplonska, B.; Gabryelewicz, T.; Pfeffer, A.; Chodakowska, M.; Luczywek, E.; Wasiak, B.; Stepien, K.; Golebiowski, M.; et al. Homocysteine, apolipoprotein E and methylenetetrahydrofolate reductase in Alzheimer's disease and mild cognitive impairment. *Dement. Geriatr. Cogn. Disord.* **2003**, *16*, 64–70. [CrossRef] [PubMed]

81. Smith, D.A.D.; Warren, M.J.; Refsum, H. Chapter Six—Vitamin B$_{12}$. *Adv. Food Nutr. Res.* **2018**, *83*, 215–279. [CrossRef] [PubMed]

82. Stabler, S.P. Vitamin B$_{12}$ deficiency. *N. Engl. J. Med.* **2013**, *368*, 149–160. [CrossRef] [PubMed]

83. Valente, E.; Scott, J.M.; Ueland, P.-M.; Cunningham, C.; Casey, M.; Molloy, A.M. Diagnostic accuracy of holotranscobalamin, methylmalonic acid, serum cobalamin, and other indicators of tissue vitamin B$_{12}$ status in the elderly. *Clin. Chem.* **2011**, *57*, 856–863. [CrossRef] [PubMed]
84. Cho, H.S.; Huang, L.K.; Lee, Y.T.; Chan, L.; Hong, C.T. Suboptimal baseline serum vitamin B$_{12}$ is associated with cognitive decline in people with Alzheimer's disease undergoing cholinesterase inhibitor treatment. *Front. Neurol.* **2018**, *9*, 325. [CrossRef] [PubMed]
85. Garcia, A.; Haron, Y.; Evans, L.; Smith, M.; Freedman, M.; Román, G. Metabolic markers of cobalamin deficiency and cognitive function in normal older adults. *J. Am. Geriatr. Soc.* **2004**, *52*, 66–71. [CrossRef]
86. Garcia, A.; Zanibbi, K. Homocysteine and cognitive function in elderly people. *CMAJ* **2004**, *171*, 897–904. [CrossRef] [PubMed]
87. Román, G.C.; Ruiz, P.J. Neurologic complications of Sjögren syndrome. *MedLink Neurol.* **2010**, *11*, 1034.
88. Andrès, E.; Loukili, N.H.; Noel, E.; Kaltenbach, G.; Abdelgheni, M.B.; Perrin, A.E.; Noblet-Dick, M.; Maloisel, F.; Schlienger, J.L.; Blicklé, J.F. Vitamin B$_{12}$ (cobalamin) deficiency in elderly patients. *CMAJ* **2004**, *171*, 251–259. [CrossRef] [PubMed]
89. Spence, D. Mechanisms of thrombogenesis in atrial fibrillation. *Lancet* **2009**, *373*, 1006. [CrossRef]
90. Wallin, A.; Román, G.C.; Esiri, M.; Kettunen, P.; Svensson, J.; Paraskevas, G.P.; Kapaki, E. Update on vascular cognitive impairment associated with subcortical small-vessel disease. *J. Alzheimers Dis.* **2018**, *62*, 1417–1441. [CrossRef] [PubMed]
91. Mohajeri, M.H.; Troesch, B.; Weber, P. Inadequate supply of vitamins and DHA in the elderly: Implications for brain aging and Alzheimer-type dementia. *Nutrition* **2015**, *31*, 261–275. [CrossRef]
92. Fuso, A.; Nicolia, V.; Cavallaro, R.A.; Ricceri, L.; D'Anselmi, F.; Coluccia, P.; Calamandrei, G.; Scarpa, S. B-vitamin deprivation induces hyperhomocysteinemia and brain S-adenosylhomocysteine, depletes brain S-adenosylmethionine, and enhances PS1 and BACE expression and amyloid-β deposition in mice. *Mol. Cell. Neurosci.* **2008**, *37*, 731–746. [CrossRef] [PubMed]
93. McCleery, J.; Abraham, R.P.; Denton, D.A.; Rutjes, A.W.; Chong, L.Y.; Al-Assaf, A.S.; Griffith, D.J.; Rafeeq, S.; Yaman, H.; Malik, M.A.; et al. Vitamin and mineral supplementation for preventing dementia or delaying cognitive decline in people with mild cognitive impairment. *Cochrane Database Syst. Rev.* **2018**, *11*, CD011905. [CrossRef]
94. Smith, A.D.; Smith, S.M.; de Jager, C.A.; Whitbread, P.; Johnston, C.; Agacinski, G.; Oulhaj, A.; Bradley, K.M.; Jacoby, R.; Refsum, H. Homocysteine-lowering by B vitamins slows the rate of accelerated brain atrophy in mild cognitive impairment. A randomized controlled trial. *PLoS ONE* **2010**, *5*, e12244. [CrossRef] [PubMed]
95. Jager, C.A.; Oulhaj, A.; Jacoby, R.; Refsum, H.; Smith, A.D. Cognitive and clinical outcomes of homocysteine-lowering B-vitamin treatment in mild cognitive impairment: A randomized controlled trial. *Int. J. Geriatr. Psychiatry* **2012**, *27*, 592–600. [CrossRef] [PubMed]
96. Douauda, G.; Refsum, H.; de Jager, C.A.; Jacoby, R.; Nichols, T.E.; Smith, S.M.; Smith, A.D. Preventing Alzheimer's disease-related gray matter atrophy by B-vitamin treatment. *PNAS Proc. Natl. Acad. Sci. USA* **2013**, *110*, 9523–9528. [CrossRef] [PubMed]
97. Sharma, A.; Gerbarg, P.; Bottiglieri, T.; Massoumi, L.; Carpenter, L.L.; Lavretsky, H.; Muskin, P.R.; Brown, R.P.; Mischoulon, D. As Work Group of the American Psychiatric Association Council on Research S-Adenosylmethionine (SAMe) for Neuropsychiatric Disorders: A Clinician-Oriented Review of Research. *J. Clin. Psychiatry* **2017**, *78*, e656–e667. [CrossRef] [PubMed]
98. Dayon, L.; Guiraud, S.P.; Corthésy, J.; Da Silva, L.; Migliavacca, E.; Tautvydaitė, D.; Oikonomidi, A.; Moullet, B.; Henry, H.; Métairon, S.; et al. One-carbon metabolism, cognitive impairment and CSF measures of Alzheimer pathology: Homocysteine and beyond. *Alzheimers. Res. Ther.* **2017**, *9*, 43. [CrossRef]
99. Galizia, I.; Oldani, L.; Macritchie, K.; Amari, E.; Dougall, D.; Jones, T.N.; Lam, R.W.; Massei, G.J.; Yatham, L.N.; Young, A.H. S-adenosyl methionine (SAMe) for depression in adults. *Cochrane Database Syst. Rev.* **2016**. [CrossRef]

International Journal of
Molecular Sciences

MDPI

Review

Recent Insights on Alzheimer's Disease Originating from Yeast Models

David Seynnaeve [1], Mara Del Vecchio [1], Gernot Fruhmann [1], Joke Verelst [1], Melody Cools [1], Jimmy Beckers [1], Daniel P. Mulvihill [2], Joris Winderickx [1] and Vanessa Franssens [1,*]

[1] Functional Biology, KU Leuven, Kasteelpark Arenberg 31, 3000 Leuven, Belgium;
david.seynnaeve@kuleuven.be (D.S.); mara.delvecchio@kuleuven.be (M.D.V.);
gernot.fruhmann@kuleuven.be (G.F.); joke.verelst@kuleuven.be (J.V.); melody.cools@kuleuven.be (M.C.);
jimmy.beckers@student.kuleuven.be (J.B.); joris.winderickx@kuleuven.be (J.W.)
[2] School of Biosciences, University of Kent, Canterbury CT2 7NJ, Kent, UK; D.P.Mulvihill@kent.ac.uk
* Correspondence: vanessa.franssens@kuleuven.be; Tel.: +32-16320381; Fax: +32-16321967

Received: 14 June 2018; Accepted: 30 June 2018; Published: 3 July 2018

Abstract: In this review article, yeast model-based research advances regarding the role of Amyloid-β (Aβ), Tau and frameshift Ubiquitin UBB^{+1} in Alzheimer's disease (AD) are discussed. Despite having limitations with regard to intercellular and cognitive AD aspects, these models have clearly shown their added value as complementary models for the study of the molecular aspects of these proteins, including their interplay with AD-related cellular processes such as mitochondrial dysfunction and altered proteostasis. Moreover, these yeast models have also shown their importance in translational research, e.g., in compound screenings and for AD diagnostics development. In addition to well-established *Saccharomyces cerevisiae* models, new upcoming *Schizosaccharomyces pombe*, *Candida glabrata* and *Kluyveromyces lactis* yeast models for Aβ and Tau are briefly described. Finally, traditional and more innovative research methodologies, e.g., for studying protein oligomerization/aggregation, are highlighted.

Keywords: Alzheimer's disease; yeast; Tau; amyloid β; ubiquitin; aggregation; oligomerization; prion

1. Introduction

1.1. Alzheimer's Disease (AD)

Alzheimer's disease (AD) is the most common neurodegenerative disease worldwide. It accounts for approximately 60–70% of all dementia cases and affects about 6% of the population aged over 65 (late-onset AD), whereas 2–10% of patients suffer from early-onset AD [1,2]. Currently, around 50 million individuals live with this devastating chronic disease and it has been estimated that the number will increase up to approximately 106 million people by 2050 due to an increasing aging population [2,3]. At the cellular level, AD is characterized by an irreversible and progressive loss of neuronal structure and function within certain regions of the brain including the hippocampus and neocortical brain, leading to cognitive dysfunction and dementia [4]. Widespread experimental evidence also suggests that AD is characterized by synaptic dysfunction early on in the disease process, disrupting communication within neural circuits important for memory formation and other cognitive functions such as intellectuality and comprehensive capacity [5–7].

Therefore, damage to these brain structures results in memory loss, language difficulties and learning deficits that are typically observed within early stages of clinical manifestation of AD. In addition, upon disease progression, a decline in other cognitive domains occurs which will result in the complete inability to function independently in basic daily activities [7]. Besides AD having a profound impact on the life quality of patients, this chronic disease also imposes a huge economic burden on healthcare systems globally with an associated cost which is estimated will exceed $1 trillion by 2050 [8].

The neuronal damage is related to the accumulation of misfolded proteins into extracellular and intracellular aggregates, consisting of Aβ peptides or protein Tau, respectively [9,10]. It is not yet clear whether the presence of these two hallmarks is the cause of AD or mainly the result of a cascade of cellular events including oxidative stress, mitochondrial dysfunction and apoptosis. Either way, the exact mechanism by which these proteins damage neurons is still unknown.

1.2. Yeast as a Model Organism to Study AD: Advantages

Studies to gain more insights on AD primarily make use of human cell lines and transgenic mouse models. However, yeast cell models are playing an increasingly important role in unravelling the fundamental disease aspects of AD. In fact, the yeast *S. cerevisiae* is a very widely studied single-celled model organism. With more than 6000 genes distributed on 16 chromosomes, its genome was the first eukaryotic genome to be completely sequenced in 1996 [11]. Since then, it has been estimated that nearly 31% of yeast genes have human orthologues [12]. Beyond the laboratory yeast strains, many different natural, brewery and clinical isolates exist and all have a core genome of about 5000 shared genes [12]. Yeast reproduction is through mitosis of either a haploid or a diploid cell. Haploids are of 2 different mating type (a or α) and a haploid cell can only mate with a cell of the opposite mating type. Mating leads to the formation of a diploid cell that can either continue to exist and bud as a diploid or, under conditions of stress, produce spores by meiosis. Spores can then later give rise to haploid cells [12]. Haploidy implies that gene-knockout strains can easily be obtained. In 2001, a collection of isogenic yeast strains, each deleted for one of the 6000 putative open reading frames (ORFs), was created [12]. This allowed for the easy phenotypic analysis of mutants, paving the way to determining gene function. In addition, yeast cells share many conserved biological processes such as cell cycle progression, protein turnover, vesicular trafficking and signal transduction with cells of higher eukaryotes [13], including human neurons. Its short generation time (1.5 h on rich medium), means that it can be very easily cultured. Thanks to its susceptibility to simple genetic and environmental manipulations, this model organism has become a valuable tool to shed more light on the complex and fundamental intracellular mechanisms underlying neurodegenerative diseases.

So-called "humanized yeast model systems" have been constructed and used as a tool to investigate the molecular mechanisms involved in several neurological disorders [14,15]. The main advantage of using yeast is its reduced complexity compared to the mammalian models. On the contrary, Tau and Aβ have no functional yeast orthologues. Heterologous expression of Tau and Aβ can be highly informative and provides useful new insights into the pathobiology of these proteins in vivo. At the same time, yeast is an excellent screening tool for compounds that may be useful in treatment and/or prevention of AD.

1.3. Yeast as a Model Organism to Study AD: Limitations

Despite being a powerful and simplified model system, yeast also has its natural limitations. As a unicellular organism, the most important limitation for neurodegenerative disease research is the analysis of disease aspects that focus on multicellularity and cell–cell interactions. These interactions include synaptic transmissions, axonal transport, glial-neuronal interactions, immune and inflammatory responses and many neuronal specializations that are likely to play an important role in neurodegeneration, but cannot be recapitulated in yeast [16]. Moreover, it is also impossible to study the cognitive aspects of AD in yeast cells.

This review discusses the findings of more recent studies on neurodegenerative disorders conducted using different yeast species.

2. Humanized Yeast Models to Study Tau Biology

2.1. Protein Tau: Structure, Functions and Modifications

Protein Tau, encoded by the 16 exon long microtubule (MT)-associated protein Tau (*MAPT*) gene located on chromosome 17q21.31, is present in neuronal and glial axons, but has also been detected outside of cells [17–22]. Tau is natively unfolded and has the tendency to adapt a paperclip-like shape, in which the N- and C-terminal domains and repeat regions are all closely located to each other (see below) [23]. It is a MT associated protein, susceptive to dynamic (de-) phosphorylation. These modifications influence its main cellular function which consists of regulating MT dynamic instability i.e., the process of polymerization (rescue) and depolymerization (catastrophe) [24–27]. Besides being involved in the regulation of MT dynamics, Tau has functions in regulating axonal transport/elongation/maturation, synaptic plasticity and maintaining DNA and RNA integrity [28–37]. It is clear that Tau is involved in numerous processes and that loss of Tau function can initiate neurotoxicity through disruption of various processes in which it is involved. For a more complete overview of physiological and pathological Tau functions, we refer to [38].

A first mechanism by which Tau function is regulated is alternative splicing. Due to alternative splicing of exon 2, 3 and 10 of the *MAPT* gene, Tau can be present in 6 different isoforms differing in the number of N-terminal inserts (1 or 2) and conserved 18 amino acid long repeats (3 to 4) in the MT binding region (C-terminal region or assembly domain) [39,40]. Tau isoforms with 4 repeat regions show a stronger interaction with MT and are more efficient in MT assembly [41–43]. The N-terminal projection region of the protein is located adjacent to a proline rich region and has a role in MT spacing and stabilization [44,45]. In addition, it was proposed that this domain, which projects away from MT, interacts with cell organelles such as the plasma membrane, mitochondria and actin filaments [46–51]. This binding could be facilitated via an interaction between the PXXP motifs in the proline-rich region and the SH3 domains of the src-family non-receptor tyrosine kinases (e.g., kinase FYN) [52–54]. Note that this plasma membrane interaction might play an important role in vesicle-mediated secretion and therefore impact the cell-to-cell spreading of protein Tau. It was proposed that the Tau-FYN interaction may regulate the post-synaptic targeting of FYN, and thereby mediate Aβ-induced excitotoxicity [23]. Additional proposed pathways for cell-to-cell transfer are tunneling nanotubes and trans-synaptic spreading [55–58]. This spreading process is still ill-defined and it still needs to be proven if spreading of (a) toxic Tau species is sufficient or necessary for the induction of a tauopathy. Guo and colleague published a comprehensive article on this emerging field within Tau biology [59]. Finally, the proline-rich domain of Tau has a role in facilitating the binding of the MT binding region to the MT [60,61].

Tau function is also regulated by several post-translational modifications including phosphorylation, glycosylation, truncation, nitration, isomerisation, acetylation, glycation, ubiquitination, deamidation, methylation, sumoylation and oxidation [38]. Tau phosphorylation has been studied extensively. The protein contains 80 putative serine/threonine and 5 potential tyrosine phosphorylation sites, of which the majority is phosphorylated in vitro, on the 2N/4R (2 N-terminal inserts and 4 amino acid repeat regions) isoform. Tau is phosphorylated by numerous kinases, grouped in 4 different classes [62,63]. More recently, GSK3α, GSK3β, MAPK13 and AMP-activated protein kinase were found to play an actual role in in vivo Tau phosphorylation using different cell lines [64,65]. Tau (de-) phosphorylation is an important factor influencing Tau's affinity for MT, thereby regulating its role in MT (de-) polymerization. On the other hand, aberrant phosphorylation (so-called hyperphosphorylation) on several epitopes (e.g., Thr181, Thr231, Ser202, Ser205, Ser214, Ser396, Ser404, Ser409, and Ser422), which severely affects Tau's MT binding capacity and stabilizing properties [66–68] can lead to an increased propensity of Tau to subsequently oligomerize and aggregate into paired helical filaments (PHF) and neurofibrillary tangles (NFT) [66]. These NFT are characteristic for a group of neurodegenerative diseases called tauopathies including AD.

The aggregation is due to a redistribution of mainly MT bound to unbound Tau, which facilitates Tau-Tau interactions made possible by 2 hexapeptide motives in repeat regions 2 and 3, which can adapt β-sheet structures [69]. While the repeat domain makes up the PHF core, the N- and C-terminal Tau region form a "coat" around this core [23].

Hyperphosphorylation can also induce pathology through other mechanisms. It can first of all lead to Tau missorting from axons to the somatodendritic compartment, which might cause synaptic dysfunction. Another consequence is affected substrate recognition, which leads to an altered proteasomal degradation [70].

Phosphorylation is, however, only one potential covalent modification Tau can undergo and it should be noted that this modification alone is not sufficient to cause aggregation. Phosphorylation at some sites (e.g., Ser214 and Ser262) in the repeat domain can even protect against aggregation [69]. Thus, it is suggested that phosphorylation might facilitate this process, and therefore serves as an indirect aggregation inducer, and that other factors are involved as well. Indeed, other modifications, and especially truncation, can be equally important for disease development. Tau truncated at Glu391 and Asp421, for example, has been identified as an event following phosphorylation and facilitating Tau filament formation. Tau truncation can even induce neurodegeneration independently of Tau aggregation through the formation of specific Tau fragments [69]. Tau truncation disrupts the paperclip-like structure, thereby promoting Tau aggregation. Tau ubiquitination, on the other hand, is considered a protective strategy of the cell to get rid of toxic Tau intermediates and accumulations of hyperphosphorylated Tau are mainly found in cells with a defective or malfunctioning ubiquitin/proteasome system. The latter can be caused by oxidative stress due to mitochondrial malfunctioning, illustrating the complex cellular pathways involved in the induction of Tau-mediated toxicity. For a more complete overview of Tau post-translational modifications and their consequences on Tau pathology, we refer to [38,71].

The Tau aggregation process itself seems to be a requirement for Tau-induced toxicity and although recent papers are pointing towards the soluble mono- or oligomeric hyperphosphorylated Tau species as being the toxic Tau forms, their relative contribution remains largely unclear. The insoluble aggregated structures are thought to act as protective structures by sequestering the toxic species [72–76].

Numerous Tau mutations, either causing AD or other tauopathies such as frontotemporal dementia and parkinsonism linked to chromosome 17 (FTDP-17), have been documented over the years and can either be missense, silent or causing a deletion. Depending on the mutation's nature and its gene location, the mutation can directly disturb Tau's MT binding capacity, thereby increasing Tau's tendency for aggregation, or indirectly by affecting the 4R:3R ratio by influencing Tau splicing [77–85]. Most of these mutations are nicely documented on the "Alzforum" website.

2.2. From Complementary Disease Models to AD Diagnostics

Historically, the AD field has been dominated by research supporting Aβ having the main role in pathogenesis. Only after the discovery of several *MAPT* mutations in FTDP-17 did Tau research receive a significant and rightful boost. Indeed, both in vitro and in vivo studies show evidence that Tau is required for Aβ-mediated neurotoxicity [86]. Therefore, only a limited amount of research articles on the pathological aspects of protein Tau using the yeast *S. cerevisiae* as a model organism have been published to date [87–89].

Figure 1 gives a visual overview of human Tau processes and modifications in *S. cerevisiae*. These studies have already been extensively reviewed, so a brief summary will be given. Upon overexpression in *S. cerevisiae*, Tau becomes hyperphosphorylated and acquires several pathological phospho-epitopes (AD2 (p-Ser396/p-Ser404), AT8 (p-Ser202/p-Ser205), AT270 (p-Thr181), AT180 (p-Thr231/p-Ser235) AT100 (p-Thr212/p-Ser214) and PG5 (p-Ser409)). Moreover, it was possible to detect the disease-relevant conformational epitope recognized by the MC1 antibody.

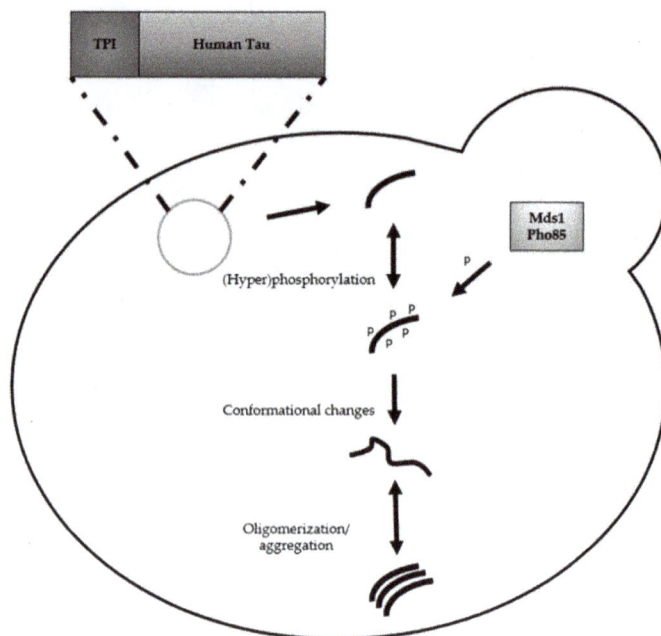

Figure 1. Humanized yeast model expressing human protein Tau: overview of Tau processes and modifications in *S. cerevisiae*. Double arrows indicate a bidirectional/reversible reaction and dashed lines specify the promoter and expressed human *Tau* gene on the plasmid. 'TPI'; Triosephosphate isomerase promoter, 'P'; phosphate group.

Pho85 and Mds1 protein kinases, yeast orthologues of human Tau kinases Cdk5 and GSK3β, respectively, were shown to play a key role in modulating Tau phosphorylation. It was suggested that Pho85 may have a direct or indirect inhibitory effect on the activity of Mds1. Upon deletion of Pho85, phosphorylation of Tau was enriched on the AD2 and PG5 epitopes. Accordingly, the MC1-reactive Tau fraction was also higher. Tau aggregation in this *pho85Δ* strain was assessed by measuring the sarkosyl-insoluble Tau (SinT) fraction and it was proposed that Tau epitopes PG5 and AT100 might play a crucial role in the accumulation of SinT aggregates, since these epitopes were especially enriched in the insoluble fraction [87]. The importance of phosphorylation of the PG5 epitope for Tau aggregation was also confirmed in a follow-up study in which aggregation of several Tau mutants was assessed. In addition, PG5 epitope phosphorylation is detrimental for Tau's MAPT function, illustrated by lack of Tau binding to taxol-stabilized MT from porcine Tubulin in vitro [88,89].

Despite all of this, Tau 2N/4R and 2N/3R expression does not induce an impaired growth phenotype in *S. cerevisiae* [87,88,90]. The latter is not necessarily expected since little attention was paid to the extent of formation of early stage, presumably toxic, soluble (oligomeric) Tau species in these studies. The possibility exists that these oligomeric Tau species are rapidly sequestered in inert aggregates as a cell protection mechanism.

As described above, several other post-translational modifications are expected to contribute to tauopathy development, besides phosphorylation. Therefore, it might be highly interesting to verify if these modifications are also recapitulated in yeast, and if so, to what extend they could offer an explanation for the (lack of) aggregation/toxicity in yeast cells.

On top of that, it was found that oxidative stress and mitochondrial dysfunction, independently of Tau phosphorylation, also strongly induce Tau aggregation in yeast cells [88]. It is also worth mentioning that inducing oxidative stress resulted in Tau dephosphorylation, in accordance with other results obtained from human, rat and mice neuronal cells [91,92]. One potential mechanism is oxidative stress-induced Pin1 activation. Pin1, a peptidyl prolyl *cis/trans* isomerase, can then subsequently activate Phosphatase 2a. De Vos and colleagues reported, in accordance with this finding, that a dysfunctional Ess1, Pin1's yeast orthologue, increases Tau hyperphosphorylation [93].

On the other hand, other studies point out that oxidative stress does not alter Tau phosphorylation or even induces Tau hyperphosphorylation. These studies were performed using a *Dropsophila melanogaster* model and human neuronal cells [94], respectively. The interplay between oxidative stress and Tau phosphorylation, therefore, needs more attention in future studies to further elucidate their relationship.

Tau was also purified, using anion-exchange chromatography, from the previously mentioned *pho85Δ S. cerevisiae* strain, maintaining its hyperphosphorylated MC1-reactive state, and could subsequently seed aggregation of wt 2N/4R Tau protein purified from a wt strain in vitro [87]. The possibility of purifying these stable, pathologically-relevant, Tau structures from *S. cerevisiae* cells paved the way for using yeast-purified Tau as an antigen source for mice immunization [95]. This strategy offers a significant benefit over *E. coli* based Tau purifications and antibody generation [76,96, 97], since Tau is not post-translationally modified in bacterial cells.

Although oligomerization can be induced by use of, for example, arachidonic acid or heparin, there is no evidence that these artificially formed oligomers/aggregates are the actual toxic species and, therefore, that the produced monoclonal antibodies recognize pathologically relevant Tau species. ADx215, an antibody developed by immunizing mice with 2N/4R Tau purified from a *pho85Δ S. cerevisiae* strain, is capable of detecting both mono- and oligomeric Tau protein [95]. This antibody was recently successfully implemented in a digital enzyme-linked immunosorbent assay (ELISA) platform and able to detect attomolar concentrations of Tau protein, thereby unlocking the potential of Tau as a serum-based AD biomarker [98]. So, over the last decade, yeast has developed from a reliable model organism, merely used to gain more understanding of pathological Tau features such as aggregation and phosphorylation, to a highly suitable platform model for disease-relevant antigen production.

In the NFT of transgenic mice models, α-synuclein can co-localize with Tau and it has been shown that α-synuclein can seed Tau aggregation in vitro and in vivo and even enhance Tau's toxicity in mice models [99–103]. It is, therefore, clear that the interplay between both proteins is important. Yeast models have been developed that enable the study of both proteins and the resulting effects on toxicity and aggregation [104,105]. Episomal expression of wt and A53T α-synuclein and wt and P301L Tau resulted in increased phosphorylation on the AD2 epitope and Tau aggregation, but no growth-inhibiting effect was detected [105]. The latter was in contrast to a previously reported study [104], where synergistic toxicity was observed upon stable genomic integration of plasmids expressing wt α-synuclein and wt Tau. This is in accordance with increased α-synuclein inclusion formation and Tau phosphorylation/aggregation [105].

2.3. Future Perspectives

Baker's yeast has been of interest to humans since the existence of brewing and bread-making. Since these two activities have been subject to continuous improvement, research on *S. cerevisiae*'s physiology was mainly application-driven. In contrast, focus on the fission yeast *S. pombe* was mainly interested-driven and initial studies were performed to gain more insights in its mating type system and sexual and cell division cycle [106]. Nevertheless, *S. pombe* might offer great potential as a complementary model to study Tau biology since several features such as the cell division machinery, cell polarity and cytoskeleton organization are more closely related to higher eukaryotes compared to *S. cerevisiae* [106–108]. This organism could, therefore, be advantageous to study Tau characteristics such as in vivo MT binding, which has not been observed so far in budding yeast models most likely due to critical gene sequence differences. So far, binding to porcine MT has only been shown for yeast extracted Tau [88]. Indeed, fluorescence microscopy studies indicate potential binding of protein Tau to MT in vivo in *S. pombe* cells. Moreover, preliminary data points out that several Tau epitopes also become phosphorylated in *S. pombe* (data not shown). The precise role(s) of each of the fission yeast Tau kinase orthologues remains unresolved.

Heinisch and colleague also proposed several arguments why the milk yeast *K. lactis* could serve as a useful model to study Tau biology, more specifically the effects of energy signaling and oxidative stress on Tau aggregation [109]. *K. lactis* has several advantages over the traditional baker's yeast model. For example, a respiratory metabolism more resembling that of mammalian cells. Moreover, *K. lactis* did not undergo a whole genome duplication throughout evolution which limits the number of redundant gene functions. This ensures a more easily trackable phenotype upon single gene deletions [110].

Experimental methodologies used in the aforementioned reports, e.g., SinT assay or fluorescence microscopy using a green fluorescent protein (GFP)-tagged Tau protein, are well suited for the study of Tau aggregation, but lack applicability for analysis of Tau oligomerization. Since the current consensus is that oligomeric, rather than aggregated, Tau is the toxic Tau species, neat technologies to study oligomerization could enhance yeast's value as a model for the study of neurodegenerative diseases such as AD. An example is the use of a split-GFP sensor system. Several split-GFP technologies are described in [111–115].

3. Humanized Yeast Models to Study Aβ Biology

3.1. Protein Aβ: Structure, Function and Aggregation

Glycoprotein amyloid precursor protein (APP) plays an important role in numerous biological activities, ranging from neuronal development and homeostasis to signaling and intracellular transport [116–119]. After synthesis in the endoplasmatic reticulum (ER), the protein is subsequently transported from the Golgi apparatus to the plasma membrane where it is cleaved by α- and γ-secretase or β- and γ-secretase following the non-amyloidogenic or amyloidogenic pathway, respectively [120,121]. This cleavage yields several Aβ species with amino acid sequences varying from 40 to 51 amino acids with $A\beta_{40}$ and $A\beta_{42}$ being the final fragments [122,123]. While β-secretase activity is primarily mediated by BACE1, γ-secretase activity actually requires the presence of 4 proteins; Presenilin 1 or 2, Nicastrin, Presenilin enhancer 2 and Anterior pharynx defective 1 (Aph1) [124]. These peptides can then be released in the extracellular space where they can bind to a variety of receptors or they remain associated with the plasma membrane and lipid raft structures [119]. The amyloidogenic pathway is central in the so-called "amyloid cascade hypothesis", which states that the formed Aβ structures sequentially oligomerize and aggregate thereby causing neurotoxicity and dementia [125]. Aβ peptides can aggregate in different structural forms i.e., soluble oligomers, protofibrils, but also insoluble amyloid fibrils and all of them feature β-sheet structures [126,127]. While the oligomers may spread throughout the brain, the fibrils can further assemble into plaques, which are commonly found in the neocortex of the AD patient brains [128]. However, there is no direct correlation between amyloid plaques and the loss of synapses and neurons in AD patient brains [129–131].

In fact, cognitive deficits appear before plaque deposition or the deposition of insoluble amyloid fibrils. Similarly to protein Tau, it is suggested that the Aβ oligomers trigger synapse dysfunction and memory impairment [132,133]. Extracellular receptor-bound Aβ oligomers were proposed to induce neurotoxic effects by causing mitochondrial dysfunction and oxidative stress in neuronal cells, which can cause a massive calcium influx [134]. This can then impair the ability of cells to conduct normal physiological functions [135].

It should be mentioned, however, that it is highly possible that different Aβ forms may contribute to neurodegeneration at different disease stages [135]. A proposed link between Aβ and Tau pathology is the Aβ aggregation-mediated kinase activation, which results in Tau hyperphosphorylation (see above) that in turn results in NFT formation. This is, however, only one possible hypothesis linking these two proteins and their interplay is presumed to be more complex [135]. Other secondary, toxicity inducing, effects of Aβ aggregation are the involvement of the innate immune system and inflammatory responses [136–138].

To maintain Aβ protein homeostasis both in the brain and in plasma, production of Aβ is counterbalanced by mechanisms such as proteolytic degradation [139–141], active transport via the blood brain barrier [142–147] and deposition of Aβ in insoluble aggregates [148,149]. This does not only involve neurons, but also other cells of the neurovascular unit, such as astrocytes [150–152]. Disruption of any of these processes might result in neuropathology. Cathepsin B, for example, was identified as a major Aβ-degrading enzyme and its expression level is altered in the brain of AD patients [153].

3.2. From Heterologously Expressed APP to Secretory Pathway-Targeted Aβ Peptides

Since Aβ peptides are generated by Secretase cleavage of APP, modelling of Aβ pathology in yeast cells can be done via a number of approaches. The APP, or the Aβ peptides can be heterologously expressed in yeast. Although there are no orthologues of the human Secretases in yeast present, both α- and β-secretase activity has been reported [154,155], with the yeast proteases Yap3 and Mkc7 suggested to exhibit α-secretase activity [156,157]. γ-secretase activity was successfully reconstituted in *S. cerevisiae* upon combined expression of APP-based substrates and human γ- secretase, resulting in the production of Aβ$_{40}$, Aβ$_{42}$, and Aβ$_{43}$ [158,159] (Figure 2).

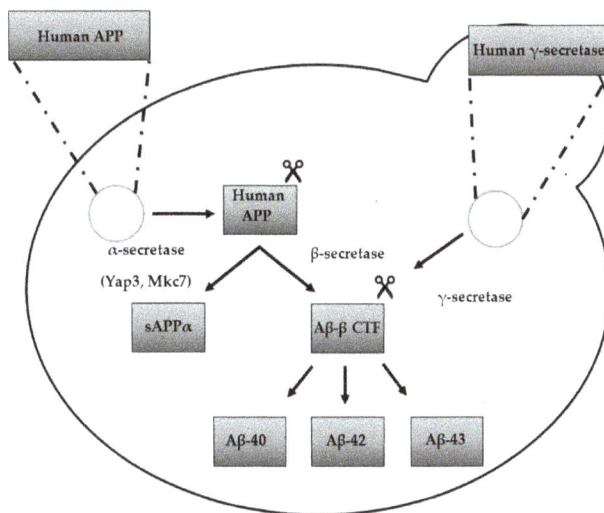

Figure 2. Humanized yeast model expressing human proteins, APP and γ-secretase: overview of Secretase-mediated APP processing and Aβ peptide production. The scissors icon indicates cleavage of the respective proteins.

As mentioned above, the γ-secretase complex consists of 4 different components and the influence of different combinations of Presenilin and Aph1 proteins on function and substrate specificity of the γ-secretase was tested in a yeast system [160,161]. Although differences were observed, the results were not well in line with parallel studies performed in mammalian cells, which, they reasoned, could be explained by the lack of additional proteins (e.g., GSAP and CD147) that affect γ-secretase function and that are not present in yeast. Moreover, the same group found that Nicastrin can be dispensable for protease activity of double-mutated γ-secretase, e.g., F411Y/S438P [162]. This first group of yeast models, in which pre-Aβ components (i.e., APP-like substrates and Secretases) are expressed, offers the possibility to screen for components and drugs that interfere with Aβ peptide generation and, therefore, have therapeutic potential.

The focus, however, has shifted more towards the expression of the actual Aβ peptides in yeast cells. This has one major benefit since it limits the amount of heterologously expressed proteins in yeast and, therefore, also any potential side reactions which result in non-Aβ pathology associated phenotypes. In this context, the prion-forming capability of *S. cerevisiae* has also been exploited to gain more insight in Aβ's oligomerization and aggregation capability.

Sup35, for example, is a translation termination factor and has the natural propensity to form self-propagating infectious amyloid aggregates which results in a prion phenotype "[PSI+]" [163].

Bagriantsev and colleague fused the MRF (middle- and release factor domain) of the protein to Aβ$_{42}$ and screened for the ability of an *ade1-14* strain to grown on medium lacking adenine. They showed that fusion of Aβ$_{42}$ to this MRF domain resulted in a similar phenotype as did the N (N-terminal domain) MRF protein, which makes up the entire amino acid sequence of the Sup35 protein. The N-terminal domain is required and sufficient for induction of the prion properties. Aβ$_{42}$ induced oligomerization, resulting in the inability of Sup35 to terminate translation, enabled growth of the *ade1-14* strain on medium lacking adenine by restoring adenine prototrophy [164]. This setup offers a neat in vivo system to screen for modulators of oligomerization. A second, independent study yielded a similar result in that the Aβ$_{42}$-Sup35 fusion was able to form aggregates, although less stable compared to Sup35 aggregates, and restore the [PSI+] phenotype of Sup35 lacking the prion forming domain [165].

To investigate Aβ$_{42}$'s location and interactions, it was fused to GFP. Apart from inducing a growth defect, Aβ$_{42}$ also induced a heat shock response [166]. The latter is in correspondence with data obtained from AD patients that indicate that heat shock protein expression is upregulated in AD as a protective measure [167]. A study on oligomerization/aggregation modifiers using this Aβ$_{42}$-GFP construct suggested that folinic acid might assist in preventing Aβ$_{42}$ misfolding and aggregation [168].

Recently, yeast was also used as a model to screen for rationally designed compounds [169]. More specifically, Thioflavin assays, circular dichroism measurements and transmission electron microscopy were used to assess the efficiency of peptidomimetic inhibitors to inhibit Aβ$_{42}$ aggregation by targeting non-covalent interactions (Table 1). This way, two compounds were able to rescue yeast from Aβ$_{42}$-induced toxicity. Yeast also served as an excellent tool to shed more light on the mechanism of action of the anti-histamine latrepirdine (Dimebon™) [170], which showed promising aggregate clearing activity in vivo (Table 1). The compound was suggested to upregulate the sequestering of aggregated GFP-Aβ$_{42}$ into autophagic-like vesicles which get targeted for degradation. Autophagy plays a crucial role in the removal of aggregated or misfolded proteins, such as Aβ, in neurodegenerative diseases [171–173] and impaired clearance of autophagic vesicles is also observed in the brains of AD mice models and patients [171–173]. Highly similar results were obtained in other cell and animal models: Steele and colleagues reported that the treatment of cultured mammalian cells with latrepirdine led to enhanced mTOR- and Atg5-dependent autophagy. Moreover, latrepirdine treatment of TgCRND8 transgenic mice was associated with improved learning behavior and with a reduction in accumulation of Aβ$_{42}$ [174].

Finally, a yeast-based screen identified clioquinol and dihydropyrimidine-thiones as compounds being able to ameliorate Aβ toxicity in a synergistic, metal-dependent, way via different mechanisms such as increasing Aβ turnover, restoring vesicle trafficking and oxidative stress protection [175,176] (Table 1). Again, also in transgenic mice models, treatment with clioquinol (analogue) compounds inhibited Aβ accumulation [177] and resulted in a dramatic improvement in learning and memory, accompanied by marked inhibition of AD-like neuropathology [178]. Finally, a study assessed the clinical effect of clioquinol analogue PBT2 using human patient cohorts. Compared to the placebo group, Aβ CSF concentration was reduced upon treating AD patients with PBT2. In addition, some cognitive test results indicated an improvement in AD patients treated with the clioquinol analogue [179]. These research and clinical studies highlight the fact that yeast-based compound screenings are extremely valuable to identify promising molecules that ameliorate Aβ pathology. Secondly, in several cases, the proposed mechanism of action of a compound, based on insights obtained from yeast research, was confirmed by other, often more comprehensive, pathological AD models [174].

Aβ peptides are generated at the plasma membrane and can subsequently be secreted and re-uptaken in the cell and eventually be found in the cytosol, mitochondria, secretory pathway and autophagosomes [180]. To recapitulate Aβ's multi-compartment trafficking, Treusch and colleagues fused a Kar2 sequence to the N-terminus of Aβ42, targeting the peptide to the ER [181] (Figure 3). After cleavage of this sequence, Aβ42 is released in the secretory pathway. The presence of a cell wall prevents diffusing of the peptides in the medium, thereby allowing interaction with the plasma membrane and endocytosis. Cell growth was decreased after expression of Aβ42 using a multicopy plasmid and galactose-inducible promoter and this in contrast to Aβ40. Screening of an overexpression library consisting of >5000 ORFs yielded several suppressors and enhancers of Aβ42 toxicity.

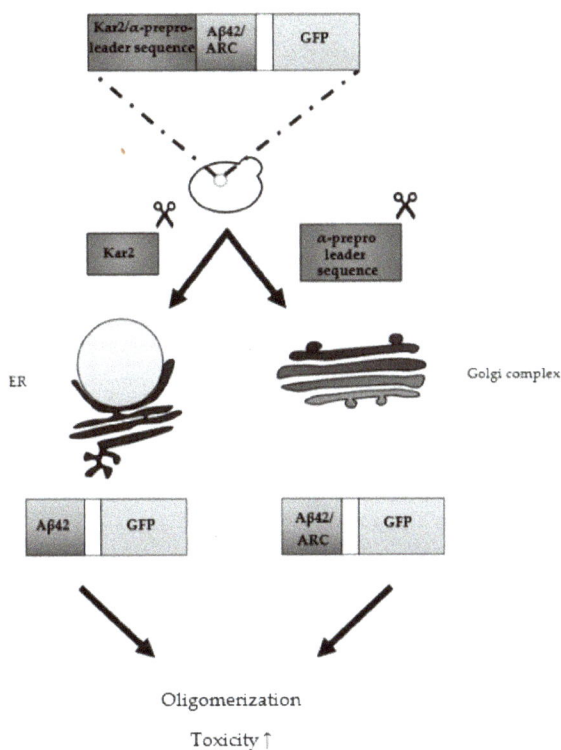

Figure 3. Humanized yeast model expressing GFP-fused Aβ peptides tagged with an endoplasmatic reticulum (ER) or Golgi complex targeting sequence. Treusch and colleagues expressed Aβ42 N-terminally tagged with the Kar2 sequence, while D'Angelo and colleagues expressed Aβ42/ARC N-terminally tagged with the α prepro-leader sequence (with and without a C-terminal GFP tag). The scissors icon indicates cleavage of the respective proteins.

PICALM (phosphatidylinositol-binding Clathrin assembly protein), of which Yap1801 and Yap1802 are the yeast homologues [182], was one of the toxicity suppressor hits and is one the most highly validated AD risk factors. The exact role of PICALM in AD is unknown, but it is thought to play a role in APP trafficking [183]. Since Aβ perturbs endocytotic trafficking, it was suggested that

PICALM has a role in restoring this process. These findings were backed up by data obtained from rat cortical neurons, in which Aβ-induced cell death was partly prevented upon PICALM expression [181]. A more recent study also reported on the beneficial role of PICALM, since it was able to reduce Aβ$_{42}$ oligomerization [183] (Table 1). In another article [184], the same yeast model was used to study the effects of native Aβ and in addition to previously shown lower growth rate, a reduced respiratory rate and elevated levels of reactive oxygen species (ROS) were exhibited. These are hallmarks of mitochondrial and ubiquitin-proteasome system dysfunction, which also occur in neurons and peripheral tissues of AD patients [185], and nicely illustrate the applicability of such a yeast model to study the role of Aβ in cell stress and damage. In fact, these results were in accordance with findings obtained from a yeast model after prolonged exposure to cytosolic Aβ$_{42}$. Several signs of mitochondrial dysfunction were observed, including increased ROS production, decreased mitochondrial membrane potential and reduced oxygen consumption [184]. A major question that remains is how Aβ peptides actually are taken up in the mitochondria.

In a follow-up study, using a more systems biology approach, the interplay between ER stress and the unfolded protein response (UPR) were studied upon constitutive expression of Aβ$_{40}$ and Aβ$_{42}$ [186]. In comparison to Aβ$_{40}$ which only induced mild stress, Aβ$_{42}$ expression resulted in prolonged high stress and an UPR failing to cope with the unfolded protein load resulting in cellular dysfunction, a shorter chronological lifespan and deregulation of lipid metabolism. These results are highly relevant for other diseases as well, especially cancer and diabetes due to the emerging role of the UPR in these diseases.

Using a similar strategy by fusing the mating type factor α prepro-leader sequence to Aβ, another group also showed that targeting Aβ in the secretory pathway is essential for toxicity in yeast [182] (Figure 3). The researchers tested both native and C-terminally GFP-tagged Aβ$_{42}$ and Aβ$_{ARC}$ and detected aggregate formation and a more profound toxic effect in case of the prepro-Aβ-linker-GFP constructs, especially Aβ$_{ARC}$ for which they measured a decrease in respiratory rate. They also suggested that Hsp104 could play a role in mediating this toxicity by favoring the conversion of large aggregates into smaller oligomeric species. Western blot data showed a decreased protein level in the case of native Aβ peptides, which indicated that the GFP moiety might have a stabilizing effect. These results were not in line with the results published by Treusch and colleagues, since in their research native Aβ$_{42}$ expression resulted in significant toxicity when ending up in the secretory pathway. Another difference was the presumed role of PICALM. In the paper published by D'Angelo and colleagues, deletion of Yap1801 and Yap1802 resulted in a decrease in Aβ-induced toxicity. Upon expression of PICALM, this toxic effect was partly restored. Therefore, more research is necessary to shed light on the actual role of PICALM in Aβ-induced toxicity.

Since it is clear that AD, mitochondrial dysfunction and altered proteostasis are linked to one another, two studies also more closely investigated the interplay between the Pitrilysin Metallopeptidase 1 (PITRM1), an oligopeptide-digesting mitochondrial matrix enzyme, and Aβ. In addition to its role in cleaving the mitochondrial targeting sequence (MTS) of proteins imported across the inner mitochondrial membrane, it also disposes mitochondrial Aβ [187–190]. However, in a first study, it was shown that the accumulation of Aβ peptides inhibits the activity of Cym1, which is the yeast PITRM1 orthologue, leading to impaired MTS processing and accumulation of precursor proteins [191]. In a second study, the effect of a missense mutation in this enzyme was documented using yeast by modelling the R183Q mutation in the Cym1 protein [192]. This resulted in a reduced Aβ$_{42}$ degradation compared to wt Cym1, suggesting a pathogenic role of this mutated protein, displaying similar behavior as in human beings.

Cenini and colleagues reported that Aβ peptides, especially Aβ$_{42}$, inhibited mitochondrial protein import by affecting an early process step when newly synthesized mitochondrial polypeptides are exposed to the cytosolic environment, rather than affecting mitochondrial membrane potential, TOM and TIM (Translocase of the outer and inner membrane, respectively) or respiratory chain metabolic protein complex composition [193]. These findings are in contrast to the study described earlier in this paragraph [191], in which Aβ peptides indirectly interfered with the processing of imported precursor proteins to the mature and active forms, which is a late step of the mitochondrial import reaction.

Finally, a *C. glabrata* model was used to assess toxicity of extracellular chemically-synthesized Aβ [194] by determining the viable colony count, using a water-based assay. It was shown that Aβ did bind the plasma membrane of *C. glabrata*, but the exact mechanism by which Aβ kills *C. glabrata* remains to be determined. Interestingly, upon oligomerization Aβ loses its toxic effect, while Aβ has a protective function against sodium hydroxide toxicity [195].

Table 1. Overview of Tau and Aβ toxicity modifiers identified using yeast-based screens.

Protein	Toxicity Modifiers	Description	Other Models	References
Tau	Pin1 (yeast homologue Ess1)	Depletion of Pin1 isomerase activity results in reduced growth of Tau expressing yeast cells.	mouse model	[93,196,197]
Aβ	peptidomimetic inhibitors	Inhibition of Aβ$_{42}$ aggregation by peptidomimetics.	-	[169]
Aβ	latrepirdine (Dimebon™)	Latrepirdine induces autophagy and decreases the intracellular GFP-Aβ$_{42}$ levels in yeast.	Hela cells, mouse model	[170,174]
Aβ	clioquinol	Small molecule screen identified several 8-hydroxyquinolines, including clioquinol, that ameliorate Aβ toxicity.	mouse model, nematode model	[175–179]
Aβ	dihydropyrimidine-thiones	Phenotypic small molecule yeast screen identified dihydropyrimidine-thiones that rescue Aβ-induced toxicity in a metal dependent manner.	nematode model	[176]
Aβ	PICALM (yeast homologues Yap1801, Yap1802)	Screening of overexpression library yielded suppressors and enhancers of Aβ$_{42}$ toxicity, including the PICALM suppressor.	rat cortical neurons	[181–183]

4. Humanized Yeast Models to Study Frameshift Ubiquitin Mutant UBB^{+1} Biology

Yeast models expressing the frameshift Ubiquitin mutant UBB^{+1} have also been developed. UBB^{+1} accumulation is found in neurons of all AD patients, but absent in those of Parkinson's disease patients, and co-localizes with the MC1 marker, i.e., NFT [198]. How UBB^{+1} is related to aberrant and phosphorylated Tau protein, both spatially and temporally, still needs to be elucidated [198]. The authors suggested that these mutant proteins may be responsible for the lack of multi-ubiquitination of the hyperphosphorylated Tau fraction found in the NFT [199]. These UBB^{+1} molecules are unable to bind to lysine residues in target molecules, since they lack the COOH-terminal glycine residue in the first repeat region, which is essential for subsequent multi-ubiquitination and activation of the proteasomal machinery [200]. Upon expression of UBB^{+1} in yeast, the protein becomes a substrate of the UPR and accumulated UBB^{+1} impairs the UPR both in yeast and mammalian cells [201–204]. This results in an accumulation of polyubiquitinated substrates which do not get degraded, partially accomplished by the inhibition of deubiquitinating enzymes [205].

Despite this impairment, no toxicity is observed. By contrast, upon prolonged expression of high levels of UBB^{+1}, cell death and mitochondrial dysfunction were observed in neuronal cells and yeast models [203,206,207]. Interestingly to keep in mind here is the fact that UBB^{+1} can be a toxic protein by itself, but it could also act as a potent modifier of toxicity of other neurotoxic proteins, such as Tau and Aβ. Therefore, yeast models combining expression of these proteins in combination with UBB^{+1} could unravel molecular mechanisms important in AD, such as UPS dysfunction and mitochondrial activity [208].

5. Studying Prion Characteristics of Aβ and Tau in Yeast

Prions are self-propagating infection protein species. They were first discovered as causative agents in mammalian diseases like Creutzfeldt-Jakob or Scrapie [209,210]. There, a normal protein (PrP^C) conformationally changes into a malicious and infectious PrP^{Sc} prion protein [211]. Besides these disease-causing prions, a plethora of prions with mostly unknown function has been discovered also in yeast [212,213]. Except for *Podospora anserina's* [Het-s] prion, most functions of all these prions are unclear and all are toxic, or at least growth-inhibitory, but still are supposed to be beneficial for the survival of cells under stress conditions [214–216]. To ensure optimal survival but limit malicious effects, prion formation has to be tightly controlled and carefully balanced. Several factors promote or inhibit prion formation like Hsp104, Hsp70, Sse1, Cur1, Btn2 [217–221]. Interestingly, some of these factors, like Hsp104, promote and inhibit prion formation depending on co-factors and expression levels [222,223].

Many human diseases are described or at least suspected to be prion diseases such as type 2 diabetes mellitus, AD or Huntington's disease [209,224,225]. Most prions are amyloid and, as its name indicates, Aβ for example stacks β-sheets to form toxic amyloid oligomers which were shown to be transmittable and infectious in mice [226–228]. If Aβ is a prion or not is still to be discussed but more and more hints are pointing towards it [227,229]. It is almost accepted that the second key-player in AD, Tau, might be a prion as well. Several studies point towards the MT binding domain of Tau being responsible for aggregation and prionization [230,231]. Several of the yeast models to study AD discussed above [181,182,191,232] are not only used to study the impact of Aβ or Tau on biochemical pathways and on organelles, but also the prion characteristics of these proteins are the focus of research. Evidence for Aβ and Tau being prions were found in mice or other higher eukaryotic model organisms but not in yeast. Tau is hyperphosphorylated and forms aggregates but it is hardly toxic in yeast models [95]. Also, transmission of neither aggregated Tau nor Aβ from affected yeast to healthy strains has been shown so far. But still, there are excellent and robust yeast in vivo techniques to study prion domains and push this field towards greater success. Brachmann and colleagues extended a model developed by Schlumperger and colleagues which makes use of the Ure2 prion system in yeast [233,234]. By replacing promoters of reporter genes by the Ure2 suppressed DAL5 promotor (pDAL5) it is possible to track Ure2-prion strength. If Ure2 is in a non-prion state it binds Gln3, the transcription factor activating pDAL5. When Ure2 forms its prion, [URE3], it releases Gln3 and, thus, induces the reporter gene expression through pDAL5. By replacing Ure2's own prion domain by any protein domain, one can easily test if it is a prion domain. By making use of different reporter genes it is not only possible to check for a domain to be a prion domain in a black-white manner, like with the URA reporter, but it is also possible to measure the strength of a prion domain by using ADE2 as a reporter. The "redness" of the reporter strain indicates the strength of prion formation and thus the release of Gln3 from unprionized Ure2.

Another technique based on a similar principle is the recently developed yTRAP [235]. Here, suspected prions are fused to a synthetic transcription factor, the synTA. When the protein is soluble and thus not prionized, it allows the synTA to bind the promotor and induce expression of a fluorescent protein, in this case mNeonGreen or mKate2. When aggregated, the transcription factor cannot reach its promotor and the expression of the reporter gene is suppressed. An overview of several traditional and more innovative yeast techniques that play(ed) a crucial role in unravelling Tau and Aβ functions such as protein–protein interaction and prion formation can be found in Table 2.

Table 2. Summary of yeast-based techniques applicable in studies on proteins involved in neurodegenerative diseases.

Technique	Used for	Description
Split-GFP system [111–115,236]	Protein–protein interaction	GFP fluorescence is reconstituted when its two subunits are in close proximity.
Synthetic genetic array [237]	Synthetic lethality	Approach for the systematic construction of double mutants for large-scale mapping of synthetic genetic interactions.
Yeast two-hybrid [238]	Protein–protein interaction	Protein interaction leads to reporter gene expression.
Prion-forming assay [233]	Prion forming	The prion domain of the yeast Ure2 prion is replaced by a potential prion domain of any protein. Reporter gene expression is induced if this domain can complement for the Ure2 prion domain.
Yeast transcriptional reporting aggregating proteins (yTRAP) [235]	Prion forming	High-throughput quantitative prion forming assay. Uses fluorescence as quantifiable reporter.

6. Conclusions

It is clear from the above discussed articles that yeast has proven its value in modern AD research. Although *S. cerevisiae* models are most prevalent, it is inspiring to see that alternative models such as *S. pombe* or *C. glabrata* are gaining popularity. The pathobiology of proteins such as Tau and Aβ is robustly recapitulated in yeast. Research using these models has shed more light on the oligomerization/aggregation and prion properties of these proteins, including their role in mitochondrial dysfunction and altered proteostasis, which are two important pathological AD-related cellular processes. This, in combination with the intrinsic benefits of using yeast such as speed and lower costs of research, puts these humanized yeast models in a unique position as a complementary model organism. Therefore, yeast may play a crucial role in overcoming the major future challenges in AD research, including identifying the relationship between all these different pathological AD-related processes.

Author Contributions: D.S. wrote the majority of this manuscript. Other listed authors provided figures/tables, wrote specific sections and/or reviewed the content.

Funding: David Seynnaeve is supported by an Aspirant fellowship (1101317N) of Research Foundation-Flanders (FWO) and Vanessa Franssens by a post-doctoral fellowship (1287517N) of FWO. Joke Verelst is supported by an FWO SBO Fellowship (1S80418N). Mara Del Vecchio, Gernot Fruhmann and Melody Cools are financed through FWO SBO (S006617N) and KU Leuven funding granted to Joris Winderickx. Work in Daniel P. Mulvihill's lab is supported by the University of Kent and by funding from the Biotechnology and Biological Sciences Research Council

Acknowledgments: Human head and scissors icon made by Freepik from www.flaticon.com.

Conflicts of Interest: The authors declare no conflicts of interest.

References

1. Prince, M.; Guerchet, M.; Prina, M. The Epidemiology and Impact of Dementia: Current State and Future Trends. Available online: http://www.who.int/mental_health/neurology/en/ (accessed on 26 March 2015).
2. Prince, M.; Bryce, R.; Albanese, E.; Wimo, A.; Ribeiro, W.; Ferri, C.P. The global prevalence of dementia: A systematic review and metaanalysis. *Alzheimer's Dement.* **2013**, *9*, 63–75. [CrossRef] [PubMed]
3. Abbott, A. Dementia: A problem for our age. *Nature* **2011**, *475*, S2–S4. [CrossRef] [PubMed]
4. Norfray, J.F.; Provenzale, J.M. Alzheimer's Disease: Neuropathologic Findings and Recent Advances in Imaging. *Am. J. Roentgenol.* **2004**, *182*, 3–13. [CrossRef] [PubMed]
5. Marsh, J.; Alifragis, P. Synaptic dysfunction in Alzheimer's disease: The effects of amyloid beta on synaptic vesicle dynamics as a novel target for therapeutic intervention. *Neural Regen. Res.* **2018**, *13*, 616–623. [CrossRef] [PubMed]
6. Selkoe, D.J. Alzheimer's Disease Is a Synaptic Failure. *Science* **2002**, *298*, 789–791. [CrossRef] [PubMed]
7. Backman, L.; Jones, S.; Berger, A.-K.; Laukka, E.J.; Small, B.J. Multiple cognitive deficits during the transition to Alzheimer's disease. *J. Intern. Med.* **2004**, *256*, 195–204. [CrossRef] [PubMed]
8. Lang, A.E. Clinical trials of disease-modifying therapies for neurodegenerative diseases: The challenges and the future. *Nat. Med.* **2010**, *16*, 1223–1226. [CrossRef] [PubMed]
9. Irie, K.; Murakami, K.; Masuda, Y.; Morimoto, A.; Ohigashi, H.; Ohashi, R.; Takegoshi, K.; Nagao, M.; Shimizu, T.; Shirasawa, T. Structure of β-amyloid fibrils and its relevance to their neurotoxicity: Implications for the pathogenesis of Alzheimer's disease. *J. Biosci. Bioeng.* **2005**, *99*, 437–447. [CrossRef] [PubMed]
10. Ittner, L.M.; Götz, J. Amyloid-β and tau—A toxic pas de deux in Alzheimer's disease. *Nat. Rev. Neurosci.* **2011**, *12*, 67–72. [CrossRef] [PubMed]
11. Goffeau, A.; Barrell, B.G.; Bussey, H.; Davis, R.W.; Dujon, B.; Feldmann, H.; Galibert, F.; Hoheisel, J.D.; Jacq, C.; Johnston, M.; et al. Life with 6000 genes. *Science* **1996**, *274*, 563–567. [CrossRef]
12. Botstein, D.; Chervitz, S.A.; Cherry, J.M. Yeast as a model organism. *Science* **1997**, *277*, 1259–1260. [CrossRef] [PubMed]
13. Petranovic, D.; Tyo, K.; Vemuri, G.N.; Nielsen, J. Prospects of yeast systems biology for human health: Integrating lipid, protein and energy metabolism. *FEMS Yeast Res.* **2010**, *10*, 1046–1059. [CrossRef] [PubMed]

14. Winderickx, J.; Delay, C.; De Vos, A.; Klinger, H.; Pellens, K.; Vanhelmont, T.; Van Leuven, F.; Zabrocki, P. Protein folding diseases and neurodegeneration: Lessons learned from yeast. *Biochim. Biophys. Acta Mol. Cell Res.* **2008**, *1783*, 1381–1395. [CrossRef] [PubMed]

15. Tenreiro, S.; Munder, M.C.; Alberti, S.; Outeiro, T.F. Harnessing the power of yeast to unravel the molecular basis of neurodegeneration. *J. Neurochem.* **2013**, *127*, 438–452. [CrossRef] [PubMed]

16. Mohammadi, S.; Saberidokht, B.; Subramaniam, S.; Grama, A. Scope and limitations of yeast as a model organism for studying human tissue-specific pathways. *BMC Syst. Biol.* **2015**, *9*, 96. [CrossRef] [PubMed]

17. Migheli, A.; Butler, M.; Brown, K.; Shelanski, M. Light and electron microscope localization of the microtubule-associated tau protein in rat brain. *J. Neurosci.* **1988**, *8*, 1846–1851. [CrossRef] [PubMed]

18. Papasozomenos, S.C.; Binder, L.I. Phosphorylation determines two distinct species of tau in the central nervous system. *Cell Motil. Cytoskelet.* **1987**, *8*, 210–226. [CrossRef] [PubMed]

19. Binder, L.I.; Frankfurter, A.; Rebhun, L.I. The distribution of tau in the mammalian central nervous system. *J. Cell Biol.* **1985**, *101*, 1371–1378. [CrossRef] [PubMed]

20. Wilhelmsen, K.C.; Lynch, T.; Pavlou, E.; Higgins, M.; Nygaard, T.G. Localization of disinhibition-dementia-parkinsonism-amyotrophy complex to 17q21-22. *Am. J. Hum. Genet.* **1994**, *55*, 1159–1165. [PubMed]

21. Pooler, A.M.; Phillips, E.C.; Lau, D.H.W.; Noble, W.; Hanger, D.P. Physiological release of endogenous tau is stimulated by neuronal activity. *EMBO Rep.* **2013**, *14*, 389–394. [CrossRef] [PubMed]

22. Yamada, K. Extracellular Tau and Its Potential Role in the Propagation of Tau Pathology. *Front. Neurosci.* **2017**, *11*. [CrossRef] [PubMed]

23. Ittner, L.M.; Ke, Y.D.; Delerue, F.; Bi, M.; Gladbach, A.; van Eersel, J.; Wölfing, H.; Chieng, B.C.; Christie, M.J.; Napier, I.A.; et al. Dendritic Function of Tau Mediates Amyloid-β Toxicity in Alzheimer's Disease Mouse Models. *Cell* **2010**, *142*, 387–397. [CrossRef] [PubMed]

24. Drechsel, D.N.; Hyman, A.A.; Cobb, M.H.; Kirschner, M.W. Modulation of the dynamic instability of tubulin assembly by the microtubule-associated protein tau. *Mol. Biol. Cell* **1992**, *3*, 1141–1154. [CrossRef] [PubMed]

25. Crowther, R.A. Straight and paired helical filaments in Alzheimer disease have a common structural unit. *Proc. Natl. Acad. Sci. USA* **1991**, *88*, 2288–2292. [CrossRef] [PubMed]

26. Wischik, C.M.; Crowther, R.A.; Stewart, M.; Roth, M. Subunit structure of paired helical filaments in Alzheimer's disease. *J. Cell Biol.* **1985**, *100*, 1905–1912. [CrossRef] [PubMed]

27. Ichihara, K.; Kitazawa, H.; Iguchi, Y.; Hotani, H.; Itoh, T.J. Visualization of the stop of microtubule depolymerization that occurs at the high-density region of microtubule-associated protein 2 (MAP2). *J. Mol. Biol.* **2001**, *312*, 107–118. [CrossRef] [PubMed]

28. Violet, M.; Delattre, L.; Tardivel, M.; Sultan, A.; Chauderlier, A.; Caillierez, R.; Talahari, S.; Nesslany, F.; Lefebvre, B.; Bonnefoy, E.; et al. A major role for Tau in neuronal DNA and RNA protection in vivo under physiological and hyperthermic conditions. *Front. Cell. Neurosci.* **2014**, *8*. [CrossRef] [PubMed]

29. Frandemiche, M.L.; De Seranno, S.; Rush, T.; Borel, E.; Elie, A.; Arnal, I.; Lante, F.; Buisson, A. Activity-Dependent Tau Protein Translocation to Excitatory Synapse Is Disrupted by Exposure to Amyloid-Beta Oligomers. *J. Neurosci.* **2014**, *34*, 6084–6097. [CrossRef] [PubMed]

30. Stamer, K.; Vogel, R.; Thies, E.; Mandelkow, E.; Mandelkow, E.-M. Tau blocks traffic of organelles, neurofilaments, and APP vesicles in neurons and enhances oxidative stress. *J. Cell Biol.* **2002**, *156*, 1051–1063. [CrossRef] [PubMed]

31. Dixit, R.; Ross, J.L.; Goldman, Y.E.; Holzbaur, E.L.F. Differential Regulation of Dynein and Kinesin Motor Proteins by Tau. *Science* **2008**, *319*, 1086–1089. [CrossRef] [PubMed]

32. Konzack, S.; Thies, E.; Marx, A.; Mandelkow, E.-M.; Mandelkow, E. Swimming against the Tide: Mobility of the Microtubule-Associated Protein Tau in Neurons. *J. Neurosci.* **2007**, *27*, 9916–9927. [CrossRef] [PubMed]

33. Vershinin, M.; Carter, B.C.; Razafsky, D.S.; King, S.J.; Gross, S.P. Multiple-motor based transport and its regulation by Tau. *Proc. Natl. Acad. Sci. USA* **2007**, *104*, 87–92. [CrossRef] [PubMed]

34. Utton, M.A.; Noble, W.J.; Hill, J.E.; Anderton, B.H.; Hanger, D.P. Molecular motors implicated in the axonal transport of tau and alpha-synuclein. *J. Cell Sci.* **2005**, *118*, 4645–4654. [CrossRef] [PubMed]

35. Kanaan, N.M.; Morfini, G.A.; LaPointe, N.E.; Pigino, G.F.; Patterson, K.R.; Song, Y.; Andreadis, A.; Fu, Y.; Brady, S.T.; Binder, L.I. Pathogenic Forms of Tau Inhibit Kinesin-Dependent Axonal Transport through a Mechanism Involving Activation of Axonal Phosphotransferases. *J. Neurosci.* **2011**, *31*, 9858–9868. [CrossRef] [PubMed]

36. Magnani, E.; Fan, J.; Gasparini, L.; Golding, M.; Williams, M.; Schiavo, G.; Goedert, M.; Amos, L.A.; Spillantini, M.G. Interaction of tau protein with the dynactin complex. *EMBO J.* **2007**, *26*, 4546–4554. [CrossRef] [PubMed]

37. Caceres, A.; Kosik, K.S. Inhibition of neurite polarity by tau antisense oligonucleotides in primary cerebellar neurons. *Nature* **1990**, *343*, 461–463. [CrossRef] [PubMed]

38. Wang, Y.; Mandelkow, E. Tau in physiology and pathology. *Nat. Rev. Neurosci.* **2016**, *17*, 22–35. [CrossRef] [PubMed]

39. Goedert, M.; Spillantini, M.G.; Potier, M.C.; Ulrich, J.; Crowther, R.A. Cloning and sequencing of the cDNA encoding an isoform of microtubule-associated protein tau containing four tandem repeats: Differential expression of tau protein mRNAs in human brain. *EMBO J.* **1989**, *8*, 393–399. [PubMed]

40. Himmler, A.; Drechsel, D.; Kirschner, M.W.; Martin, D.W. Tau consists of a set of proteins with repeated C-terminal microtubule-binding domains and variable N-terminal domains. *Mol. Cell. Biol.* **1989**, *9*, 1381–1388. [CrossRef] [PubMed]

41. Rosenberg, K.J.; Ross, J.L.; Feinstein, H.E.; Feinstein, S.C.; Israelachvili, J. Complementary dimerization of microtubule-associated tau protein: Implications for microtubule bundling and tau-mediated pathogenesis. *Proc. Natl. Acad. Sci. USA* **2008**, *105*, 7445–7450. [CrossRef] [PubMed]

42. Bunker, J.M.; Wilson, L.; Jordan, M.A.; Feinstein, S.C. Modulation of microtubule dynamics by tau in living cells: Implications for development and neurodegeneration. *Mol. Biol. Cell* **2004**, *15*, 2720–2728. [CrossRef] [PubMed]

43. Lu, M.; Kosik, K.S. Competition for microtubule-binding with dual expression of tau missense and splice isoforms. *Mol. Biol. Cell* **2001**, *12*, 171–184. [CrossRef] [PubMed]

44. Derisbourg, M.; Leghay, C.; Chiappetta, G.; Fernandez-Gomez, F.-J.; Laurent, C.; Demeyer, D.; Carrier, S.; Buée-Scherrer, V.; Blum, D.; Vinh, J.; et al. Role of the Tau N-terminal region in microtubule stabilization revealed by new endogenous truncated forms. *Sci. Rep.* **2015**, *5*, 9659. [CrossRef] [PubMed]

45. Chen, J.; Kanai, Y.; Cowan, N.J.; Hirokawa, N. Projection domains of MAP2 and tau determine spacings between microtubules in dendrites and axons. *Nature* **1992**, *360*, 674–677. [CrossRef] [PubMed]

46. Sattilaro, R.F.; Dentler, W.L.; LeCluyse, E.L. Microtubule-associated proteins (MAPs) and the organization of actin filaments in vitro. *J. Cell Biol.* **1981**, *90*, 467–473. [CrossRef] [PubMed]

47. Correas, I.; Padilla, R.; Avila, J. The tubulin-binding sequence of brain microtubule-associated proteins, tau and MAP-2, is also involved in actin binding. *Biochem. J.* **1990**, *269*, 61–64. [CrossRef] [PubMed]

48. Rendon, A.; Jung, D.; Jancsik, V. Interaction of microtubules and microtubule-associated proteins (MAPs) with rat brain mitochondria. *Biochem. J.* **1990**, *269*, 555–556. [CrossRef] [PubMed]

49. Brandt, R.; Léger, J.; Lee, G. Interaction of tau with the neural plasma membrane mediated by tau's amino-terminal projection domain. *J. Cell Biol.* **1995**, *131*, 1327–1340. [CrossRef] [PubMed]

50. Zmuda, J.F.; Rivas, R.J. Actin disruption alters the localization of tau in the growth cones of cerebellar granule neurons. *J. Cell Sci.* **2000**, *113*, 2797–2809. [PubMed]

51. Manczak, M.; Reddy, P.H. Abnormal interaction between the mitochondrial fission protein Drp1 and hyperphosphorylated tau in Alzheimer's disease neurons: Implications for mitochondrial dysfunction and neuronal damage. *Hum. Mol. Genet.* **2012**, *21*, 2538–2547. [CrossRef] [PubMed]

52. Bhaskar, K.; Hobbs, G.A.; Yen, S.-H.; Lee, G. Tyrosine phosphorylation of tau accompanies disease progression in transgenic mouse models of tauopathy. *Neuropathol. Appl. Neurobiol.* **2010**, *36*, 462–477. [CrossRef] [PubMed]

53. Lee, G.; Newman, S.T.; Gard, D.L.; Band, H.; Panchamoorthy, G. Tau interacts with src-family non-receptor tyrosine kinases. *J. Cell Sci.* **1998**, *111*, 3167–3177. [PubMed]

54. Lee, G.; Thangavel, R.; Sharma, V.M.; Litersky, J.M.; Bhaskar, K.; Fang, S.M.; Do, L.H.; Andreadis, A.; Van Hoesen, G.; Ksiezak-Reding, H. Phosphorylation of Tau by Fyn: Implications for Alzheimer's Disease. *J. Neurosci.* **2004**, *24*, 2304–2312. [CrossRef] [PubMed]

55. Dujardin, S.; Bégard, S.; Caillierez, R.; Lachaud, C.; Delattre, L.; Carrier, S.; Loyens, A.; Galas, M.-C.; Bousset, L.; Melki, R.; et al. Ectosomes: A New Mechanism for Non-Exosomal Secretion of Tau Protein. *PLoS ONE* **2014**, *9*, e100760. [CrossRef] [PubMed]

56. Calafate, S.; Buist, A.; Miskiewicz, K.; Vijayan, V.; Daneels, G.; de Strooper, B.; de Wit, J.; Verstreken, P.; Moechars, D. Synaptic Contacts Enhance Cell-to-Cell Tau Pathology Propagation. *Cell Rep.* **2015**, *11*, 1176–1183. [CrossRef] [PubMed]

57. Tardivel, M.; Bégard, S.; Bousset, L.; Dujardin, S.; Coens, A.; Melki, R.; Buée, L.; Colin, M. Tunneling nanotube (TNT)-mediated neuron-to neuron transfer of pathological Tau protein assemblies. *Acta Neuropathol. Commun.* **2016**, *4*, 117. [CrossRef] [PubMed]

58. Wang, Y.; Balaji, V.; Kaniyappan, S.; Krüger, L.; Irsen, S.; Tepper, K.; Chandupatla, R.; Maetzler, W.; Schneider, A.; Mandelkow, E.; Mandelkow, E.-M. The release and trans-synaptic transmission of Tau via exosomes. *Mol. Neurodegener.* **2017**, *12*, 5. [CrossRef] [PubMed]

59. Guo, J.L.; Lee, V.M.Y. Cell-to-cell transmission of pathogenic proteins in neurodegenerative diseases. *Nat. Med.* **2014**, *20*, 130–138. [CrossRef] [PubMed]

60. Goode, B.L.; Denis, P.E.; Panda, D.; Radeke, M.J.; Miller, H.P.; Wilson, L.; Feinstein, S.C. Functional interactions between the proline-rich and repeat regions of tau enhance microtubule binding and assembly. *Mol. Biol. Cell* **1997**, *8*, 353–365. [CrossRef] [PubMed]

61. Sillen, A.; Barbier, P.; Landrieu, I.; Lefebvre, S.; Wieruszeski, J.-M.; Leroy, A.; Peyrot, V.; Lippens, G. NMR Investigation of the Interaction between the Neuronal Protein Tau and the Microtubules. *Biochemistry* **2007**. [CrossRef] [PubMed]

62. Hanger, D.P.; Seereeram, A.; Noble, W. Mediators of tau phosphorylation in the pathogenesis of Alzheimer's disease. *Expert Rev. Ther.* **2009**, *9*, 1647–1666. [CrossRef] [PubMed]

63. Sergeant, N.; Bretteville, A.; Hamdane, M.; Caillet-Boudin, M.-L.; Grognet, P.; Bombois, S.; Blum, D.; Delacourte, A.; Pasquier, F.; Vanmechelen, E.; et al. Biochemistry of Tau in Alzheimer's disease and related neurological disorders. *Expert Rev. Proteom.* **2008**, *5*, 207–224. [CrossRef] [PubMed]

64. Domise, M.; Didier, S.; Marinangeli, C.; Zhao, H.; Chandakkar, P.; Buée, L.; Viollet, B.; Davies, P.; Marambaud, P.; Vingtdeux, V. AMP-activated protein kinase modulates tau phosphorylation and tau pathology in vivo. *Sci. Rep.* **2016**, *6*, 26758. [CrossRef] [PubMed]

65. Cavallini, A.; Brewerton, S.; Bell, A.; Sargent, S.; Glover, S.; Hardy, C.; Moore, R.; Calley, J.; Ramachandran, D.; Poidinger, M.; et al. An unbiased approach to identifying tau kinases that phosphorylate tau at sites associated with alzheimer disease. *J. Biol. Chem.* **2013**, *288*, 23331–23347. [CrossRef] [PubMed]

66. Trinczek, B.; Biernat, J.; Baumann, K.; Mandelkow, E.M.; Mandelkow, E. Domains of tau protein, differential phosphorylation, and dynamic instability of microtubules. *Mol. Biol. Cell* **1995**, *6*, 1887–1902. [CrossRef] [PubMed]

67. Hanger, D.P.; Anderton, B.H.; Noble, W. Tau phosphorylation: The therapeutic challenge for neurodegenerative disease. *Trends Mol. Med.* **2009**, *15*, 112–119. [CrossRef] [PubMed]

68. Lu, P.-J.; Wulf, G.; Zhou, X.Z.; Davies, P.; Lu, K.P. The prolyl isomerase Pin1 restores the function of Alzheimer-associated phosphorylated tau protein. *Nature* **1999**, *399*, 784–788. [CrossRef] [PubMed]

69. von Bergen, M.; Friedhoff, P.; Biernat, J.; Heberle, J.; Mandelkow, E.M.; Mandelkow, E. Assembly of tau protein into Alzheimer paired helical filaments depends on a local sequence motif ((306)VQIVYK(311)) forming beta structure. *Proc. Natl. Acad. Sci. USA* **2000**, *97*, 5129–5134. [CrossRef] [PubMed]

70. Dickey, C.A.; Kamal, A.; Lundgren, K.; Klosak, N.; Bailey, R.M.; Dunmore, J.; Ash, P.; Shoraka, S.; Zlatkovic, J.; Eckman, C.B.; et al. The high-affinity HSP90-CHIP complex recognizes and selectively degrades phosphorylated tau client proteins. *J. Clin. Investig.* **2007**, *117*, 648–658. [CrossRef] [PubMed]

71. Šimić, G.; Babić Leko, M.; Wray, S.; Harrington, C.; Delalle, I.; Jovanov-Milošević, N.; Bažadona, D.; Buée, L.; de Silva, R.; Giovanni, G.D.; et al. Tau protein hyperphosphorylation and aggregation in alzheimer's disease and other tauopathies, and possible neuroprotective strategies. *Biomolecules* **2016**, *6*, 2–28. [CrossRef] [PubMed]

72. Berger, Z.; Roder, H.; Hanna, A.; Carlson, A.; Rangachari, V.; Yue, M.; Wszolek, Z.; Ashe, K.; Knight, J.; Dickson, D.; et al. Accumulation of Pathological Tau Species and Memory Loss in a Conditional Model of Tauopathy. *J. Neurosci.* **2007**, *27*, 3650–3662. [CrossRef] [PubMed]

73. Bretteville, A.; Planel, E. Tau aggregates: Toxic, inert, or protective species? *J. Alzheimer's Dis.* **2008**, *14*, 431–436. [CrossRef]

74. Cowan, C.M.; Mudher, A. Are tau aggregates toxic or protective in tauopathies? *Front. Neurol.* **2013**, *4*, 1–13. [CrossRef] [PubMed]

75. Maeda, S.; Sahara, N.; Saito, Y.; Murayama, M.; Yoshiike, Y.; Kim, H.; Miyasaka, T.; Murayama, S.; Ikai, A.; Takashima, A. Granular Tau Oligomers as Intermediates of Tau Filaments. *Biochemistry* **2007**, *46*, 3856–3861. [CrossRef] [PubMed]

76. Lasagna-Reeves, C.A.; Castillo-Carranza, D.L.; Sengupta, U.; Sarmiento, J.; Troncoso, J.; Jackson, G.R.; Kayed, R. Identification of oligomers at early stages of tau aggregation in Alzheimer's disease. *FASEB J.* **2012**, *26*, 1946–1959. [CrossRef] [PubMed]

77. Hutton, M.; Lendon, C.L.; Rizzu, P.; Baker, M.; Froelich, S.; Houlden, H.; Pickering-Brown, S.; Chakraverty, S.; Isaacs, A.; Grover, A.; et al. Association of missense and 5′-splice-site mutations in tau with the inherited dementia FTDP-17. *Nature* **1998**, *393*, 702–705. [CrossRef] [PubMed]

78. Poorkaj, P.; Bird, T.D.; Wijsman, E.; Nemens, E.; Garruto, R.M.; Anderson, L.; Andreadis, A.; Wiederholt, W.C.; Raskind, M.; Schellenberg, G.D. Tau is a candidate gene for chromosome 17 frontotemporal dementia. *Ann. Neurol.* **1998**, *43*, 815–825. [CrossRef] [PubMed]

79. Spillantini, M.G.; Murrell, J.R.; Goedert, M.; Farlow, M.R.; Klug, A.; Ghetti, B. Mutation in the tau gene in familial multiple system tauopathy with presenile dementia. *Proc. Natl. Acad. Sci. USA* **1998**, *95*, 7737–7741. [CrossRef] [PubMed]

80. Spillantini, M.G.; Bird, T.D.; Ghetti, B. Frontotemporal dementia and Parkinsonism linked to chromosome 17: A new group of tauopathies. *Brain Pathol.* **1998**, *8*, 387–402. [CrossRef] [PubMed]

81. Rademakers, R.; Cruts, M.; van Broeckhoven, C. The role of tau (MAPT) in frontotemporal dementia and related tauopathies. *Hum. Mutat.* **2004**, *24*, 277–295. [CrossRef] [PubMed]

82. Dermaut, B.; Kumar-Singh, S.; Rademakers, R.; Theuns, J.; Cruts, M.; Van Broeckhoven, C. Tau is central in the genetic Alzheimer–frontotemporal dementia spectrum. *Trends Genet.* **2005**, *21*, 664–672. [CrossRef] [PubMed]

83. D'Souza, I.; Poorkaj, P.; Hong, M.; Nochlin, D.; Lee, V.M.; Bird, T.D.; Schellenberg, G.D. Missense and silent tau gene mutations cause frontotemporal dementia with parkinsonism-chromosome 17 type, by affecting multiple alternative RNA splicing regulatory elements. *Proc. Natl. Acad. Sci. USA* **1999**, *96*, 5598–5603. [CrossRef] [PubMed]

84. Barghorn, S.; Zheng-Fischhöfer, Q.; Ackmann, M.; Biernat, J.; von Bergen, M.; Mandelkow, E.M.; Mandelkow, E. Structure, microtubule interactions, and paired helical filament aggregation by tau mutants of frontotemporal dementias. *Biochemistry* **2000**, *39*, 11714–11721. [CrossRef] [PubMed]

85. Hong, M.; Zhukareva, V.; Vogelsberg-Ragaglia, V.; Wszolek, Z.; Reed, L.; Miller, B.I.; Geschwind, D.H.; Bird, T.D.; McKeel, D.; Goate, A.; et al. Mutation-specific functional impairments in distinct tau isoforms of hereditary FTDP-17. *Science* **1998**, *282*, 1914–1917. [CrossRef] [PubMed]

86. Roberson, E.D.; Scearce-Levie, K.; Palop, J.J.; Yan, F.; Cheng, I.H.; Wu, T.; Gerstein, H.; Yu, G.-Q.; Mucke, L. Reducing Endogenous Tau Ameliorates Amyloid β-Induced Deficits in an Alzheimer's Disease Mouse Model. *Science* **2007**, *316*, 750–754. [CrossRef] [PubMed]

87. Vandebroek, T.; Vanhelmont, T.; Terwel, D.; Borghgraef, P.; Lemaire, K.; Snauwaert, J.; Wera, S.; Van Leuven, F.; Winderickx, J. Identification and isolation of a hyperphosphorylated, conformationally changed intermediate of human protein tau expressed in yeast. *Biochemistry* **2005**, *44*, 11466–11475. [CrossRef] [PubMed]

88. Vanhelmont, T.; Vandebroek, T.; De Vos, A.; Terwel, D.; Lemaire, K.; Anandhakumar, J.; Franssens, V.; Swinnen, E.; Van Leuven, F.; Winderickx, J. Serine-409 phosphorylation and oxidative damage define aggregation of human protein tau in yeast. *FEMS Yeast Res.* **2010**, *10*, 992–1005. [CrossRef] [PubMed]

89. Vandebroek, T.; Terwel, D.; Vanhelmont, T.; Gysemans, M.; Van Haesendonck, C.; Engelborghs, Y.; Winderickx, J.; Van Leuven, F. Microtubule binding and clustering of human Tau-4R and Tau-P301L proteins isolated from yeast deficient in orthologues of glycogen synthase kinase-3beta or cdk5. *J. Biol. Chem.* **2006**, *281*, 25388–25397. [CrossRef] [PubMed]

90. De Vos, A.; Anandhakumar, J.; Van den Brande, J.; Verduyckt, M.; Franssens, V.; Winderickx, J.; Swinnen, E. Yeast as a model system to study tau biology. *Int. J. Alzheimers Dis.* **2011**, *2011*, 428970. [CrossRef] [PubMed]

91. Zambrano, C.A.; Egaña, J.T.; Núñez, M.T.; Maccioni, R.B.; González-Billault, C. Oxidative stress promotes tau dephosphorylation in neuronal cells: The roles of cdk5 and PP1. *Free Radic. Biol. Med.* **2004**, *36*, 1393–1402. [CrossRef] [PubMed]

92. Kang, S.W.; Kim, S.J.; Kim, M.S. Oxidative stress with tau hyperphosphorylation in memory impaired 1,2-diacetylbenzene-treated mice. *Toxicol. Lett.* **2017**, *279*, 53–59. [CrossRef] [PubMed]

93. De Vos, A.; Bynens, T.; Rosseels, J.; Coun, C.; Ring, J.; Madeo, F.; Galas, M.-C.; Winderickx, J.; Franssens, V. The peptidyl prolyl cis/trans isomerase Pin1/Ess1 inhibits phosphorylation and toxicity of tau in a yeast model for Alzheimer's disease. *AIMS Mol. Sci.* **2015**, *2*, 144–160. [CrossRef]

94. Su, B.; Wang, X.; Lee, H.; Tabaton, M.; Perry, G.; Smith, M.A.; Zhu, X. Chronic oxidative stress causes increased tau phosphorylation in M17 neuroblastoma cells. *Neurosci. Lett.* **2010**, *468*, 267–271. [CrossRef] [PubMed]

95. Rosseels, J.; Van Den Brande, J.; Violet, M.; Jacobs, D.; Grognet, P.; Lopez, J.; Huvent, I.; Caldara, M.; Swinnen, E.; Papegaey, A.; et al. Tau monoclonal antibody generation based on humanized yeast models: Impact on tau oligomerization and diagnostics. *J. Biol. Chem.* **2015**, *290*, 4059–4074. [CrossRef] [PubMed]

96. Castillo-Carranza, D.L.; Gerson, J.E.; Sengupta, U.; Guerrero-Muñoz, M.J.; Lasagna-Reeves, C.A.; Kayed, R. Specific Targeting of Tau Oligomers in Htau Mice Prevents Cognitive Impairment and Tau Toxicity Following Injection with Brain-Derived Tau Oligomeric Seeds. *J. Alzheimer Dis.* **2014**, *40*, S97–S111. [CrossRef] [PubMed]

97. Patterson, K.R.; Remmers, C.; Fu, Y.; Brooker, S.; Kanaan, N.M.; Vana, L.; Ward, S.; Reyes, J.F.; Philibert, K.; Glucksman, M.J.; et al. Characterization of prefibrillar Tau oligomers in vitro and in Alzheimer disease. *J. Biol. Chem.* **2011**, *286*, 23063–23076. [CrossRef] [PubMed]

98. Pérez-Ruiz, E.; Decrop, D.; Ven, K.; Tripodi, L.; Leirs, K.; Rosseels, J.; van de Wouwer, M.; Geukens, N.; De Vos, A.; Vanmechelen, E.; et al. Digital ELISA for the quantification of attomolar concentrations of Alzheimer's disease biomarker protein Tau in biological samples. *Anal. Chim. Acta* **2018**, *1015*, 74–81. [CrossRef] [PubMed]

99. Giasson, B.I.; Forman, M.S.; Higuchi, M.; Golbe, L.I.; Graves, C.L.; Kotzbauer, P.T.; Trojanowski, J.Q.; Lee, V.M.Y. Initiation and synergistic fibrillization of tau and alpha-synuclein. *Science* **2003**, *300*, 636–640. [CrossRef] [PubMed]

100. Jensen, P.H.; Hager, H.; Nielsen, M.S.; Hojrup, P.; Gliemann, J.; Jakes, R. alpha-synuclein binds to Tau and stimulates the protein kinase A-catalyzed tau phosphorylation of serine residues 262 and 356. *J. Biol. Chem.* **1999**, *274*, 25481–25489. [CrossRef] [PubMed]

101. Duka, T.; Rusnak, M.; Drolet, R.E.; Duka, V.; Wersinger, C.; Goudreau, J.L.; Sidhu, A. Alpha-Synuclein Induces Hyperphosphorylation of *Au* in the Mptp Model of Parkinsonism. *FASEB J.* **2006**, *20*, 2302–2312. [CrossRef] [PubMed]

102. Waxman, E.A.; Giasson, B.I. Induction of Intracellular Tau Aggregation Is Promoted by α-Synuclein Seeds and Provides Novel Insights into the Hyperphosphorylation of Tau. *J. Neurosci.* **2011**, *31*, 7604–7618. [CrossRef] [PubMed]

103. Frasier, M.; Walzer, M.; McCarthy, L.; Magnuson, D.; Lee, J.M.; Haas, C.; Kahle, P.; Wolozin, B. Tau phosphorylation increases in symptomatic mice overexpressing A30P α-synuclein. *Exp. Neurol.* **2005**, *192*, 274–287. [CrossRef] [PubMed]

104. Zabrocki, P.; Pellens, K.; Vanhelmont, T.; Vandebroek, T.; Griffioen, G.; Wera, S.; Van Leuven, F.; Winderickx, J. Characterization of α-synuclein aggregation and synergistic toxicity with protein tau in yeast. *FEBS J.* **2005**, *272*, 1386–1400. [CrossRef] [PubMed]

105. Ciaccioli, G.; Martins, A.; Rodrigues, C.; Vieira, H.; Calado, P. A powerful yeast model to investigate the synergistic interaction of α-synuclein and tau in neurodegeneration. *PLoS ONE* **2013**, *8*, e55848. [CrossRef] [PubMed]

106. Hoffman, C.S.; Wood, V.; Fantes, P.A. An Ancient Yeast for Young Geneticists: A Primer on the Schizosaccharomyces pombe Model System. *Genetics* **2015**, *201*, 403–423. [CrossRef] [PubMed]

107. Balasubramanian, M.K.; Bi, E.; Glotzer, M. Comparative analysis of cytokinesis in budding yeast, fission yeast and animal cells. *Curr. Biol.* **2004**, *14*, R806–R818. [CrossRef] [PubMed]

108. Humphrey, T.; Pearce, A. Cell cycle molecules and mechanisms of the budding and fission yeasts. *Methods Mol. Biol.* **2005**, *296*, 3–29. [PubMed]

109. Heinisch, J.J.; Brandt, R. Signaling pathways and posttranslational modifications of tau in Alzheimer's disease: The humanization of yeast cells. *Microb. Cell* **2016**, *3*, 135–146. [CrossRef] [PubMed]

110. Rodicio, R.; Heinisch, J.J. Yeast on the milky way: Genetics, physiology and biotechnology of *Kluyveromyces lactis*. *Yeast* **2013**, *30*, 165–177. [CrossRef] [PubMed]

111. Barnard, E.; McFerran, N.V.; Trudgett, A.; Nelson, J.; Timson, D.J. Development and implementation of split-GFP-based bimolecular fluorescence complementation (BiFC) assays in yeast. *Biochem. Soc. Trans.* **2008**, *36*, 479–482. [CrossRef] [PubMed]

112. Cabantous, S.; Nguyen, H.B.; Pedelacq, J.D.; Koraïchi, F.; Chaudhary, A.; Ganguly, K.; Lockard, M.A.; Favre, G.; Terwilliger, T.C.; Waldo, G.S. A new protein-protein interaction sensor based on tripartite split-GFP association. *Sci. Rep.* **2013**, *3*, 02854. [CrossRef] [PubMed]

113. Cabantous, S.; Terwilliger, T.C.; Waldo, G.S. Protein tagging and detection with engineered self-assembling fragments of green fluorescent protein. *Nat. Biotechnol.* **2005**, *23*, 102–107. [CrossRef] [PubMed]

114. Cabantous, S.; Waldo, G.S. In vivo and in vitro protein solubility assays using split GFP. *Nat. Methods* **2006**, *3*, 845–854. [CrossRef] [PubMed]

115. Foglieni, C.; Papin, S.; Salvadè, A.; Afroz, T.; Pinton, S.; Pedrioli, G.; Ulrich, G.; Polymenidou, M.; Paganetti, P. Split GFP technologies to structurally characterize and quantify functional biomolecular interactions of FTD-related proteins. *Sci. Rep.* **2017**, *7*, 14013. [CrossRef] [PubMed]

116. Priller, C.; Bauer, T.; Mitteregger, G.; Krebs, B.; Kretzschmar, H.A.; Herms, J. Synapse Formation and Function Is Modulated by the Amyloid Precursor Protein. *J. Neurosci.* **2006**, *26*, 7212–7221. [CrossRef] [PubMed]

117. Turner, P.R.; O'Connor, K.; Tate, W.P.; Abraham, W.C. Roles of amyloid precursor protein and its fragments in regulating neural activity, plasticity and memory. *Prog. Neurobiol.* **2003**, *70*, 1–32. [CrossRef]

118. Duce, J.A.; Tsatsanis, A.; Cater, M.A.; James, S.A.; Robb, E.; Wikhe, K.; Leong, S.L.; Perez, K.; Johanssen, T.; Greenough, M.A.; et al. Iron-Export Ferroxidase Activity of β-Amyloid Precursor Protein Is Inhibited by Zinc in Alzheimer's Disease. *Cell* **2010**, *142*, 857–867. [CrossRef] [PubMed]

119. Chen, G.; Xu, T.; Yan, Y.; Zhou, Y.; Jiang, Y.; Melcher, K.; Xu, H.E. Amyloid beta: Structure, biology and structure-based therapeutic development. *Acta Pharmacol. Sin.* **2017**, *38*, 1205–1235. [CrossRef] [PubMed]

120. Haass, C.; Koo, E.H.; Mellon, A.; Hung, A.Y.; Selkoe, D.J. Targeting of cell-surface β-amyloid precursor protein to lysosomes: Alternative processing into amyloid-bearing fragments. *Nature* **1992**, *357*, 500–503. [CrossRef] [PubMed]

121. Joshi, G.; Wang, Y. Golgi defects enhance APP amyloidogenic processing in Alzheimer's disease. *BioEssays* **2015**, *37*, 240–247. [CrossRef] [PubMed]

122. Olsson, F.; Schmidt, S.; Althoff, V.; Munter, L.M.; Jin, S.; Rosqvist, S.; Lendahl, U.; Multhaup, G.; Lundkvist, J. Characterization of Intermediate Steps in Amyloid Beta (Aβ) Production under Near-native Conditions. *J. Biol. Chem.* **2014**, *289*, 1540–1550. [CrossRef] [PubMed]

123. Takami, M.; Nagashima, Y.; Sano, Y.; Ishihara, S.; Morishima-Kawashima, M.; Funamoto, S.; Ihara, Y. γ-Secretase: Successive Tripeptide and Tetrapeptide Release from the Transmembrane Domain of β-Carboxyl Terminal Fragment. *J. Neurosci.* **2009**, *29*, 13042–13052. [CrossRef] [PubMed]

124. Wolfe, M.S. Inhibition and modulation of γ-secretase for Alzheimer's disease. *Neurotherapeutics* **2008**, *5*, 391–398. [CrossRef] [PubMed]

125. Hensley, K.; Carney, J.M.; Mattson, M.P.; Aksenova, M.; Harris, M.; Wu, J.F.; Floyd, R.A.; Butterfield, D.A. A model for beta-amyloid aggregation and neurotoxicity based on free radical generation by the peptide: Relevance to Alzheimer disease. *Proc. Natl. Acad. Sci. USA* **1994**, *91*, 3270–3274. [CrossRef] [PubMed]

126. .Benzinger, T.L.; Gregory, D.M.; Burkoth, T.S.; Miller-Auer, H.; Lynn, D.G.; Botto, R.E.; Meredith, S.C. Propagating structure of Alzheimer's beta-amyloid(10-35) is parallel beta-sheet with residues in exact register. *Proc. Natl. Acad. Sci. USA* **1998**, *95*, 13407–13412. [CrossRef] [PubMed]

127. Yu, L.; Edalji, R.; Harlan, J.E.; Holzman, T.F.; Lopez, A.P.; Labkovsky, B.; Hillen, H.; Barghorn, S.; Ebert, U.; Richardson, P.L.; et al. Structural Characterization of a Soluble Amyloid β-Peptide Oligomer. *Biochemistry* **2009**, *48*, 1870–1877. [CrossRef] [PubMed]

128. Haass, C.; Schlossmacher, M.G.; Hung, A.Y.; Vigo-Pelfrey, C.; Mellon, A.; Ostaszewski, B.L.; Lieberburg, I.; Koo, E.H.; Schenk, D.; Teplow, D.B.; et al. Amyloid β-peptide is produced by cultured cells during normal metabolism. *Nature* **1992**, *359*, 322–325. [CrossRef] [PubMed]

129. Terry, R.D.; Masliah, E.; Salmon, D.P.; Butters, N.; DeTeresa, R.; Hill, R.; Hansen, L.A.; Katzman, R. Physical basis of cognitive alterations in alzheimer's disease: Synapse loss is the major correlate of cognitive impairment. *Ann. Neurol.* **1991**, *30*, 572–580. [CrossRef] [PubMed]

130. Masliah, E.; Terry, R.D.; Alford, M.; DeTeresa, R.; Hansen, L.A. Cortical and subcortical patterns of synaptophysinlike immunoreactivity in Alzheimer's disease. *Am. J. Pathol.* **1991**, *138*, 235–246. [PubMed]

131. Masliah, E.; Achim, C.L.; Ge, N.; DeTeresa, R.; Terry, R.D.; Wiley, C.A. Spectrum of human immunodeficiency virus-associated neocortical damage. *Ann. Neurol.* **1992**, *32*, 321–329. [CrossRef] [PubMed]

132. Mucke, L.; Masliah, E.; Yu, G.Q.; Mallory, M.; Rockenstein, E.M.; Tatsuno, G.; Hu, K.; Kholodenko, D.; Johnson-Wood, K.; McConlogue, L. High-level neuronal expression of abeta 1-42 in wild-type human amyloid protein precursor transgenic mice: Synaptotoxicity without plaque formation. *J. Neurosci.* **2000**, *20*, 4050–4058. [CrossRef] [PubMed]

133. Hsia, A.Y.; Masliah, E.; McConlogue, L.; Yu, G.Q.; Tatsuno, G.; Hu, K.; Kholodenko, D.; Malenka, R.C.; Nicoll, R.A.; Mucke, L. Plaque-independent disruption of neural circuits in Alzheimer's disease mouse models. *Proc. Natl. Acad. Sci. USA* **1999**, *96*, 3228–3233. [CrossRef] [PubMed]

134. Canevari, L.; Abramov, A.Y.; Duchen, M.R. Toxicity of amyloid beta peptide: Tales of calcium, mitochondria, and oxidative stress. *Neurochem. Res.* **2004**, *29*, 637–650. [CrossRef] [PubMed]

135. Mattson, M.P. Pathways towards and away from Alzheimer's disease. *Nature* **2004**, *430*, 631–639. [CrossRef] [PubMed]

136. Weggen, S.; Eriksen, J.L.; Das, P.; Sagi, S.A.; Wang, R.; Pietrzik, C.U.; Findlay, K.A.; Smith, T.E.; Murphy, M.P.; Bulter, T.; et al. A subset of NSAIDs lower amyloidogenic Aβ42 independently of cyclooxygenase activity. *Nature* **2001**, *414*, 212–216. [CrossRef] [PubMed]

137. Neniskyte, U.; Neher, J.J.; Brown, G.C. Neuronal Death Induced by Nanomolar Amyloid β Is Mediated by Primary Phagocytosis of Neurons by Microglia. *J. Biol. Chem.* **2011**, *286*, 39904–39913. [CrossRef] [PubMed]

138. Heneka, M.T.; Golenbock, D.T.; Latz, E. Innate immunity in Alzheimer's disease. *Nat. Immunol.* **2015**, *16*, 229–236. [CrossRef] [PubMed]

139. Iwata, N.; Tsubuki, S.; Takaki, Y.; Watanabe, K.; Sekiguchi, M.; Hosoki, E.; Kawashima-Morishima, M.; Lee, H.-J.; Hama, E.; Sekine-Aizawa, Y.; et al. Identification of the major Aβ_{1-42}-degrading catabolic pathway in brain parenchyma: Suppression leads to biochemical and pathological deposition. *Nat. Med.* **2000**, *6*, 143–150. [CrossRef] [PubMed]

140. Iwata, N.; Tsubuki, S.; Takaki, Y.; Shirotani, K.; Lu, B.; Gerard, N.P.; Gerard, C.; Hama, E.; Lee, H.J.; Saido, T.C. Metabolic Regulation of Brain Abeta by Neprilysin. *Science* **2001**, *292*, 1550–1552. [CrossRef] [PubMed]

141. Selkoe, D.J. Clearing the brain's amyloid cobwebs. *Neuron* **2001**, *32*, 177–180. [CrossRef]

142. DeMattos, R.B.; Bales, K.R.; Cummins, D.J.; Dodart, J.-C.; Paul, S.M.; Holtzman, D.M. Peripheral anti-Aβ antibody alters CNS and plasma Aβ clearance and decreases brain Aβ burden in a mouse model of Alzheimer's disease. *Proc. Natl. Acad. Sci. USA* **2001**, *98*, 8850–8855. [CrossRef] [PubMed]

143. Zlokovic, B.V.; Yamada, S.; Holtzman, D.; Ghiso, J.; Frangione, B. Clearance of amyloid beta-peptide from brain: Transport or metabolism? *Nat. Med.* **2000**, *6*, 718. [CrossRef] [PubMed]

144. Martel, C.L.; Mackic, J.B.; Matsubara, E.; Governale, S.; Miguel, C.; Miao, W.; McComb, J.G.; Frangione, B.; Ghiso, J.; Zlokovic, B. V Isoform-specific effects of apolipoproteins E2, E3, and E4 on cerebral capillary sequestration and blood-brain barrier transport of circulating Alzheimer's amyloid beta. *J. Neurochem.* **1997**, *69*, 1995–2004. [CrossRef] [PubMed]

145. Ghersi-Egea, J.F.; Gorevic, P.D.; Ghiso, J.; Frangione, B.; Patlak, C.S.; Fenstermacher, J.D. Fate of cerebrospinal fluid-borne amyloid beta-peptide: Rapid clearance into blood and appreciable accumulation by cerebral arteries. *J. Neurochem.* **1996**, *67*, 880–883. [CrossRef] [PubMed]

146. Maness, L.M.; Banks, W.A.; Podlisny, M.B.; Selkoe, D.J.; Kastin, A.J. Passage of human amyloid beta-protein 1–40 across the murine blood-brain barrier. *Life Sci.* **1994**, *55*, 1643–1650. [CrossRef]

147. Zlokovic, B.V.; Martel, C.L.; Mackic, J.B.; Matsubara, E.; Wisniewski, T.; Mccomb, J.G.; Frangione, B.; Ghiso, J. Brain Uptake of Circulating Apolipoproteins J and E Complexed to Alzheimer's Amyloid β. *Biochem. Biophys. Res. Commun.* **1994**, *205*, 1431–1437. [CrossRef] [PubMed]

148. Hardy, J.; Selkoe, D.J. The Amyloid Hypothesis of Alzheimer's Disease: Progress and Problems on the Road to Therapeutics. *Science* **2002**, *297*, 353–356. [CrossRef] [PubMed]

149. Tanzi, R.E.; Moir, R.D.; Wagner, S.L. Clearance of Alzheimer's Aβ Peptide. *Neuron* **2004**, *43*, 605–608. [CrossRef] [PubMed]

150. Wyss-Coray, T.; Loike, J.D.; Brionne, T.C.; Lu, E.; Anankov, R.; Yan, F.; Silverstein, S.C.; Husemann, J. Adult mouse astrocytes degrade amyloid-β in vitro and in situ. *Nat. Med.* **2003**, *9*, 453–457. [CrossRef] [PubMed]

151. Koistinaho, M.; Lin, S.; Wu, X.; Esterman, M.; Koger, D.; Hanson, J.; Higgs, R.; Liu, F.; Malkani, S.; Bales, K.R.; et al. Apolipoprotein E promotes astrocyte colocalization and degradation of deposited amyloid-β peptides. *Nat. Med.* **2004**, *10*, 719–726. [CrossRef] [PubMed]

152. Zlokovic, B.V.; Deane, R.; Sallstrom, J.; Chow, N.; Miano, J.M. Neurovascular pathways and Alzheimer amyloid beta-peptide. *Brain Pathol.* **2005**, *15*, 78–83. [CrossRef] [PubMed]

153. Cataldo, A.M.; Hamilton, D.J.; Barnett, J.L.; Paskevich, P.A.; Nixon, R.A. Properties of the endosomal-lysosomal system in the human central nervous system: Disturbances mark most neurons in populations at risk to degenerate in Alzheimer's disease. *J. Neurosci.* **1996**, *16*, 186–199. [CrossRef] [PubMed]

154. Zhang, H.; Komano, H.; Fuller, R.S.; Gandy, S.E.; Frail, D.E. Proteolytic processing and secretion of human beta-amyloid precursor protein in yeast. Evidence for a yeast secretase activity. *J. Biol. Chem.* **1994**, *269*, 27799–27802. [PubMed]

155. Hines, V.; Zhang, W.; Ramakrishna, N.; Styles, J.; Mehta, P.; Kim, K.S.; Innis, M.; Miller, D.L. The expression and processing of human beta-amyloid peptide precursors in Saccharomyces cerevisiae: Evidence for a novel endopeptidase in the yeast secretory system. *Cell. Mol. Biol. Res.* **1994**, *40*, 273–284. [PubMed]

156. Zhang, W.; Espinoza, D.; Hines, V.; Innis, M.; Mehta, P.; Miller, D.L. Characterization of β-amyloid peptide precursor processing by the yeast Yap3 and Mkc7 proteases. *Biochim. Biophys. Acta Mol. Cell Res.* **1997**, *1359*, 110–122. [CrossRef]

157. Komano, H.; Seeger, M.; Gandy, S.; Wang, G.T.; Krafft, G.A.; Fuller, R.S. Involvement of cell surface glycosyl-phosphatidylinositol-linked aspartyl proteases in alpha-secretase-type cleavage and ectodomain solubilization of human Alzheimer beta-amyloid precursor protein in yeast. *J. Biol. Chem.* **1998**, *273*, 31648–31651. [CrossRef] [PubMed]

158. Edbauer, D.; Winkler, E.; Regula, J.T.; Pesold, B.; Steiner, H.; Haass, C. Reconstitution of γ-secretase activity. *Nat. Cell Biol.* **2003**, *5*, 486–488. [CrossRef] [PubMed]

159. Yagishita, S.; Futai, E.; Ishiura, S. In vitro reconstitution of gamma-secretase activity using yeast microsomes. *Biochem. Biophys. Res. Commun.* **2008**, *377*, 141–145. [CrossRef] [PubMed]

160. Yonemura, Y.; Futai, E.; Yagishita, S.; Suo, S.; Tomita, T.; Iwatsubo, T.; Ishiura, S. Comparison of presenilin 1 and presenilin 2 γ-secretase activities using a yeast reconstitution system. *J. Biol. Chem.* **2011**, *286*, 44569–44575. [CrossRef] [PubMed]

161. Yonemura, Y.; Futai, E.; Yagishita, S.; Kaether, C.; Ishiura, S. Specific combinations of presenilins and Aph1s affect the substrate specificity and activity of γ-secretase. *Biochem. Biophys. Res. Commun.* **2016**, *478*, 1751–1757. [CrossRef] [PubMed]

162. Futai, E.; Yagishita, S.; Ishiura, S. Nicastrin is dispensable for gamma-secretase protease activity in the presence of specific presenilin mutations. *J. Biol. Chem.* **2009**, *284*, 13013–13022. [CrossRef] [PubMed]

163. Wickner, R.B.; Edskes, H.K.; Kryndushkin, D.; McGlinchey, R.; Bateman, D.; Kelly, A. Prion diseases of yeast: Amyloid structure and biology. *Semin. Cell Dev. Biol.* **2011**, *22*, 469–475. [CrossRef] [PubMed]

164. Bagriantsev, S.; Liebman, S. Modulation of Abeta42 low-n oligomerization using a novel yeast reporter system. *BMC Biol.* **2006**, *4*, 32. [CrossRef] [PubMed]

165. von der Haar, T.; Jossé, L.; Wright, P.; Zenthon, J.; Tuite, M.F. Development of a novel yeast cell-based system for studying the aggregation of Alzheimer's disease-associated Abeta peptides in vivo. *Neurodegener. Dis.* **2007**, *4*, 136–147. [CrossRef] [PubMed]

166. Caine, J.; Sankovich, S.; Antony, H.; Waddington, L.; Macreadie, P.; Varghese, J.; Macreadie, I. Alzheimer's Abeta fused to green fluorescent protein induces growth stress and a heat shock response. *FEMS Yeast Res.* **2007**, *7*, 1230–1236. [CrossRef] [PubMed]

167. Hamos, J.E.; Oblas, B.; Pulaski-Salo, D.; Welch, W.J.; Bole, D.G.; Drachman, D.A. Expression of heat shock proteins in Alzheimer's disease. *Neurology* **1991**, *41*, 345–350. [CrossRef] [PubMed]

168. Macreadie, I.; Lotfi-Miri, M.; Mohotti, S.; Shapira, D.; Bennett, L.; Varghese, J. Validation of folate in a convenient yeast assay suited for identification of inhibitors of Alzheimer's amyloid-beta aggregation. *J. Alzheimer's Dis.* **2008**, *15*, 391–396. [CrossRef]

169. Rajasekhar, K.; Suresh, S.N.; Manjithaya, R.; Govindaraju, T. Rationally Designed Peptidomimetic Modulators of Aβ Toxicity in Alzheimer's Disease. *Sci. Rep.* **2015**, *5*, 8139. [CrossRef] [PubMed]

170. Bharadwaj, P.R.; Verdile, G.; Barr, R.K.; Gupta, V.; Steele, J.W.; Lachenmayer, M.L.; Yue, Z.; Ehrlich, M.E.; Petsko, G.; Ju, S.; et al. Latrepirdine (dimebon) enhances autophagy and reduces intracellular GFP-Aβ42 levels in yeast. *J. Alzheimer's Dis.* **2012**, *32*, 949–967. [CrossRef] [PubMed]

171. .Nixon, R.A.; Wegiel, J.; Kumar, A.; Yu, W.H.; Peterhoff, C.; Cataldo, A.; Cuervo, A.M. Extensive involvement of autophagy in Alzheimer disease: An immuno-electron microscopy study. *J. Neuropathol. Exp. Neurol.* **2005**, *64*, 113–122. [CrossRef] [PubMed]

172. Yu, W.H.; Cuervo, A.M.; Kumar, A.; Peterhoff, C.M.; Schmidt, S.D.; Lee, J.-H.; Mohan, P.S.; Mercken, M.; Farmery, M.R.; Tjernberg, L.O.; et al. Macroautophagy—A novel β-amyloid peptide-generating pathway activated in Alzheimer's disease. *J. Cell Biol.* **2005**, *171*, 87–98. [CrossRef] [PubMed]

173. Lee, J.-H.; Yu, W.H.; Kumar, A.; Lee, S.; Mohan, P.S.; Peterhoff, C.M.; Wolfe, D.M.; Martinez-Vicente, M.; Massey, A.C.; Sovak, G.; et al. Lysosomal Proteolysis and Autophagy Require Presenilin 1 and Are Disrupted by Alzheimer-Related PS1 Mutations. *Cell* **2010**, *141*, 1146–1158. [CrossRef] [PubMed]

174. Steele, J.W.; Lachenmayer, M.L.; Ju, S.; Stock, A.; Liken, J.; Kim, S.H.; Delgado, L.M.; Alfaro, I.E.; Bernales, S.; Verdile, G.; et al. Latrepirdine improves cognition and arrests progression of neuropathology in an Alzheimer's mouse model. *Mol. Psychiatry* **2013**, *18*, 889–897. [CrossRef] [PubMed]

175. Matlack, K.E.S.; Tardiff, D.F.; Narayan, P.; Hamamichi, S.; Caldwell, K.A.; Caldwell, G.A.; Lindquist, S. Clioquinol promotes the degradation of metal-dependent amyloid-β (Aβ) oligomers to restore endocytosis and ameliorate Aβ toxicity. *Proc. Natl. Acad. Sci. USA* **2014**, *111*, 4013–4018. [CrossRef] [PubMed]

176. Tardiff, D.F.; Brown, L.E.; Yan, X.; Trilles, R.; Jui, N.T.; Barrasa, M.I.; Caldwell, K.A.; Caldwell, G.A.; Schaus, S.E.; Lindquist, S. Dihydropyrimidine-Thiones and Clioquinol Synergize To Target β-Amyloid Cellular Pathologies through a Metal-Dependent Mechanism. *ACS Chem. Neurosci.* **2017**, *8*, 2039–2055. [CrossRef] [PubMed]

177. Cherny, R.A.; Atwood, C.S.; Xilinas, M.E.; Gray, D.N.; Jones, W.D.; McLean, C.A.; Barnham, K.J.; Volitakis, I.; Fraser, F.W.; Kim, Y.-S.; et al. Treatment with a Copper-Zinc Chelator Markedly and Rapidly Inhibits β-Amyloid Accumulation in Alzheimer's Disease Transgenic Mice. *Neuron* **2001**, *30*, 665–676. [CrossRef]

178. Adlard, P.A.; Cherny, R.A.; Finkelstein, D.I.; Gautier, E.; Robb, E.; Cortes, M.; Volitakis, I.; Liu, X.; Smith, J.P.; Perez, K.; et al. Rapid Restoration of Cognition in Alzheimer's Transgenic Mice with 8-Hydroxy Quinoline Analogs Is Associated with Decreased Interstitial Aβ. *Neuron* **2008**, *59*, 43–55. [CrossRef] [PubMed]

179. Lannfelt, L.; Blennow, K.; Zetterberg, H.; Batsman, S.; Ames, D.; Harrison, J.; Masters, C.L.; Targum, S.; Bush, A.I.; Murdoch, R.; et al. Safety, efficacy, and biomarker findings of PBT2 in targeting Aβ as a modifying therapy for Alzheimer's disease: A phase IIa, double-blind, randomised, placebo-controlled trial. *Lancet Neurol.* **2008**, *7*, 779–786. [CrossRef]

180. LaFerla, F.M.; Green, K.N.; Oddo, S. Intracellular amyloid-β in Alzheimer's disease. *Nat. Rev. Neurosci.* **2007**, *8*, 499–509. [CrossRef] [PubMed]

181. .Treusch, S.; Hamamichi, S.; Goodman, J.L.; Matlack, K.E.S.; Chung, C.Y.; Baru, V.; Shulman, J.M.; Parrado, A.; Bevis, B.J.; Valastyan, J.S.; et al. Functional links between Aβ toxicity, endocytic trafficking, and Alzheimer's disease risk factors in yeast. *Science* **2011**, *334*, 1241–1245. [CrossRef] [PubMed]

182. D'Angelo, F.; Vignaud, H.; Di Martino, J.; Salin, B.; Devin, A.; Cullin, C.; Marchal, C. A yeast model for amyloid-β aggregation exemplifies the role of membrane trafficking and PICALM in cytotoxicity. *Dis. Model. Mech.* **2013**, *6*, 206–216. [CrossRef] [PubMed]

183. Park, S.-K.; Ratia, K.; Ba, M.; Valencik, M.; Liebman, S.W. Inhibition of Aβ$_{42}$ oligomerization in yeast by a PICALM ortholog and certain FDA approved drugs. *Microb. Cell* **2016**, *3*, 53–64. [CrossRef] [PubMed]

184. Chen, X.; Petranovic, D. Amyloid-β peptide-induced cytotoxicity and mitochondrial dysfunction in yeast. *FEMS Yeast Res.* **2015**, *15*, fov061. [CrossRef] [PubMed]

185. Fukui, H.; Moraes, C.T. The mitochondrial impairment, oxidative stress and neurodegeneration connection: Reality or just an attractive hypothesis? *Trends Neurosci.* **2008**, *31*, 251–256. [CrossRef] [PubMed]

186. Chen, X.; Bisschops, M.M.M.; Agarwal, N.R.; Ji, B.; Shanmugavel, K.P.; Petranovic, D. Interplay of Energetics and ER Stress Exacerbates Alzheimer's Amyloid-β (Aβ) Toxicity in Yeast. *Front. Mol. Neurosci.* **2017**, *10*, 232. [CrossRef] [PubMed]

187. Stahl, A.; Moberg, P.; Ytterberg, J.; Panfilov, O.; Brockenhuus Von Lowenhielm, H.; Nilsson, F.; Glaser, E. Isolation and identification of a novel mitochondrial metalloprotease (PreP) that degrades targeting presequences in plants. *J. Biol. Chem.* **2002**, *277*, 41931–41939. [CrossRef] [PubMed]

188. Alikhani, N.; Berglund, A.-K.; Engmann, T.; Spånning, E.; Vögtle, F.-N.; Pavlov, P.; Meisinger, C.; Langer, T.; Glaser, E. Targeting Capacity and Conservation of PreP Homologues Localization in Mitochondria of Different Species. *J. Mol. Biol.* **2011**, *410*, 400–410. [CrossRef] [PubMed]

189. Alikhani, N.; Guo, L.; Yan, S.; Du, H.; Pinho, C.M.; Chen, J.X.; Glaser, E.; Yan, S.S. Decreased proteolytic activity of the mitochondrial amyloid-β degrading enzyme, PreP peptidasome, in Alzheimer's disease brain mitochondria. *J. Alzheimer's Dis.* **2011**, *27*, 75–87. [CrossRef] [PubMed]

190. Teixeira, P.F.; Glaser, E. Processing peptidases in mitochondria and chloroplasts. *Biochim. Biophys. Acta-Mol. Cell Res.* **2013**, *1833*, 360–370. [CrossRef] [PubMed]

191. Mossmann, D.; Vögtle, F.-N.; Taskin, A.A.; Teixeira, P.F.; Ring, J.; Burkhart, J.M.; Burger, N.; Pinho, C.M.; Tadic, J.; Loreth, D.; et al. Amyloid-β Peptide Induces Mitochondrial Dysfunction by Inhibition of Preprotein Maturation. *Cell Metab.* **2014**, *20*, 662–669. [CrossRef] [PubMed]

192. Brunetti, D.; Torsvik, J.; Dallabona, C.; Teixeira, P.; Sztromwasser, P.; Fernandez-Vizarra, E.; Cerutti, R.; Reyes, A.; Preziuso, C.; D'Amati, G.; et al. Defective PITRM1 mitochondrial peptidase is associated with A amyloidotic neurodegeneration. *EMBO Mol. Med.* **2016**, *8*, 176–190. [CrossRef] [PubMed]

193. Cenini, G.; Rub, C.; Bruderek, M.; Voos, W. Amyloid β-peptides interfere with mitochondrial preprotein import competence by a coaggregation process. *Mol. Biol. Cell* **2016**, *27*, 3257–3272. [CrossRef] [PubMed]

194. Bharadwaj, P.; Waddington, L.; Varghese, J.; Macreadie, I.G. A new method to measure cellular toxicity of non-fibrillar and fibrillar Alzheimer's Abeta using yeast. *J. Alzheimer's Dis.* **2008**, *13*, 147–150. [CrossRef]

195. Dubey, A.K.; Bharadwaj, P.R.; Varghese, J.N.; Macreadie, I.G. Alzheimer's Amyloid-β Rescues Yeast from Hydroxide Toxicity. *J. Alzheimer's Dis.* **2009**, *18*, 31–33. [CrossRef] [PubMed]

196. Liou, Y.-C.; Sun, A.; Ryo, A.; Zhou, X. Z.; Yu, Z.-X.; Huang, H.-K.; Uchida, T.; Bronson, R.; Bing, G.; Li, X.; et al. Role of the prolyl isomerase Pin1 in protecting against age-dependent neurodegeneration. *Nature* **2003**, *424*, 556–561. [CrossRef] [PubMed]

197. Kondo, A.; Albayram, O.; Zhou, X.Z.; Lu, K.P. Pin1 Knockout Mice: A Model for the Study of Tau Pathology in Alzheimer's Disease. *Methods Mol Biol.* **2017**, *1523*, 415–425. [PubMed]

198. van Leeuwen, F.W.; de Kleijn, D.P.; van den Hurk, H.H.; Neubauer, A.; Sonnemans, M.A.; Sluijs, J.A.; Köycü, S.; Ramdjielal, R.D.; Salehi, A.; Martens, G.J.; et al. Frameshift mutants of beta amyloid precursor protein and ubiquitin-B in Alzheimer's and Down patients. *Science* **1998**, *279*, 242–247. [CrossRef] [PubMed]

199. Morishima-Kawashima, M.; Hasegawa, M.; Takio, K.; Suzuki, M.; Titani, K.; Ihara, Y. Ubiquitin is conjugated with amino-terminally processed tau in paired helical filaments. *Neuron* **1993**, *10*, 1151–1160. [CrossRef]

200. Hilt, W.; Wolf, D.H. Proteasomes: Destruction as a programme. *Trends Biochem. Sci.* **1996**, *21*, 96–102. [CrossRef]

201. van Tijn, P.; de Vrij, F.M.S.; Schuurman, K.G.; Dantuma, N.P.; Fischer, D.F.; van Leeuwen, F.W.; Hol, E.M. Dose-dependent inhibition of proteasome activity by a mutant ubiquitin associated with neurodegenerative disease. *J. Cell Sci.* **2007**, *120*, 1615–1623. [CrossRef] [PubMed]

202. Lindsten, K.; de Vrij, F.M.S.; Verhoef, L.G.G.C.; Fischer, D.F.; van Leeuwen, F.W.; Hol, E.M.; Masucci, M.G.; Dantuma, N.P. Mutant ubiquitin found in neurodegenerative disorders is a ubiquitin fusion degradation substrate that blocks proteasomal degradation. *J. Cell Biol.* **2002**, *157*, 417–427. [CrossRef] [PubMed]

203. Braun, R.J.; Sommer, C.; Leibiger, C.; Gentier, R.J.G.; Dumit, V.I.; Paduch, K.; Eisenberg, T.; Habernig, L.; Trausinger, G.; Magnes, C.; et al. Accumulation of Basic Amino Acids at Mitochondria Dictates the Cytotoxicity of Aberrant Ubiquitin. *Cell Rep.* **2015**, *10*, 1557–1571. [CrossRef] [PubMed]

204. Tank, E.M.H.; True, H.L. Disease-Associated Mutant Ubiquitin Causes Proteasomal Impairment and Enhances the Toxicity of Protein Aggregates. *PLoS Genet.* **2009**, *5*, e1000382. [CrossRef] [PubMed]

205. Krutauz, D.; Reis, N.; Nakasone, M.A.; Siman, P.; Zhang, D.; Kirkpatrick, D.S.; Gygi, S.P.; Brik, A.; Fushman, D.; Glickman, M.H. Extended ubiquitin species are protein-based DUB inhibitors. *Nat. Chem. Biol.* **2014**, *10*, 664–670. [CrossRef] [PubMed]

206. De Vrij, F.M.S.; Sluijs, J.A.; Gregori, L.; Fischer, D.F.; Hermens, W.T.J.M.C.; Goldgaber, D.; Verhaagen, J.; Van Leeuwen, F.W.; Hol, E.M. Mutant ubiquitin expressed in Alzheimer's disease causes neuronal death. *FASEB J.* **2001**, *15*, 2680–2688. [CrossRef] [PubMed]

207. Tan, Z.; Sun, X.; Hou, F.-S.; Oh, H.-W.; Hilgenberg, L.G.W.; Hol, E.M.; van Leeuwen, F.W.; Smith, M.A.; O'Dowd, D.K.; Schreiber, S.S. Mutant ubiquitin found in Alzheimer's disease causes neuritic beading of mitochondria in association with neuronal degeneration. *Cell Death Differ.* **2007**, *14*, 1721–1732. [CrossRef] [PubMed]

208. Braun, R.J. Ubiquitin-dependent proteolysis in yeast cells expressing neurotoxic proteins. *Front. Mol. Neurosci.* **2015**, *8*, 8. [CrossRef] [PubMed]

209. Colby, D.W.; Prusiner, S.B. Prions. *Cold Spring Harb. Perspect. Biol.* **2011**, *3*, a006833. [CrossRef] [PubMed]

210. Prusiner, S.B. Creutzfeldt-Jakob disease and scrapie prions. *Alzheimer Dis. Assoc. Disord.* **1989**, *3*, 52–78. [CrossRef] [PubMed]

211. Safar, J.G. Molecular pathogenesis of sporadic prion diseases in man. *Prion* **2012**, *6*, 108–115. [CrossRef] [PubMed]

212. Wickner, R.B.; Shewmaker, F.P.; Bateman, D.A.; Edskes, H.K.; Gorkovskiy, A.; Dayani, Y.; Bezsonov, E.E. Yeast Prions: Structure, Biology, and Prion-Handling Systems. *Microbiol. Mol. Biol. Rev.* **2015**, *79*, 1–17. [CrossRef] [PubMed]
213. Wickner, R.B.; Edskes, H.K.; Son, M.; Bezsonov, E.E.; DeWilde, M.; Ducatez, M. Yeast Prions Compared to Functional Prions and Amyloids. *J. Mol. Biol.* **2018**. [CrossRef] [PubMed]
214. Riek, R.; Saupe, S.J. The HET-S/s Prion Motif in the Control of Programmed Cell Death. *Cold Spring Harb. Perspect. Biol.* **2016**, *8*, a023515. [CrossRef] [PubMed]
215. Coustou, V.; Deleu, C.; Saupe, S.; Begueret, J. The protein product of the het-s heterokaryon incompatibility gene of the fungus Podospora anserina behaves as a prion analog. *Proc. Natl. Acad. Sci. USA* **1997**, *94*, 9773–9778. [CrossRef] [PubMed]
216. Wickner, R.B.; Bezsonov, E.E.; Son, M.; Ducatez, M.; DeWilde, M.; Edskes, H.K. Anti-Prion Systems in Yeast and Inositol Polyphosphates. *Biochemistry* **2018**, *57*, 1285–1292. [CrossRef] [PubMed]
217. Wu, Y.-X.; Greene, L.E.; Masison, D.C.; Eisenberg, E. Curing of yeast [PSI+] prion by guanidine inactivation of Hsp104 does not require cell division. *Proc. Natl. Acad. Sci. USA* **2005**, *102*, 12789–12794. [CrossRef] [PubMed]
218. Wegrzyn, R.D.; Bapat, K.; Newnam, G.P.; Zink, A.D.; Chernoff, Y.O. Mechanism of prion loss after Hsp104 inactivation in yeast. *Mol. Cell. Biol.* **2001**, *21*, 4656–4669. [CrossRef] [PubMed]
219. Moriyama, H.; Edskes, H.K.; Wickner, R.B. [URE3] prion propagation in Saccharomyces cerevisiae: Requirement for chaperone Hsp104 and curing by overexpressed chaperone Ydj1p. *Mol. Cell. Biol.* **2000**, *20*, 8916–8922. [CrossRef] [PubMed]
220. Kryndushkin, D.S.; Shewmaker, F.; Wickner, R.B. Curing of the [URE3] prion by Btn2p, a Batten disease-related protein. *EMBO J.* **2008**, *27*, 2725–2735. [CrossRef] [PubMed]
221. Wickner, R.B.; Bezsonov, E.; Bateman, D.A. Normal levels of the antiprion proteins Btn2 and Cur1 cure most newly formed [URE3] prion variants. *Proc. Natl. Acad. Sci. USA* **2014**, *111*, E2711–E2720. [CrossRef] [PubMed]
222. O'Driscoll, J.; Clare, D.; Saibil, H. Prion aggregate structure in yeast cells is determined by the Hsp104-Hsp110 disaggregase machinery. *J. Cell Biol.* **2015**, *211*, 145–158. [CrossRef] [PubMed]
223. Romanova, N.V.; Chernoff, Y.O. Hsp104 and prion propagation. *Protein Pept. Lett.* **2009**, *16*, 598–605. [CrossRef] [PubMed]
224. Mukherjee, A.; Soto, C. Prion-Like Protein Aggregates and Type 2 Diabetes. *Cold Spring Harb. Perspect. Med.* **2017**, *7*, a024315. [CrossRef] [PubMed]
225. Moore, R.C.; Xiang, F.; Monaghan, J.; Han, D.; Zhang, Z.; Edström, L.; Anvret, M.; Prusiner, S.B. Huntington Disease Phenocopy Is a Familial Prion Disease. *Am. J. Hum. Genet.* **2001**, *69*, 1385–1388. [CrossRef] [PubMed]
226. Abbott, A. The red-hot debate about transmissible Alzheimer's. *Nature* **2016**, *531*, 294–297. [CrossRef] [PubMed]
227. Olsson, T.T.; Klementieva, O.; Gouras, G.K. Prion-like seeding and nucleation of intracellular amyloid-β. *Neurobiol. Dis.* **2018**, *113*, 1–10. [CrossRef] [PubMed]
228. Lu, J.-X.; Qiang, W.; Yau, W.-M.; Schwieters, C.D.; Meredith, S.C.; Tycko, R. Molecular Structure of β-Amyloid Fibrils in Alzheimer's Disease Brain Tissue. *Cell* **2013**, *154*, 1257–1268. [CrossRef] [PubMed]
229. Coalier, K.A.; Paranjape, G.S.; Karki, S.; Nichols, M.R. Stability of early-stage amyloid-β(1–42) aggregation species. *Biochim. Biophys. Acta Proteins Proteom.* **2013**, *1834*, 65–70. [CrossRef] [PubMed]
230. Holmes, B.B.; Diamond, M.I. Prion-like properties of Tau protein: The importance of extracellular Tau as a therapeutic target. *J. Biol. Chem.* **2014**, *289*, 19855–19861. [CrossRef] [PubMed]
231. .Kaufman, S.K.; Sanders, D.W.; Thomas, T.L.; Ruchinskas, A.J.; Vaquer-Alicea, J.; Sharma, A.M.; Miller, T.M.; Diamond, M.I. Tau Prion Strains Dictate Patterns of Cell Pathology, Progression Rate, and Regional Vulnerability In Vivo. *Neuron* **2016**, *92*, 796–812. [CrossRef] [PubMed]
232. Fruhmann, G.; Seynnaeve, D.; Zheng, J.; Ven, K.; Molenberghs, S.; Wilms, T.; Liu, B.; Winderickx, J.; Franssens, V. Yeast buddies helping to unravel the complexity of neurodegenerative disorders. *Mech. Ageing Dev.* **2017**, *161*, 288–305. [CrossRef] [PubMed]
233. Brachmann, A.; Toombs, J.A.; Ross, E.D. Reporter assay systems for [URE3] detection and analysis. *Methods* **2006**, *39*, 35–42. [CrossRef] [PubMed]
234. Schlumberger, M.; Prusiner, S.B.; Herskowitz, I. Induction of distinct [URE3] yeast prion strains. *Mol. Cell. Biol.* **2001**, *21*, 7035–7046. [CrossRef] [PubMed]

235. Newby, G.A.; Kiriakov, S.; Hallacli, E.; Kayatekin, C.; Tsvetkov, P.; Mancuso, C.P.; Bonner, J.M.; Hesse, W.R.; Chakrabortee, S.; Manogaran, A.L.; et al. A Genetic Tool to Track Protein Aggregates and Control Prion Inheritance. *Cell* **2017**, *171*, 966–979.e18. [CrossRef] [PubMed]

236. Chun, W.; Waldo, G.S.; Johnson, G.V.W. Split GFP complementation assay: a novel approach to quantitatively measure aggregation of tau in situ: Effects of GSK3β activation and caspase 3 cleavage. *J. Neurochem.* **2007**, *103*, 2529–2539. [CrossRef] [PubMed]

237. Zhao, L.; Yang, Q.; Zheng, J.; Zhu, X.; Hao, X.; Song, J.; Lebacq, T.; Franssens, V.; Winderickx, J.; Nystrom, T.; et al. A genome-wide imaging-based screening to identify genes involved in synphilin-1 inclusion formation in Saccharomyces cerevisiae. *Sci. Rep.* **2016**, *6*, 30134. [CrossRef] [PubMed]

238. Yu, Y.; Li, Y.; Zhang, Y. Screening of APP interaction proteins by DUALmembrane yeast two-hybrid system. *Int. J. Clin. Exp. Pathol.* **2015**, *8*, 2802–2808. [PubMed]

International Journal of
Molecular Sciences

MDPI

Article

Development of Convenient System for Detecting Yeast Cell Stress, Including That of Amyloid Beta

Yen Nhi Luu and Ian Macreadie *

School of Science, RMIT University, Bundoora, VIC 3083, Australia; s3691025@student.rmit.edu.au
* Correspondence: ian.macreadie@rmit.edu.au; Tel.: +61-3-9925-6627

Received: 19 June 2018; Accepted: 21 July 2018; Published: 23 July 2018

Abstract: (1) Background: As a model eukaryote, the study of stress responses in yeast can be employed for studying human health and disease, and the effects of various drugs that may impact health. "Reporting" of stress in yeast has frequently utilised enzymes like β-galactosidase that require laborious assays for quantitative results. The use of a stress reporter that can be measured quantitatively and with high sensitivity in living cells in a multi-well plate reader is a more desirable approach; (2) Methods: A multi-copy yeast-*Escherichia coli* shuttle plasmid containing the *HSP42* promoter upstream of the mCherry reporter, along with the *URA3* selectable marker was constructed and tested; (3) Results: Under certain stress conditions inducing the heat shock response, transformants containing the plasmid produced red fluorescence that could be readily quantitated in a microtitre plate reader. Stresses that produced red fluorescence included exposure to heat shock, copper ions, oligomeric amyloid beta ($A\beta_{42}$) and fibrillar $A\beta_{42}$; (4) Conclusions: Being able to conveniently and quantitatively monitor stresses in whole live populations of yeast offers great opportunities to screen compounds and conditions that cause stress, as well as conditions that alleviate stress. While freshly prepared oligomeric amyloid beta has previously been shown to exhibit high toxicity, fibrils have been generally considered to be non-toxic or of low toxicity. In this study, fibrillar amyloid beta has also been shown to induce stress.

Keywords: heat shock response; heat shock protein; Alzheimer's disease; beta amyloid; yeast

1. Introduction

Heat shock proteins (HSPs) are ubiquitously expressed and conserved in both yeast and humans [1]. Low level, constitutive expression of HSPs perform housekeeping functions, assisting in maintenance of proteostasis [2]. HSPs target up to 3% of the total number of genes in yeast, with some acting as molecular chaperones to assist in binding of and folding of proteins and sequester misfolded polypeptides towards proteolytic pathways, while others are involved in intracellular transport, cell wall maintenance, and oxidative stress mechanisms [3–5]. While HSPs are always present within cells, their expression may be upregulated during the heat shock response (HSR) in response to cellular stress which may include changes in environment, such as elevated temperature [6,7], misfolding or aggregation of proteins [2,3], and reactive oxygen species (ROS) production [8,9]. The HSR is mediated by activity of heat response factors (HRFs) that bind to a 5 bp heat shock element (HSE) in the promotor regions of heat shock genes to initiate transcription [4,10,11].

As a defense mechanism against misfolded and aggregated proteins, HSPs are vital in the response against the toxic Alzheimer's disease (AD) protein Aβ. Aβ is present in many forms in an AD-affected brain including monomers, toxic oligomeric intermediates and fibrils. The soluble oligomeric form, particularly $A\beta_{42}$, produces cytotoxic effects that initiate a cascade of events that contribute to the development of AD due to a higher propensity to aggregate [12–14]. HSPs are found at elevated levels in AD-affected brains, activating microglial phagocytosis and degradation, inhibiting Aβ formation, and slowing down or inhibiting the rate of aggregation, thereby contributing to the clearance of Aβ [3,5,15,16].

β-galactosidase reporter assays have previously been used for measurement of the HSR towards Aβ, in which a yeast HSE was placed upstream of the *lacZ* gene and β-galactosidase levels were measured with and without the presence of Aβ [17]. The use of such an assay is somewhat inconvenient in that a preparation of cell lysate is required as well as several commercial reagents.

The aim of this study was to develop an alternative reporter assay for quick screening of the HSR using the mCherry fluorescence reporter, to measure cell stress in whole living cell populations without a need for any reagent addition. Development of this expression system could allow for quick, high throughput screening assays to determine conditions that may cause stress to cells. The demonstration of the use of this system is outlined as follows.

2. Results

2.1. Construction of the pYHSRed1 Plasmid

The schematic map of the pYHSRed1 plasmid is shown in Figure 1. It has a 2 μ *ori* for high copy replication in *S. cerevisiae*. The mCherry reporter is located downstream from the promotor of *HSP42*, the most abundant cytosolic HSP in yeast for suppression of aggregation [18]. It contains a *URA3* gene for selection in *S. cerevisiae* strains that have a *URA3* gene mutation or disruption and therefore require uracil supplementation. For propagation of the plasmid in *E. coli* it contains the pUC *ori* and an ampicillin resistance selection marker (encoding β-lactamase). Transformation of this plasmid into a *ura3* mutant *S. cerevisiae* BY4743 strain that requires uracil for growth produces a transformant that no longer requires uracil supplementation. The expression of mCherry expression and red fluorescence should be regulated by the *HSP42* promoter, so the intensity of red fluorescence should indicate the amount of the stress response in the recombinant yeast.

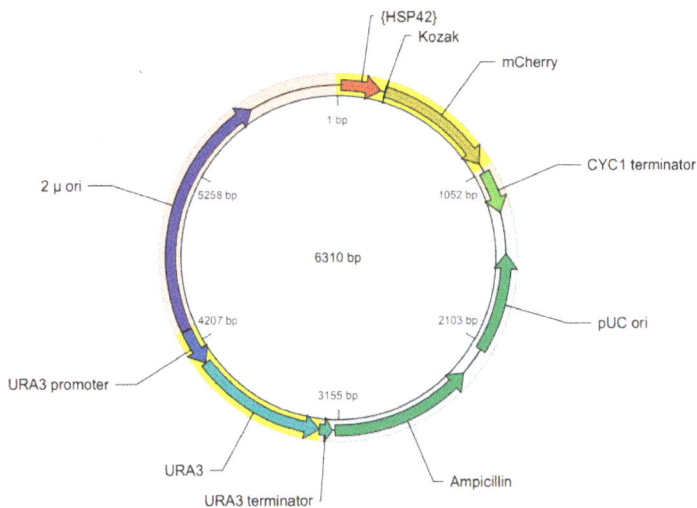

Figure 1. Schematic map of the pYHSRed1 plasmid.

2.2. Transformation of Yeast with pYHSRed1 and Basal Expression of mCherry

To examine the basal levels of red fluorescence afforded by pYHSRed1, a comparison was made between BY4743 and its transformant, BY4743 [pYHSRed1]. BY4743 and BY4743 [pYHSRed1] were grown to exponential phase in liquid minimal media with the appropriate supplementation required by each strain, incubated at 30 °C for two hours. Transformants had some production of mCherry, as indicated by the increased red fluorescence (Figure 2). This basal expression of mCherry fluorescence in BY4743 [pYHSRed1] was significantly higher ($p < 0.05$ for all comparisons at same cell density) than that of the untransformed parental BY4743 strain, indicative of the low level constitutive expression of HSPs in yeast cells when unstressed (Figure 2). The measurement of mCherry fluorescence in cultures of varying cell densities was also analysed to determine how the density of cell cultures affected fluorescence. Fluorescence was proportional to cell density, with cell cultures of larger OD_{600} readings producing higher mCherry fluorescence. Arbitrary measurements of higher mCherry fluorescence of the parental BY4743 strain were also observed at greater cell densities which may be attributed to high sensitivity of the spectrophotometer. For measurement of fluorescence in successive experiments, cell cultures of $OD_{600} \geq 0.6$ were utilised.

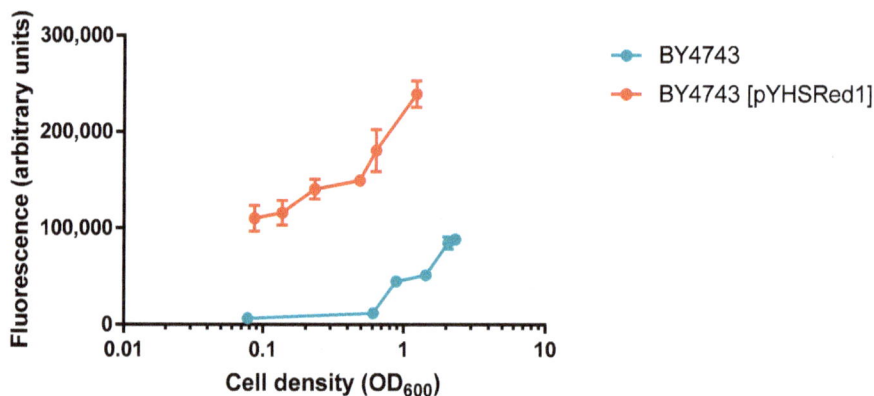

Figure 2. Comparison of basal levels of red fluorescence in BY4743 and BY4743 [pYHSRed1] in cultures of varying cell densities. Mean ±SEM of Data are shown as triplicate measurements.

2.3. Increased mCherry Fluorescence in Cells Exposed to Heat Shock and Copper Sulphate

The mCherry reporter was examined under conditions that induce HSR in cells: elevated temperatures and exposure to metal ions. BY4743 [pYHSRed1] cells were incubated at 42 °C for two hours, with control cells being incubated at 30 °C. Exposure to 42 °C resulted in a significant increase in mCherry fluorescence being measured, indicating a significant upregulation in heat shock response genes compared to the control (Figure 3a).

Yeast cells were also exposed to copper sulphate for two hours. Treatment with 0.1 and 0.3 mM $CuSO_4$ did not produce a significant increase in HSR but cells treated with 0.5 mM $CuSO_4$ produced a significant increase in mCherry fluorescence (Figure 3b).

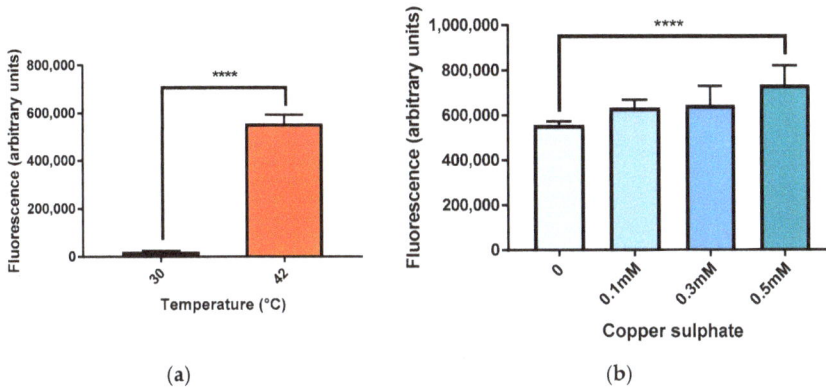

Figure 3. mCherry fluorescence in BY4743 [pYHSRed1] under heat and copper stress. (a) Heat stress; (b) Stress due to copper sulphate. Data shown as mean ± SEM of triplicate measurements; **** $p < 0.0001$.

2.4. Stress Induced by Oligomeric and Fibrillar Aβ$_{42}$ Measured by mCherry Fluorescence

Elevated mCherry levels were measured when yeast cells were treated with both oligomeric and fibrillar Aβ$_{42}$. Cell responses of both exponential and stationary phase cells were measured due to the differing vulnerability of yeast cells in different growth phases to Aβ$_{42}$ toxicity [19].

Oligomeric Aβ$_{42}$ (Figure 4) induced a dose-dependent response in mCherry fluorescence in both stationary and exponential phase cells. A significant increase in mCherry fluorescence was observed in stationary phase yeast cells at 500 nM and 1 μM Aβ$_{42}$, but not with 50 nM Aβ$_{42}$. Aβ$_{42}$ also induced significant mCherry fluorescence at 50 nM in exponential phase yeast cells, but there was no significant effect at lower levels.

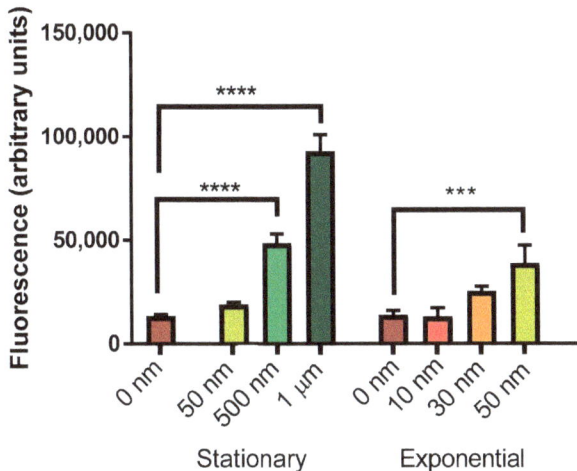

Figure 4. Measurement of mCherry fluorescence of BY4743 [pYHSRed1] cells in stationary and exponential phase growth treated with oligomeric Aβ$_{42}$. Data are shown as mean ± SEM of triplicate measurements; *** $p > 0.001$, **** $p < 0.0001$.

Fibrillar Aβ₄₂ (Figure 5) induced a significant elevation in mCherry fluorescence in stationary phase yeast cells at 50, 500 nM and 1 μM Aβ₄₂. Levels of 30 and 50 nM Aβ₄₂ also induced significant mCherry fluorescence in exponential phase yeast cells.

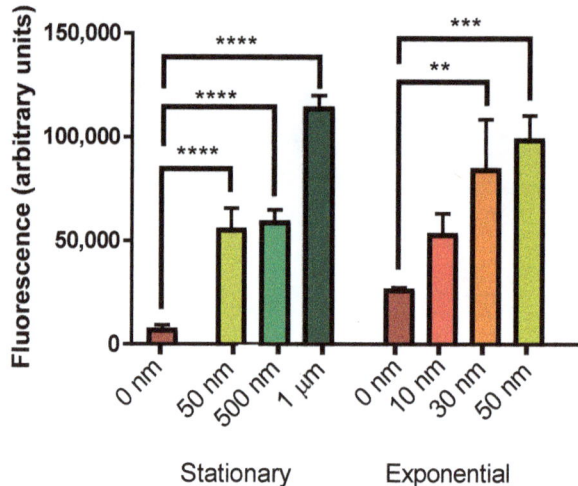

Figure 5. Measurement of mCherry fluorescence of BY4743 [pYHSRed1] cells in stationary and exponential phase growth treated with fibrillar Aβ₄₂. Data are shown as mean ± SEM of triplicate measurements; ** $p > 0.01$, *** $p > 0.001$, **** $p < 0.0001$.

3. Discussion

This study aimed to develop a convenient yeast reporter system to measure cell stress correlated with induction of the heat shock response by measuring fluorescence of the mCherry reporter, induced from the HSE of *HSP42*.

Significant basal mCherry expression of the transformant compared to the wildtype was observed, confirming functionality of pYHSRed1 in transformed yeast. Basal expression of HSPs is expected, as in unstressed conditions, HSPs perform housekeeping functions for proteostasis and regulation of protein quality control [9,20].

mCherry fluorescence in BY4743 [pYHSRed1] transformants after exposure to some known inducers of the heat shock response was measured and significant increases in red fluorescence were observed. For example, there was a significant increase in HSR at 42 °C, as this temperature is a heat shock condition in both mammalian and yeast cells and known to activate HSFs [4,21]. Likewise the HSR is also induced by heavy metal ions and oxidants [22], and at 0.5 mM levels it induced highly significant red mCherry expression. Copper can cause stress as it is a heavy metal and may promote oxidative damage at elevated levels in cells [23].

Oligomeric Aβ₄₂ is unstable and toxic, with many studies showing its effect in killing of both yeast cells and neurons [24–26]. Effects of externally supplemented Aβ₄₂ cells may differ based on growth stages, as non-quiescent cells are more susceptible to Aβ₄₂ toxicity compared to quiescent cells in the first 24 h of exposure [19]. There is reduced viability of cells exposed to Aβ₄₂ oligomers compared to fibrils [24], therefore, lower concentrations of oligomers were applied for treatment. The lower levels of mCherry fluorescence observed compared to fibrillar Aβ₄₂ treatment may be due to this cell killing, reducing the number of cells able to emit fluorescence.

In contrast to oligomers, fibrils are generally viewed as harmless and benign, contributing to the non-toxic plaques found in the brain [14]. Though HSPs can sequester oligomer aggregates, they do not cause significant changes to fibrillar Aβ, possibly accumulating on the fibrils due to their inability to process them [3,27]. However, insoluble fibrils may induce oxidative stress from fibrillization [28,29]. Ladiwala et al. [30] also found fibrils formed at elevated concentrations of Aβ were toxic. Fibrils prepared with both HFIP and NH$_4$OH pretreatment caused toxicity to *S. cerevisiae* [19]. It is possible that, while not cytotoxic like oligomers, the fibrils cause stress and HSR induction in cells through production of ROS and mild cell killing.

Further work needs to be performed to gain greater understanding of this new attribute of fibrillar Aβ$_{42}$.

4. Materials and Methods

4.1. pYHSRed1, Yeast Strain and Transformation

The pYHSRed1 was custom designed and produced by VectorBuilder (Cyagen, Santa Clara, CA, USA). It utilises a *URA3* multi-copy VectorBuilder plasmid with 238 nt of the *HSP42* promoter sequences inserted immediately upstream of the mCherry reporter.

The *Saccharomyces cerevisiae* yeast strain BY4743 (*MATa/α his3Δ1/his3Δ1, leu2Δ0/leu2Δ0 LYS2/lys2Δ0 met15Δ0/MET15 ura3Δ0/ura3Δ0*) was the host strain used in this study. The plasmid pYHSRed1 was transformed into the host strain as described by Porzoor and Macreadie [31].

4.2. Yeast Culture Protocol

Minimal media was used for growth of the BY4743 transformants. The media composition is as follows: Yeast nitrogen base without amino acids (0.67%) and dextrose (2%). For solidification of media agar (1.5%) was added. Supplementation of auxotrophic requirements of BY4743 [pYHSRed1] was performed by adding 20 mg/L histidine and 30 mg/L leucine.

Overnight fresh cultures of the transformants were obtained by inoculating one colony into 10 mL fresh selective minimal media in a 50 mL tube. The tubes were incubated at 30 °C at 250 rpm. Overnight stationary cultures were further grown to exponential phase by transferring 100 µL aliquot to fresh selective minimal media in 15 mL tubes and incubating at 30 °C at 250 rpm for a further two hours.

4.3. Preparation of Aβ$_{42}$

Aβ$_{42}$ was pretreated with NH$_4$OH as described by [19]. To obtain oligomers, Aβ$_{42}$ was solubilized in water and used immediately. To obtain fibrils, the Aβ$_{42}$ was solubilised in water and incubated at 37 °C for 24 h.

4.4. Exposure of Yeast Cells to Heat Shock

Yeast cells were analysed for the effect of exposure to heat shock conditions on mCherry levels. Yeast cells from exponential phase cultures were aliquoted into wells in 96-well microtiter plates. Cells were incubated for a further two hours at 30 and 42 °C.

4.5. Exposure of Yeast Cells to Copper Ions

Yeast cells were analysed for the effect of exposure to copper sulphate on mCherry levels. Cells from overnight cultures and exponential phase cultures were suspended in water and then were aliquoted into wells in 96-well microtiter plates. Copper sulphate was added to the diluted cell suspension to required concentrations. The final volume of each well was made up to 200 µL. The microtiter plate was incubated at 30 °C for two hours.

4.6. Effect of Exposure to Oligomeric and Fibrillar Aβ on Yeast Cells

Yeast cells were analysed for the effect of exposure to fibrillar and oligomeric Aβ$_{42}$ on mCherry levels. Cells from overnight cultures and exponential phase cultures were pelleted by centrifugation and resuspended in water and aliquoted into wells in 96-well microtiter plates. Oligomeric and fibrillar

Aβ$_{42}$ were added to the diluted cell suspension to required concentrations. The final volume of each well was made up to 200 μL. The microtiter plate was incubated at 30 °C for two hours.

4.7. Spectrophotometry

Cell density and mCherry fluorescence was measured with the POLARstar omega microplate reader and analysed with BMG Labtech Mars Data Analysis Software (Ortenberg, Germany). Cell density was measured in Corning 96 Well TC-Treated microplates at 600 nm. mCherry fluorescence was measured from a Nunclon Surface black F96 microtiter plate with top optics using an Ex584 excitation filter and 600–680 emission filter.

Raw data was blank corrected, subtracting the mCherry fluorescence reading of the liquid the cell culture was suspended in, to remove background fluorescence. This figure was then divided by the cell density (OD$_{600}$) reading for the culture.

5. Conclusions

The heat shock response is vital in both yeast and human cells for defense against various cell stressors, including misfolded and aggregated proteins associated with neurodegenerative diseases such as Alzheimer's disease. A novel outcome in this study was development of pYHSRed1, a plasmid reporting on stress, especially the HSR, in yeast cells. Coupled with measurement of mCherry fluorescence by a spectrophotometer, the level of stress in live yeast cells may be determined.

Yeast cells containing the pYHSRed1 plasmid were exposed to several conditions known to induce the heat shock response. Elevated temperatures, exposure to metal ions and the subsequent ROS production, and oligomeric and fibrillar Aβ$_{42}$ all induced significant increases in mCherry production, indicative of the upregulation of transcription of heat shock genes. The significant upregulation of mCherry observed after exposure to fibrillar Aβ, considered to be of low or no toxicity, implicates fibrils as a contributor to cellular stress by induction of the HSR.

Transformation of this plasmid into yeast provides an improved method of stress and HSR detection and may be useful for high throughput analysis of therapeutic compounds that may reduce stress caused by Aβ. Future studies could also determine the stress of mutant versions of Aβ on yeast cells and to identify therapeutic compounds that may alleviate the effects of other deleterious proteins, biochemicals or cellular states that cause cellular stress.

Author Contributions: Y.N.L. conceived, designed, and performed the experiments and wrote the paper. I.M. conceived, designed, and supervised the study.

Funding: This research received no external funding.

Conflicts of Interest: The authors declare no conflict of interest.

Abbreviations

AD	Alzheimer's disease
Aβ	Beta-amyloid
HSE	Heat shock element
HSF	Heat shock factor
HSP	Heat shock protein
HSR	Heat shock response
ROS	Reactive oxygen species

References

1. Lindquist, S. The heat-shock proteins. *Annu. Rev. Genet.* **1988**, *22*, 631–677. [CrossRef] [PubMed]
2. Fink, A.L. Chaperone-mediated protein folding. *Physiol. Rev.* **1999**, *79*, 425–449. [CrossRef] [PubMed]
3. Evans, C.G.; Wisen, S.; Gestwicki, J.E. Heat shock proteins 70 and 90 inhibit early stages of amyloid β-(1-42) aggregation in vitro. *J. Biol. Chem.* **2006**, *281*, 33182–33191. [CrossRef] [PubMed]

4. Hahn, J.S.; Hu, Z.; Thiele, D.J.; Iyer, V.R. Genome-wide analysis of the biology of stress responses through heat shock transcription factor. *Mol. Cell. Biol.* **2004**, *24*, 5249–5256. [CrossRef] [PubMed]

5. Ou, J.R.; Tan, M.S.; Xie, A.M.; Yu, J.T.; Tan, L. Heat shock protein 90 in Alzheimer's disease. *BioMed Res. Int.* **2014**, *2014*, 1–8. [CrossRef] [PubMed]

6. Nisamedtinov, I.; Lindsey, G.G.; Karreman, R.; Orumets, K.; Koplimaa, M.; Kevvai, K.; Paalme, T. The response of the yeast *Saccharomyces cerevisiae* to sudden vs. gradual changes in environmental stress monitored by expression of the stress response protein hsp12p. *FEMS Yeast Res.* **2008**, *8*, 829–838. [CrossRef] [PubMed]

7. Sanchez, Y.; Lindquist, S. Hsp104 required for induced thermotolerance. *Science* **1990**, *248*, 1112–1115. [CrossRef] [PubMed]

8. Dubey, A.K.; Bharadwaj, P.R.; Varghese, J.N.; Macreadie, I.G. Alzheimer's amyloid-β rescues yeast from hydroxide toxicity. *J. Alzheimer's Dis.* **2009**, *18*, 31–33. [CrossRef] [PubMed]

9. Kalmar, B.; Greensmith, L. Induction of heat shock proteins for protection against oxidative stress. *Adv. Drug Deliv. Rev.* **2009**, *61*, 310–318. [CrossRef] [PubMed]

10. Liu, X.D.; Liu, P.C.; Santoro, N.; Thiele, D.J. Conservation of a stress response: Human heat shock transcription factors functionally substitute for yeast hsf. *EMBO J.* **1997**, *16*, 6466–6477. [CrossRef] [PubMed]

11. Wu, C. Heat shock transcription factors: Structure and regulation. *Annu. Rev. Cell Dev. Biol.* **1995**, *11*, 441–469. [CrossRef] [PubMed]

12. Thirumalai, D.; Reddy, G.; Straub, J.E. Role of water in protein aggregation and amyloid polymorphism. *Acc. Chem. Res.* **2012**, *45*, 83–92. [CrossRef] [PubMed]

13. Hartley, D.M.; Walsh, D.M.; Ye, C.P.; Diehl, T.; Vasquez, S.; Vassilev, P.M.; Teplow, D.B.; Selkoe, D.J. Protofibrillar intermediates of amyloid β-protein induce acute electrophysiological changes and progressive neurotoxicity in cortical neurons. *J. Neurosci.* **1999**, *19*, 8876–8884. [CrossRef] [PubMed]

14. Hardy, J.; Selkoe, D.J. The amyloid hypothesis of Alzheimer's disease: Progress and problems on the road to therapeutics. *Science* **2002**, *297*, 353–356. [CrossRef] [PubMed]

15. Kakimura, J.; Kitamura, Y.; Takata, K.; Umeki, M.; Suzuki, S.; Shibagaki, K.; Taniguchi, T.; Nomura, Y.; Smith, M.A.; Gebicke-Haerter, P.J.; et al. Microglial activation and amyloid-β clearance induced by exogenous heat-shock proteins. *FASEB J.* **2002**, *16*, 601–603. [CrossRef] [PubMed]

16. Smith, R.C.; Rosen, K.M.; Pola, R.; Magrane, J. Stress proteins in Alzheimer's disease. *Int. J. Hyperther.* **2005**, *21*, 421–431. [CrossRef] [PubMed]

17. Caine, J.; Sankovich, S.; Antony, H.; Waddington, L.; Macreadie, P.; Varghese, J.; Macreadie, I. Alzheimer's Aβ fused to green fluorescent protein induces growth stress and a heat shock response. *FEMS Yeast Res.* **2007**, *7*, 1230–1236. [CrossRef] [PubMed]

18. Haslbeck, M.; Braun, N.; Stromer, T.; Richter, B.; Model, N.; Weinkauf, S.; Buchner, J. Hsp42 is the general small heat shock protein in the cytosol of *Saccharomyces cerevisiae*. *EMBO J.* **2004**, *23*, 638–649. [CrossRef] [PubMed]

19. Porzoor, A.; Caine, J.M.; Macreadie, I.G. Pretreatment of chemically-synthesized Aβ42 affects its biological activity in yeast. *Prion* **2014**, *8*, 404–410. [CrossRef] [PubMed]

20. Jakobsen, B.K.; Pelham, H.R. Constitutive binding of yeast heat shock factor to DNA in vivo. *Mol. Cell. Biol.* **1988**, *8*, 5040–5042. [CrossRef] [PubMed]

21. Mager, W.H.; De Kruijff, A.J. Stress-induced transcriptional activation. *Microbiol. Rev.* **1995**, *59*, 506–531. [PubMed]

22. Ananthan, J.; Goldberg, A.L.; Voellmy, R. Abnormal proteins serve as eukaryotic stress signals and trigger the activation of heat shock genes. *Science* **1986**, *232*, 522–524. [CrossRef] [PubMed]

23. Avery, S.V.; Howlett, N.G.; Radice, S. Copper toxicity towards *Saccharomyces cerevisiae*: Dependence on plasma membrane fatty acid composition. *Appl. Environ. Microbiol.* **1996**, *62*, 3960–3966. [PubMed]

24. Bharadwaj, P.; Waddington, L.; Varghese, J.; Macreadie, I.G. A new method to measure cellular toxicity of non-fibrillar and fibrillar Alzheimer's Aβ using yeast. *J. Alzheimer's Dis.* **2008**, *13*, 147–150. [CrossRef]

25. Dahlgren, K.N.; Manelli, A.M.; Stine, W.B., Jr.; Baker, L.K.; Krafft, G.A.; LaDu, M.J. Oligomeric and fibrillar species of amyloid-ß peptides differentially affect neuronal viability. *J. Biol. Chem.* **2002**, *277*, 32046–32053. [CrossRef] [PubMed]

26. Stefani, M. Biochemical and biophysical features of both oligomer/fibril and cell membrane in amyloid cytotoxicity. *FEBS J.* **2010**, *277*, 4602–4613. [CrossRef] [PubMed]

27. Lee, S.; Carson, K.; Rice-Ficht, A.; Good, T. Small heat shock proteins differentially affect Aβ aggregation and toxicity. *Biochem. Biophys. Res. Commun.* **2006**, *347*, 527–533. [CrossRef] [PubMed]

28. Chauhan, V.; Chauhan, A. Oxidative stress in Alzheimer's disease. *Pathophysiology* **2006**, *13*, 195–208. [CrossRef] [PubMed]

29. Varadarajan, S.; Yatin, S.; Aksenova, M.; Butterfield, D.A. Alzheimer's amyloid β-peptide-associated free radical oxidative stress and neurotoxicity. *J. Struct. Biol.* **2000**, *130*, 184–208. [CrossRef] [PubMed]

30. Ladiwala, A.R.; Litt, J.; Kane, R.S.; Aucoin, D.S.; Smith, S.O.; Ranjan, S.; Davis, J.; Van Nostrand, W.E.; Tessier, P.M. Conformational differences between two amyloid β oligomers of similar size and dissimilar toxicity. *J. Biol. Chem.* **2012**, *287*, 24765–24773. [CrossRef] [PubMed]

31. Porzoor, A.; Macreadie, I. Yeast as a model for studies on abeta aggregation toxicity in Alzheimer's disease, autophagic responses, and drug screening. *Methods Mol. Biol.* **2016**, *1303*, 217–226. [PubMed]